Dynamic Earth

Dynamic Earth presents the principles of convection in the earth's mantle in an accessible style. Mantle convection is the process underlying plate tectonics, volcanic hotspots and, hence, most geological processes. This book is one of the first to synthesise the exciting insights into the earth's basic internal mechanisms that have flowed from the plate tectonics revolution of the 1960s.

The book summarises key observations and presents the relevant physics starting from basic principles. The core of the text shows how direct inferences from observations and basic physics clarify the roles of the tectonic plates and mantle plumes. The main concepts and arguments are presented with minimal mathematics, although more mathematical versions of important aspects are included for those who desire them. The book also surveys the geochemical constraints on the mantle and discusses its dynamical evolution, with implications for changes in the surface tectonic regime.

The audience for Geoff Davies' book will be the broad range of geologists who desire a better understanding of the earth's internal dynamics, as well as graduate students and researchers working on the many aspects of mantle dynamics and its implications for geological processes on earth and other planets. It is also suitable as a text or supplementary text for upper undergraduate and postgraduate courses in geophysics, geochemistry, and tectonics.

GEOFF DAVIES is a Senior Fellow in the Research School of Earth Sciences at the Australian National University. He received B.Sc.(Hons.) and M.Sc. degrees from Monash University, Australia, and his Ph.D. from the California Institute of Technology. He was a postdoctoral fellow at Harvard University and held faculty positions at the University of Rochester and Washington University in St. Louis, before returning to his home country. He is the author of over 80 scientific papers published in leading international journals and was elected a Fellow of the American Geophysical Union in 1992.

Dynamic Earth

Plates, Plumes and Mantle Convection

GEOFFREY F. DAVIES
Australian National University

PUBLISHED BY THE PRESS SYNDICATE OF THE UNIVERSITY OF CAMBRIDGE
The Pitt Building, Trumpington Street, Cambridge, United Kingdom

CAMBRIDGE UNIVERSITY PRESS
The Edinburgh Building, Cambridge CB2 2RU, UK www.cup.cam.ac.uk
40 West 20th Street, New York, NY 10011-4211, USA www.cup.org
10 Stamford Road, Oakleigh, Melbourne 3166, Australia
Ruiz de Alarcón 13, 28014 Madrid, Spain

First published 1999

Printed in the United Kingdom at the University Press, Cambridge

Typeset in Times $10\frac{1}{2}$/13pt, in 3B2 [KW]

A catalogue record for this book is available from the British Library

Library of Congress Cataloguing in Publication data

Davies, Geoffrey F. (Geoffrey Frederick)
Dynamic earth : plates, plumes, and mantle convection/ Geoffrey F. Davies.
p. cm.
Includes bibliographical references.
ISBN 0 521 59067 1 (hbk.). – ISBN 0 521 59933 4 (pbk.)
1. Earth–Mantle. 2. Geodynamics. I. Title.
QE509.4.D38 1999
551.1′16–dc21 98–51722 CIP

ISBN 0 521 59067 1 hardback
ISBN 0 521 59933 4 paperback

Contents

PART 1

ORIGINS

There is a central group of ideas that underlies our understanding of the process of convection in the earth's solid mantle. These ideas are that the earth is very old, that temperatures and pressures are high in the earth's interior, and that given high temperature, high pressure and sufficient time, solid rock can flow like a fluid. As well there is the idea that the earth's crust has been repeatedly and often profoundly deformed and transmuted. This idea is a central product of several centuries' practice of the science of geology. It is the perceived deformations of the crust that ultimately have led to the development of the idea of mantle convection, as their explanation. Our subject thus connects directly to more than two centuries' development of geological thought, especially through crustal deformation, heat, time and the age of the earth.

I think we scientists should more often examine the origins of our discipline. In doing so we gain respect for our scientific forebears and we may encounter important neglected ideas. We will usually gain a perspective that will make us more effective and productive scientists. Looking at our history also helps us to understand the way science is done, which is very differently from the hoary stereotype of cold logic, objectivity, 'deduction' from observations, and inexorable progress towards 'truth'.

We may be reminded also that science has profoundly changed our view of the world and we may feel some humility regarding the place of humans in the world. The deformation processes that are the subject of this book are only very marginally a part of immediate human experience, even though they are not as exotic as, for example, quantum physics or relativity. Partly because of this, understanding of them emerged only gradually over a long period, through the efforts of a great many scientists. It is easy to take for granted the magnitude of the accumulated shifts in concepts that have resulted.

Finally, there are many people in our society who are very ignorant of the earth and its workings, or who actively resist ideas such as that the earth is billions of years old. If we are to give our society the benefit of our insights without sounding authoritarian, we must be very clear about where those ideas derive from.

For these reasons, and because there is a fascinating story to be told, Chapters 2 and 3 present a short account of the emergence of the central ideas that have engendered the theories of plate tectonics and mantle convection. Chapter 1 outlines the rationale of the book.

CHAPTER 1

Introduction

1.1 Objectives

The purpose of this book is to present the principles of convection, to show how those principles apply in the peculiar conditions of the earth's mantle, and to present the most direct and robust inferences about mantle convection that can be drawn from observations. The main arguments are presented in as simple a form as possible, with a minimum of mathematics (though more mathematical versions are also included). Where there are controversies about mantle convection I give my own assessment, but I have tried to keep these assessments separate from the presentation of principles, main observations and direct inferences. My decision to write this book arose from my judgement that the broad picture of how mantle convection works was becoming reasonably settled. There are many secondary aspects that remain to be clarified.

There are many connections between mantle convection and geology, using the term 'geology' in the broadest sense: the study of the earth's crust and interior. The connections arise because mantle convection is the source of all tectonic motions, and because it controls the thermal regime in the mantle and through it the flow of heat into the crust. Some of these connections are noted along the way, but there are three aspects that are discussed more fully. The first is in Part 1, where the historical origins of the ideas that fed into the conception of mantle convection are described. Especially in Chapter 2 those historical connections are with geology. Another major connection is through Chapter 13, in which the relationship between mantle chemistry and mantle convection is considered. The third respect arises in the last chapter, where the broad tectonic implications of hypothetical past mantle regimes are discussed.

A theory of mantle convection is a *dynamical* theory of geology, in that it describes the *forces* that give rise to the motions apparent in the deformation of the earth's crust and in earthquakes and to the magmatism and metamorphism that has repeatedly affected the crust. Such a dynamical theory is a more fundamental one than plate tectonics, which is a *kinematic* theory: it describes the *motions* of plates but not the forces that move them. Also plate tectonics does not encompass mantle plumes, which comprise a distinct mode of mantle convection. It is this fundamental dynamical theory that I wish to portray here.

This book is focused on those arguments that derive most directly from observations and the laws of physics, with a minimum of assumption and inference, and that weigh most strongly in telling us how the mantle works. These arguments are developed from a level of mathematics and physics that a first or second year undergraduate should be familiar with, and this should make them accessible not just to geophysicists, but to most others engaged in the study of geology, in the broad sense. To maximise their accessibility to all geologists, I have tried to present them in terms of simple physical concepts and in words, before moving to more mathematical versions.

For some time now there has been an imperative for geologists to become less specialised. This has been true especially since the advent of the theory of plate tectonics, which has already had a great unifying effect on geology. I hope my presentation here is sufficiently accessible that specialists in other branches of geology will be able to make their own informed judgements of the validity and implications of the main ideas.

Whether my judgement is correct, that the main ideas presented here will become and remain broadly accepted, is something that only the passage of time will reveal. Scientific consensus on major ideas only arises from a prolonged period of examination and testing. There can be no simple 'proof' of their correctness.

This point is worth elaborating a little. One often encounters the phrase 'scientifically proven'. This betrays a fundamental misconception about science. Mathematicians prove things. Scientists, on the other hand, develop models whose behaviour they compare with observations of the real world. If they do not correspond (and assuming the observations are accurate), the model is not a useful representation of the real world, and it is abandoned. If the model behaviour does correspond with observations, then we can say that it works, and we keep it and call it a theory. This does not preclude the possibility that another model will work as well or better (by corresponding with observations more accurately or in a broader context). In this case, we say that the new model is better, and usually we drop the old one.

However, the old model is not 'wrong'. It is merely less useful, but it may be simpler to use and sufficient in some situations. Thus Newton's theory of gravitation works very well in the earth's vicinity, even though Einstein's theory is better. For that matter, the old Greek two-sphere model of the universe (terrestrial and celestial) is still quite adequate for navigation (strictly, the celestial sphere works but the non-spherical shape of the earth needs to be considered). Scientists do not 'prove' things. Instead, they develop more useful models of the world. I believe the model of mantle dynamics presented here is the most useful available at present.

Mantle convection has a fundamental place in geology. There are two sources of energy that drive geological processes. The sun's energy drives the weather and ocean circulation and through them the physical and chemical weathering and transport processes that are responsible for erosion and the deposition of sediments. The sun's energy also supports life, which affects these processes.

The other energy source is the earth's internal heat. It is widely believed, and it will be so argued here, that this energy drives the dynamics of the mantle, and thus it is the fundamental energy source for all the non-surficial geological processes. In considering mantle dynamics, we are thus concerned with the fundamental mechanism of all of those geological processes. Inevitably the implications flow into many geological disciplines and the evidence for the theory that we develop is to be found widely scattered through those disciplines.

Inevitably too the present ideas connect with many ideas and great debates that have resonated through the history of our subject: the rates and mechanisms of upheavals, the ages of rocks and of the earth, the sources of heat, the means by which it escapes from the interior, the motions of continents. These connections will be related in Part 1. The historical origins of ideas are often neglected in science, but I think it is important to include them, for several reasons. First, to acknowledge the great thinkers of the past, however briefly. Second, to understand the context of ideas and theories. They do not pop out of a vacuum, but emerge from real people embedded in their own culture and history, as was portrayed so vividly by Jacob Bronowski in his television series and book *The Ascent of Man* [1]. Third, it is not uncommon for alternative possibilities to be neglected once a particular interpretation becomes established. If we returned more often to the context in which choices were made, we might be less channelled in our thinking.

1.2 Scope

The book has four parts. Part 3, Essence, presents the essential arguments that lead most directly to a broad outline of how mantle dynamics works. Part 2, Foundations, lays the foundations for Part 3, including key surface observations, the structure and physical properties of the interior, and principles and examples of viscous fluid flow and heat flow.

Parts 1 and 4 connect the core subject of mantle convection to the broader subject of geology. Part 1 looks at the origin and development of key ideas. Part 4 discusses possible implications for the chemical and thermal evolution of the mantle, the tectonic evolution and history of the continental crust. Many aspects of the latter topics are necessarily conjectural.

1.3 Audience

The book is intended for a broad geological audience as well as for more specialised audiences, including graduate students studying more general aspects of geophysics or mantle convection in particular. For the latter it should function as an introductory text and as a summary of the present state of the main arguments. I do not attempt to summarise the many types of numerical model currently being explored, nor to present the technicalities of numerical methods; these are likely to progress rapidly and it is not appropriate to try to summarise them in a book. My expectation is that the broad outlines of mantle convection given here will not change as more detailed understanding is acquired.

In order to accommodate this range of readership, the material is presented as a main narrative with more advanced or specialised items interspersed. Each point is first developed as simply as possible. Virtually all the key arguments can be appreciated through some basic physics and simple quantitative estimates. Where more advanced treatments are appropriate, they are clearly identified and separated from the main narrative. Important conclusions from the advanced sections are also included in the main narrative.

It is always preferable to understand first the qualitative arguments and simple estimates, before a more elaborate analysis or model is attempted. Otherwise a great deal of effort can be wasted on a point that turns out to be unimportant. Worse, it is sometimes true that the relevance and significance of numerical results cannot be properly evaluated because scaling behaviour and dependence on parameter values are incompletely presented. Therefore the mode of presentation used here is a model for the way theoretical models can be developed, as well as a useful way of reaching an audience with a range of levels of interest and mathematical proficiency.

1.4 Reference

1. J. Bronowski, *The Ascent of Man*, 448 pp., Little, Brown, Boston, 1973.

CHAPTER 2

Emergence

We begin with a look at some of the 'classical' questions about the earth: its age, its internal heat, and how rocks may deform. These questions are famous both because they are fundamental and because some great controversies raged during the course of their resolution. In looking at how the age of the earth was first inferred, we soon encounter the question of whether great contortions of the crust happened suddenly or slowly. The fact that the interior of the earth is hot is central, both to the occurrence of mantle convection and geological processes, but also historically because one estimate of the age of the earth was based on the rate at which it would lose internal heat.

Much of my limited knowledge of the history of geology prior to this century comes from Hallam's very readable short book *Great Geological Controversies* [1]. I make this general acknowledgement here to save undue interruption of the narrative through this chapter. My interpretations are my own responsibility.

2.1 Time

The idea that continents shift slowly about the face of the earth becomes differentiated from fantasy only with an appreciation of time. One of the most profound shifts in the history of human thought began about 200 years ago, when geologists first began to glimpse the expanse of time recorded in the earth's crust. This revolution has been less remarked upon than some others, perhaps because it occurred gradually and with much argument, and because the sources of evidence for it are less accessible to common observation than, for example, the stars and planets that measure the size of the local universe, or the living things that are the products of natural selection.

During the time since the formulation of the theory of plate tectonics, my home in Australia has moved about 1.8 m closer to the equator. Within the same period, that displacement has become accessible to direct scientific observation, but not to unaided human perception. Mostly the landscape is static to human perception. It is not an uncommon experience to see the aftermath of a landslide or rockfall, and it is occasionally possible to see a fresh fault scarp after an earthquake. Students of geology now take for granted that these are irreversible events that are part of the processes of erosion and tectonic deformation. However, the relationship of these observations to the form of the land surface and to folded and faulted rock strata is not at all immediately obvious. Indeed it is only 200 years since this connection began to be made seriously and systematically, and less than 100 years since earthquakes, fault scarps and sudden slip on buried faults were coherently related through the ideas of accumulated elastic stress and frictional fault surfaces.

One person's dawning comprehension of the expanse of geological time is recorded in the account (quoted by Hallam [1], p. 33) by the mathematician John Playfair of his visit in 1788 to Siccar Point in Britain, in the company of the geologists Hutton and Hall, to observe a famous unconformity where subhorizontal Devonian sandstones rest on near-vertical Silurian slates (which he called schistus).

> We felt ourselves necessarily carried back to the time when the schistus on which we stood was yet at the bottom of the sea, and when the sandstone before us was only beginning to be deposited, in the shape of sand and mud, from the waters of a superincumbent ocean. An epocha still more remote presented itself, when even the most ancient of these rocks, instead of standing upright in vertical beds, lay in horizontal planes at the bottom of the sea, and was not yet disturbed by that immeasurable force which has burst asunder the solid pavement of the globe. Revolutions still more remote appeared in the distance of this extraordinary perspective. The mind seemed to grow giddy by looking so far into the abyss of time...

Playfair and Hutton did not have a clear quantitative measure of the time intervals they were contemplating, but they knew they were dealing with periods vastly greater than the thousands of years commonly believed at the time. Hutton especially must have appreciated this, because he is perhaps most famous for expounding the idea of indefinite time in a famous statement from that same year [2] '... we find no vestige of a beginning, no prospect of an end.' (I said 'indefinite time' rather than 'infinite time' here because Hutton's words do not necessarily imply the latter. In modern parlance, we could say that Hutton was proposing that the earth was in a steady state, and it is characteristic of steady-state processes that information about their initial conditions has been lost.)

The work and approach of Lyell in the first half of the nineteenth century provided a basis for quantitative estimates of the elapse of time recorded in the crust. Lyell is famous for expounding and applying systematically the idea that geological structures might be explained solely by the slow action of presently observable processes. He and many others subsequently made use of observations that could be related to historical records, of erosion rates and deposition rates, and of stratigraphic relationships, to demonstrate that a great expanse of time was required. Though still rather qualitative, an eloquent example comes from an address by Lyell in 1850 (Hallam [1], p. 58; [3]).

> The imagination may well recoil from the vain effort of conceiving a succession of years sufficiently vast to allow of the accomplishment of contortions and inversions of stratified masses like those of the higher Alps; but its powers are equally incapable of comprehending the time required for grinding down the pebbles of a conglomerate 8000 feet [2650 metres] in thickness. In this case, however, there is no mode of evading the obvious conclusion, since every pebble tells its own tale. Stupendous as is the aggregate result, there is no escape from the necessity of assuming a lapse of time sufficiently enormous to allow of so tedious an operation.

According to Hallam (p. 106), it was Charles Darwin who made one of the first quantitative estimates of the lapse of geological time, in the first edition of his *Origin of Species* [4]. This was an estimate for the time to erode a particular formation in England, and Darwin's estimate, not intended to be anything more than an illustration, was 300 million years. Though it might have been only rough, Darwin's estimate conveys the idea that the time spans involved in geology, that can be characterised qualitatively only by vague terms such as 'vast', are not 300 000 years and not 300 billion years, for example.

During the middle and later years of the nineteenth century, a great debate raged amongst geologists and between geologists and physicists, particularly Lord Kelvin, about the age of the earth (which I will discuss in Section 2.4). What impresses me is not so much the magnitudes of the differences being argued as the general level of agreement and correctness, especially amongst geologists' estimates. We must realise that initially they knew only that the number must be orders of magnitude greater than the 10^4 years or so inferred from scriptures, a number that was then still commonly accepted in the non-scientific community, though even in Hutton's time this was actively doubted within the scientific community. We must also bear in mind what they were attempting to measure, which was the time necessary to accumulate the sedimentary strata which we now know as the Phanerozoic (the Cambrian to the present, the period of large fossils). Their estimates were accurate to better than an order of magnitude, and some were within a factor of two. Thus the geologists' estimates tended to be a few hundred million years (Kelvin was arguing for less than 100 million years), and the base of the Cambrian is now measured as being about 540 million years old.

Let us also acknowledge that even the estimate of a few thousand years is a good measure of the time since written records began. This we now think of as the period within which civilisation arose, rather than the age of the earth. The point is that the estimates of both the scriptural scholars and the nineteenth century geologists were quantitatively quite good. What we now disagree with is the interpretation that either of these numbers represents the age of the earth.

Even today, no-one has directly measured the age of the earth [5]. The oldest rocks known are about 4 Ga old [6], and a few grains of the mineral zircon, incorporated into younger sediment, have ages up to 4.27 Ga [7]. (Units used commonly in this book, and their multiples, are summarised in Appendix 1.) By now we do have a clear record of changes in the way the earth works, because we can see much further back than Hutton and Lyell could. We even have a 'vestige of a beginning' that allows us to estimate the age of the earth, but it is a subtle one requiring some assumptions and comparisons with meteorites for its interpretation. The ages of meteorites have been measured using the decay of two uranium isotopes into lead isotopes: ^{238}U into ^{206}Pb and ^{235}U into ^{207}Pb. These define a line in a plot of $^{207}Pb/^{204}Pb$ versus $^{206}Pb/^{204}Pb$ whose slope corresponds to an age of 4.57 Ga. It has also been demonstrated that estimates of the mean lead isotope composition of the earth fall close to this line, which is consistent with the earth having a similar age [8]. But that is not the end of the story, because different assumptions can lead to different estimates of the earth's mean lead isotope composition, and for all reasonable estimates of this the age obtained is significantly younger than that of the meteorites. Further, the event dated in this way need not be the formation of the earth.

The ages obtained from estimates of the earth's lead isotope composition range from about 4.45 to 4.52 billion years, 50 to 120 million years after the formation of the meteorites [5]. This plausibly represents the mean age of separation of the silicate mantle from the metallic core of the earth, which probably separated the uranium from some of the lead: lead is much more soluble in liquid iron than is uranium, so it is expected the uranium partitioned almost entirely into the silicate mantle, whereas some lead went into the iron core and some stayed in the mantle [9]. There are good reasons for believing that the separation of the core from the mantle material was contemporaneous with the later stages of the accretion of the earth from the cloud of material orbiting the sun, of which the meteorites are believed to be surviving chunks [10]. So, we are really still left with the *assumption* that the earth formed at about the same time as the meteorites, and we infer with somewhat more basis that it was probably substantially formed at the mean time of separation of core and mantle.

2.2 Catastrophes and increments

Most geologists have heard of the great debate between 'catastrophists' and 'uniformitarians' that raged around the beginning of the nineteenth century. As Hallam explains (p. 30), the catastrophism that was challenged by Hutton, Lyell and others was not a simplistic, theistic appeal to sudden supernatural causes. It was an expression of the genuine difficulty in connecting observations of dramatically tilted and contorted strata with any presently observable process, such as small local uplifts associated with volcanic activity.

Hutton argued for the action of slow, presently observable processes acting over indeterminate amounts of time, but his arguments were focussed on deposition. Hallam notes (p. 34) that Hutton still conceived of catastrophic disturbance, as reflected in his reference to '... that enormous force of which regular strata have been broken and displaced; ... strata, which have been formed in a regular manner at the bottom of the sea, have been violently bent, broken and removed from their original place and situation.'

Lyell went further, and argued that *all* geological observations could be explained by the prolonged action of presently observable processes. A principal influence on the development of his ideas seems to have been his observations in Sicily of progressive accumulation of volcanic deposits and of associated progressive uplifts of strata containing marine fossils. Clearly he conceived, if only vaguely, that volcanism and associated earthquakes provided a mechanism that could produce large uplifts and distortions of strata. Thus he was able to argue not only that sedimentary deposition was protracted but that 'disturbance' was also protracted.

It was evidently easier to assemble observations and arguments for prolonged deposition than for prolonged deformation. Clearly a major reason for this is that the mechanism of erosion and deposition is more accessible to common observation than is the mechanism of mountain building. Thus in the passage quoted in Section 2.1, Lyell appeals to deposition to defend his notion of prolonged process, rather than to the 'contortions and inversions' of disturbance, because 'every pebble tells its own tale'; and the tale told by the pebble is intelligible to the thoughtful lay observer. This passage is from 1850 [3], twenty years after the first publication of Lyell's *Principles of Geology*. A century later, conventional geology had little more to offer by way of a mountain building mechanism, but we are now able to redress this imbalance.

I do not mean to imply that prolonged deformation has only recently been established. It has been established, for example, by a secondary effect of mountain building, namely more erosion and deposition. By documenting the prolonged deposition of such sediments, the long duration of uplift has been definitively established. The point is that geologists could not as easily appeal directly to the mechanism for deformation as they could to the mechanism for grinding pebbles.

Charles Darwin made important contributions to geology as well as to biology, and he provided early examples of both indirect and direct arguments (Hallam, p. 55). While in South America during his famous voyage on the *Beagle*, Darwin observed a correlation between the elevation of strata in the Andes mountains and the declining proportion of extant species in the fossils contained in the strata. From this he inferred the progressive uplift of the strata and thus, indirectly, of the progressive uplift of the mountains.

Darwin's experience of a powerful earthquake in the Andes yielded a more direct argument, and also contributed to his adoption of Lyell's empirical approach. The inference he made from this bears powerfully on the question of tectonic mechanism. The earthquake elevated the coastline by several feet (1–2 metres) for as much as 100 miles (160 km). Earthquakes of such magnitude apparently occurred about once per century in that area, a frequency, Darwin pointed out, sufficient to raise a mountain range like the Andes in much less than a million years.

Darwin's arguments were influential, since they supported an alternative to a catastrophist theory of the time that the Andes had been raised in one great convulsion. In retrospect they are very significant also because they foreshadowed a profound insight that could have come from the study of earthquakes, but that came instead from another branch of geophysics. This will be covered in Sections 3.5 and 3.6.

Lyell went too far. He argued that catastrophes played no part in forming the geological record, and that the earth was in a steady state. He claimed that the record revealed no discernible change in rates of process, or kinds of process, or even in kinds of organisms (Hallam, p. 121). He maintained this position in the face of rapidly accumulating evidence from the fossil record that there had been dramatic progressive changes in life forms. I note this not to diminish his great achievements, but to illustrate that he was human like the rest of us and overstated his case, and that not everything he said was gospel.

Unfortunately, Lyell was so influential that his work has been treated almost as gospel by some geologists. His 'principle of uniformitarianism' was elevated almost to the level of natural law by some, and arcane debates can still be found on precisely at what level of method, process or 'law' his principle applies. It is really rather simpler than that. He was arguing against the natural and prevailing assumption of his time that dramatic results necessarily required dramatic causes.

Catastrophes have become more respectable in recent decades. For example, there was a well-documented series of catastrophic floods at the end of the ice age in the western United States [11] and it has become increasingly accepted that the extinction of the dinosaurs was caused either by the impact of a giant meteorite into the earth [12] or by a giant eruption of 'flood basalts' in India [13], or by a combination of the two. Also stratigraphers have recognised that many sedimentary deposits are biased towards unusually large and infrequent events like exceptionally large floods, which tend to carry a disproportionate amount of sediment [14].

Whewell [15], quoted by Hallam (p. 54), said all that was necessary not long after Lyell's first volume was published.

> *Time*, inexhaustible and ever accumulating his efficacy, can undoubtedly do much for the theorist in geology; but *Force*, whose limits we cannot measure and whose nature we cannot fathom, is also a power never to be slighted: and to call in one, to protect us from the other, is equally presumptuous to whichever of the two our superstition leans.

Pointing out that geological increments come in a great range of magnitudes and recur with a great range of frequencies, he notes elsewhere

In order to enable ourselves to represent geological causes as operating with uniform energy through all time, we must measure our time in long cycles, in which repose and violence alternate; how long must we extend this cycle of change, the repetition of which we express by the word *uniformity*?

And why must we suppose that all our experience, geological as well as historical, includes more than one such cycle? Why must we insist upon it, that man has been long enough an observer to obtain the *average* of forces which are changing through immeasurable time?

2.3 Heat

Heat enters the subject of mantle convection in several ways, each of them fundamental. Heat is the form of energy driving the mantle system. High temperatures are required for mantle rocks to flow like a fluid. The time required for the earth to cool was an important way of estimating the earth's age in the nineteenth century, as we will see in the next section. The cooling of the earth has probably had a major effect on its tectonic mechanisms through its history. Tectonic modes must, since they are driven by the earth's heat, be capable of removing that heat at a sufficient rate to keep the earth cooling. The last point is important in constraining conjectures on the tectonic modes that might have operated early in earth's history, as we will see in Chapter 14.

The idea is common and old that the earth's interior is hot. There is supporting evidence available to common experience in some places, and such evidence has become widely known. Hot magma issues from the earth's interior through volcanoes. There are hot springs in volcanic areas. The temperature in very deep mines is uncomfortably high.

This knowledge is sufficiently widespread that many people have the misconception that the interior below the crust is molten. It is not uncommon to hear people speak of the earth's molten core, and mean everything below the crust. Geologists through the first half of this century might have done well to ponder this common belief more deeply, because it is quite a reasonable inference, and it becomes more so in the light of more precise knowledge of temperatures at depth.

Many measurements have been made of temperature in mines and boreholes, and it has been long established that the temperature in the crust increases with depth at a rate usually in the range 15–25 °C/km. This directly implies quite high temperatures, even in the crust. Continental crust is typically 35–40 km thick, so by extrapolating the observed gradient we can estimate temperatures at its base to be in the range 600–1000 °C. Granites melt at temperatures of about 750 °C. The rocks of the mantle, below the crust, begin to melt at about 1200 °C. It is entirely plausible that the deep crust and the top of the mantle sometimes melt to yield volcanic magmas. It is also plausible, in the absence of contrary evidence, that the earth's interior is entirely molten below a depth of about 80 km.

The argument is not quite so simple as this, though this simple version validly establishes plausibility. A more complete argument takes into account that a significant fraction of the heat emerging from the continental crust is due to radioactivity within the upper crust. Below this, a lower gradient of temperature is sufficient to conduct the balance of the heat from greater depth. We will see in Chapter 7 that temperatures at the base of the crust are likely to be in the range 400–700 °C, and the depth at which the mantle would start to melt is about 100 km.

However, the earth's interior is not entirely molten. The evidence comes from seismology, and has been known since early this century. As we will see in Chapter 5, the crust and mantle transmit two kinds of waves, compressional and shear. Shear waves are transmitted by solids, but not by liquids, so the crust and mantle are inferred to be solid. The solid mantle extends about halfway to the centre of the earth. Below that is the core, through which shear waves are not transmitted, so the core is inferred to be liquid, and is probably composed mostly of iron. Partial melting does occur locally in the crust and near the top of the mantle, and this gives rise to volcanoes and to subterranean intrusions of magma.

How can the observations of the increase of temperature with depth be reconciled with the clear evidence from seismology that the mantle is solid through most of its vast bulk? One plausible supposition was that if the earth were molten early in its history it would have cooled relatively efficiently until it had solidified. We might then expect temperatures deep in the interior to be still close to the melting temperature. This has implications for the expected strength of the mantle, as we will see in Section 2.5. Another implication of the observed temperature increase with depth only emerged when the age of the earth was determined to be several billion years: it was not clear how the observed heat flow out of the earth's interior could be maintained for so long, but this anticipates the next section.

2.4 Cooling age of earth

The observations of the increase of temperature with depth in the crust were sufficiently well-established by the middle of the nineteenth century to be the basis of Kelvin's estimate of the age of the earth. He calculated the time it would take for the temperature gradient to decrease to its present value, assuming the earth started in a molten state. You will find the full argument in Chapter 7, in a different context. It depends on the fact that if a body starts out hot and is then cooled from the surface, the temperature gradient near the surface decreases with time in a predictable way, if the heat is escaping by thermal conduction. Knowing the thermal conductivity of rocks, and estimating the initial temperature as being close to that of molten rock, Kelvin calculated the age of the earth to be about 100 Ma (million years), with an uncertainty of a factor of about four [16].

Kelvin was very impressed that this corresponded with his estimate of the age of the sun [17], based on quite different reasoning. He had assumed that the sun's energy was derived from gravitational energy during its formation, and calculated the time for which it could sustain its present rate of heat loss. We can understand Kelvin's satisfaction with the consistency of his ages of the sun and the earth, but I can't think of any reason, in the light of our present understanding, why it is anything more than a coincidence.

I remarked earlier that age estimates by various geologists and by Kelvin were not really very different, all being of the order of 100 Ma. To understand the heated nature of the argument between Kelvin and geologists, we must appreciate that Kelvin was arguing against the extreme position of Hutton, Lyell and others that the earth was in a steady state of indeterminate age. Kelvin's fundamental point was that his estimate of age yielded a finite number, rather than an infinite or indeterminate number. His perspective was that of a physicist who was prominent in the development of the science of thermodynamics. His arguments were based on the implication of the second law of thermodynamics that the sun and the earth must be running down, and that a steady-state earth contradicted the laws of physics which he had helped to establish.

Kelvin was complaining, in effect, that Lyell's earth was a perpetual motion machine, which is a very reasonable point, but evidently Kelvin did not concede that Lyell's earth was much closer to a steady state than the catastrophists' earth that Lyell had been challenging. Even from the modern perspective, remembering that the interpreted geological record at that time was confined to the Phanerozoic, Lyell was essentially correct concerning tectonics: there has been little *discernible overall* change in modes and rates of tectonics within this period (although, acknowledging Whewell, there have been fluctuations). Unfortunately the argument degenerated into a squabble about whether the age of the earth was less than or more than one hundred million years. In pursuit of his point, Kelvin through his later life pared his upper limit progressively down, to 24 Ma in 1897 (Hallam, p. 122).

It is salutary to see how much the perspective had changed between 1780 and 1860. Clerics argued for an age of the order of 10^4 years. Early geologists tried to fit their theories into the biblical time framework, and assumed as well that the more distant past had been far more violent than the more recent past. Hutton and Lyell argued against a short, catastrophist, declining history of the earth's activity, but made no quantitative estimates. When Darwin finally made an estimate using their methods, he got a number of the order of 10^8 years. This was surely sufficiently different from the biblical estimate to justify the rhetorical steady-state position of Hutton and Lyell. Kelvin in turn reacted against their rhetorical position, justifiably in the light of emerging knowledge of physics, but he argued on the basis of a number that was also of the order of 10^8 years.

To his credit, Kelvin qualified his estimates, at least early on, with the explicit assumption that there was no unknown physical process at work. In fact there were two. One of these is well-known, the other deserves to be better known. First, the discovery of radioactivity near the end of the nineteenth century revealed an energy source that could sustain the heat of both the earth and the sun for billions of years, and Rutherford was able to proclaim in 1904, in effect, that although Kelvin's number might not have been right, 'the old boy' had never actually been wrong (Hallam, p. 123).

The second process was conjectured by the Reverend Osmond Fisher, who in 1881 published a book, entitled *The Physics of the Earth's Crust* [18], in which he expounded the view that the crust must reside on a plastic (that is, deformable) substratum. He noted that Kelvin's argument depended on the assumption that heat is conducted out of the earth's interior. Given the known thermal conductivity of rocks, this process could cool the earth only to a depth of about 100 km in 100 Ma. However, if the interior were deformable, then convection could transport heat from a much greater depth to replace that lost through the surface. In this way the presently observed temperature gradient in the crust could be maintained for much longer than Kelvin's estimate of 100 Ma.

It was not until around 1980 that the earth's internal thermal regime and thermal evolution began to be reasonably well understood. At that time several people combined an approximate expression for the rate of heat transport by convection with estimates of the radioactive heat generation rate within the mantle to calculate the thermal evolution of the mantle. (It was also important to include an expression for the strong temperature dependence of the viscosity of mantle rocks.) Such calculations plausibly resolved the apparent contradictions of previous interpretations, and they will be described in Chapter 14.

2.5 Flowing rocks

My grandfather, who was a sheep drover, would have been more familiar with the agency of heat acting on solid material than are many modern city dwellers. In an age when technology was closer to daily life, he would certainly have been aware that horseshoes need frequent adjustment and replacement, and that a blacksmith's indispensable tool is his forge. With his forge, he can heat metal and render it malleable, so that it can be bent and shaped to fit each horse's hoof.

Neither my grandfather nor we are so familiar with the idea that rocks can be malleable. If a rock is cold, it is brittle, and if it is heated sufficiently it melts and becomes magma. There is however a range of intermediate behaviour, at temperatures near but below the melting temperature, in which rocks and other solids can deform without breaking, but this behaviour is only rarely perceptible in non-metals. This is because it is slow, and we require an elapse of time before it becomes manifest. An example commonly cited is that of glass, which at normal temperatures deforms under the action of its own weight at a rate that makes its deformation observable over decades or centuries. Thus it is reported that drink bottles recovered from the desert have sagged like an object in a surrealist painting, and that windows in old cathedrals of Europe are noticeably thicker at the bottom than at the top, and we infer that the craftsmen of old did not install them in that condition. (We need not be concerned here that glass is technically not a crystalline solid, but rather a super-cooled liquid with its atoms caught in the disordered arrangements of liquids. All of our statements are equally true of crystalline solids, though they deform less rapidly as a rule.)

For solids to deform without breaking also requires that they are not stressed too greatly. It turns out that the action of pressure suppresses brittle fracture in favour of ductile deformation. In effect, the two sides of a fracture (potential or existing) are locked together under the action of pressure, and then the solid's only available response is its tendency to deform throughout its volume. Of course the pressures are very high in the earth's interior, and evidently sufficient to suppress brittle behaviour completely at depths greater than about 100 km. (This has left deep earthquakes as a puzzle. They occur in widely separated parts of the earth at depths up to nearly 700 km, well below the depth at which brittle behaviour is suppressed in most regions. However, an answer may be at hand in terms of a special mechanism involving transformations of crystal structure induced by high pressure. It is also clear that the places where deep earthquakes occur are not normal parts of the mantle in other respects.)

The result is that with the high temperatures and pressures of the earth's interior, and enough time, rocks can deform and flow, and thus be considered as fluids, even though they are solid in practical experience. The rate of deformation is extremely slow: about twenty orders of magnitude slower than liquids of common experience, for similar stress levels. Seismic waves have periods in the range of seconds to thousands of seconds, and the deformation is so small in this period that shear waves can be transmitted with little dissipation of energy. Thus to seismic waves the material effectively is a solid. On the other hand, there can be significant deformation in a few thousand years (about 10^{11} s), as we will see in Chapter 6.

The observations that first led to the inference that the mantle is deformable on geological time scales are discussed in Section 3.2. It also turns out that the viscosity of mantle rocks (which is the measure of how 'stiff' or 'runny' they are) is strongly dependent on temperature. As the melting temperature of the rock is approached, the viscosity drops by nearly an order of magnitude for every hundred degree rise in temperature. This has a major effect on the thermal evolution of the mantle. It also led to a simple and robust argument that the mantle is likely to be convecting and, by implication, for there to be large tectonic displacements of the crust. This argument was put by Tozer in 1965 [19] (see Section 3.8).

2.6 References

1. A. Hallam, *Great Geological Controversies*, 244 pp., Oxford University Press, Oxford, 1989.
2. J. Hutton, *Trans. R. Soc. Edinbugh* **1**, 217, 1788.
3. C. Lyell, *Q. J. Geol. Soc. London* **6**, xxxii, 1850.
4. C. Darwin, *On the Origin of Species*, Murray, London, 1859.
5. G. B. Dalrymple, *The Age of the Earth*, 474 pp., Stanford University Press, Stanford, CA, 1991.
6. S. A. Bowring and T. Housh, The Earth's early evolution, *Science* **269**, 1535–40, 1995.
7. D. O. Froude, T. R. Ireland, P. D. Kinny, I. S. Williams, W. Compston, I. R. Williams and J. S. Myers, Ion microprobe identifica-

tion of 4100–4200 Myr-old terrestrial zircons, *Nature* **304**, 616–18, 1983.

8. C. Patterson, Age of meteorites and of the Earth, *Geochim. Cosmochim. Acta* **10**, 230–7, 1956.
9. H. S. C. O'Neill and H. Palme, Composition of the silicate Earth: implications for accretion and core formation, in: *The Earth's Mantle: Composition, Structure and Evolution*, I. N. S. Jackson, ed., Cambridge University Press, Cambridge, 3–126, 1998.
10. D. J. Stevenson, Fluid dynamics of core formation, in: *Origin of the Earth*, H. E. Newsom and J. H. Jones, eds., Oxford University Press, New York, 231–49, 1990.
11. J. H. Bretz, The Lake Missoula floods and the channeled scabland, *J. Geol.* **77**, 505–43, 1969.
12. W. Alvarez, *T. Rex and the Crater of Doom*, 185 pp., Princeton University Press, Princeton, NJ, 1997.
13. V. Courtillot, J. Besse, D. Vandamme, R. Montigny, J.-J. Jaeger and J. Cappetta, Deccan flood basalts at the Cretaceous/Tertiary boundary?, *Earth Planet. Sci. Lett.* **80**, 361–74, 1986.
14. D. V. Ager, *The Nature of the Stratigraphic Record*, Macmillan, London, 1973.
15. W. Whewell, *History of the Inductive Sciences, from the Earliest to the Present Time*, Parker, London, 1837.
16. Lord Kelvin, *Philos. Mag.* (ser. 4) **25**, 1, 1863.
17. Lord Kelvin, On the age of the sun's heat, *Macmillans Magazine* **5**, 288, 1862.
18. O. Fisher, *The Physics of the Earth's Crust*, Murray, London, 1881.
19. D. C. Tozer, Heat transfer and convection currents, *Philos. Trans. R. Soc. London, Ser. A* **258**, 252–71, 1965.

CHAPTER 3

Mobility

The idea that parts of the earth have moved slowly relative to each other over distances comparable to the size of the globe belongs mostly to the twentieth century. There were some earlier suggestions of catastrophic global displacements, but it was in the twentieth century that large slow displacements of the continents were proposed and systematically advocated, and eventually their existence was decisively established.

Historically, the idea of mantle convection is closely entwined with the ideas of continental drift and plate tectonics. The idea that the earth's interior is mobile can be traced back at least to the middle of the nineteenth century, but it became the focus of sharp debate early in the twentieth century with the acquisition of seismological evidence that below the crust is a solid, rocky mantle extending about halfway to the centre of the earth. To many this seemed to make continental drift impossible.

After Holmes proposed mantle convection as a possible mechanism of continental drift around 1930, most thinking about continental drift and the emerging plate tectonics was strongly conditioned by expectations of how such convection would work. One can argue that this interaction of ideas actually held back the recognition of the pattern of movements on the earth's surface (Section 3.8).

Others have told the story of the theory of continental drift and its ultimate evolution into the theory of plate tectonics [1–6]. There are several reasons for recounting it here, rather than simply proceeding to a description of the plates and how they work. Mainly, I want to include the complementary development of ideas about the mobility of the mantle, which I think played a larger role in the story than has been appreciated. Also I want to highlight some aspects of the plate tectonics story that I think deserve more attention than they have received, and to continue the theme of

Chapter 2, showing some of the context from which important ideas emerged. Without this, it is easy to overlook the large amount of good science, done by a great many scientists, upon which such insights are usually founded. Finally, when a theory becomes as widely accepted as the theory of plate tectonics, it is easy to lose sight of why we think it is a good theory. I therefore describe some of the early and compelling evidence for the existence of the plates and their motions.

There is another story that developed in the shadow of plate tectonics: the story of mantle plumes. Plumes are a distinct component of mantle convection. They have played a significant role in the history of the continents, and possibly had a larger role early in earth history, though that is still quite uncertain. They may have had decisive effects on the history of life. The story of the idea of plumes is just as long but not as complex as the story of plates. Perhaps four names can be identified as principals: Darwin, Dana, Wilson and Morgan, with a footnote for Holmes.

This chapter is about how the idea of large, slow displacements became established. It is a long story, so my account here has to be selective, including only key evidence and key arguments. Because the stories of continental drift and plate tectonics have been told before in some detail, I do not provide a lot of detail nor many illustrations here, except where I want to emphasise particular points.

3.1 Drifting continents

In 1912 Alfred Wegener first spoke publicly and wrote of his idea that whole continents had undergone large, slow displacements. These were published in book form in 1915 under the title *Die Entstehung der Kontinente und Ozeane* (*The Origin of Continents and Oceans*), which went through several editions both before and after his death [7, 8].

The seed of Wegener's theory came from the similarity in map view of the shapes of the continental margins on either side of the Atlantic Ocean. This similarity is reflected more crudely by the coastline, and had been remarked upon previously. His ideas became more definite when he learned of similarities in fossils occurring on opposite sides of the Atlantic, and later of geological similarities. He developed his ideas into the proposal that all of the continents had been grouped into one supercontinent, and that this had fragmented and the pieces had drifted apart starting in the Mesozoic era, about 200 Ma ago.

In later editions he added more data, and also used evidence from various kinds of deposits that could be used to infer climate and thus to distinguish equatorial and polar regions, arguing that the distributions of palaeoclimate made more sense if the continents had moved.

The similarities in fossils were already being noted, and had led palaeontologists to postulate the past existence of land connections between the widely separated continents. Initially these connections were assumed to be continents that had later subsided under the ocean. Geologists had for decades held that continents episodically emerged from and subsided into the ocean, on the basis of the widespread occurrence of marine fossils in continental sediments. Little was known of the nature of the rocks of the sea floor at this time, which limited the possibility of direct geological tests of this idea.

Wegener argued that such large-scale vertical movements of continents were not viable, because gravity measurements had shown that the earth's crust is close to isostatic equilibrium. I will discuss this in more detail in the next section. The point is that it was already established that the continents stand higher than the seafloor because continental crust must be less dense, and that the continents in effect 'float' in a denser substratum. If large continental blocks subsided by several kilometres, there would be a large negative gravity anomaly created, and such large anomalies were excluded by the observations. One response to Wegener's isostasy argument was to assume that the land connections were smaller than continental scale – narrow land 'bridges'. This proposal was *ad hoc*, and has never had any evidence to support it.

Another argument Wegener made against the idea of rising and sinking continents was that if it occurred erratically in space and time, as the geological record suggested, one would expect the elevations of the earth's surface to be spread more evenly than they are between the highest and the lowest, with a preponderance of areas at intermediate elevations. Instead, the observation is that there are two preponderant levels of the earth's surface, one at the level of the deep sea floor and one at the level of the continental surfaces, just above sea level.

This bimodal distribution of elevation had been recognised for a long time as a first-order feature of the earth requiring explanation. Wegener's point is a very sound one, that the observed topography looks like a very improbable consequence of the older idea of rising and sinking continents. Its rhetorical weakness was that he did not have an explanation for the bimodal topography either. In the absence of explicit mechanisms for either vertical or horizontal displacements of continents, it was not possible to quantify the argument at all, and so the opponents of his theory were free to make suppositions to suit their point of view, and to note, for instance, that the particular Gaussian distribution that he assumed to make his point was no more probable, a priori, than the observed distribution.

Hallam [1] points to two factors that may have facilitated Wegener's boldness of thinking. One was that he was not a geologist by training, but a meteorologist, so he had no particular commitment to prevailing ideas in the geological community. The other was that in Germany at the time, the geophysics community embraced meteorology and climatology as well as the 'solid' earth, which perhaps made their thinking more open to mobilist ideas.

Wegener was tentative about what force or forces might cause continents to drift, writing, 'The Newton of drift theory has not yet appeared . . .' and conceding that it might be a long time before this was clarified [1]. In what can be seen in retrospect as a key tactical error, he suggested a differential rotational force and tidal forces as possible causes. (Recall that Darwin also made this tactical 'error' when he made a rough estimate of the age of an erosional episode, thereby providing a target for Kelvin to snipe at.)

By the time of his third edition (1922), which became better known in the English-speaking world, Wegener's theory began to generate very strong opposition. According to Menard [6], there were two reasons in particular that might have contributed to this. One was that Wegener (perhaps like most of us) had begun with the naive idea that his theory was so obvious that it would quickly be accepted. When this did not occur, and he observed palaeontologists failing to understand his argument against land bridges, he became more of an advocate. The other reason was that he believed that geodetic measurements showed a shift of Greenland relative to Europe, and that this was a dramatic confirmation of his theory. (The drift rate implied by the data was metres per year, but he had also correlated Pleistocene glacial moraines across the Atlantic that would have required this rate.) The data later turned out to be in error, but by this time he may have become totally convinced, and adopted a more evangelical approach. When some (but by no means all) of his arguments were found wanting (such as the correlation of moraines), his credibility dropped, and the annoyance of his detractors rose.

Prominent among the opponents of continental drift was Harold Jeffreys. Jeffreys showed that Wegener's proposed driving forces were many orders of magnitude smaller than would be required to overcome the resistance from the oceanic crust through which the continents were presumed to move. Even in 1926 Jeffreys' language reveals a reaction to Wegener's fervour. He parodied Wegener by accusing him of arguing that a small force acting for a very long time could overcome a much larger force acting for the same time, and characterised this idea as 'a very dangerous one, liable to lead to serious error' ([1], [9] p. 150).

Jeffreys' dismissal of Wegener's proposed mechanisms extended to the whole idea of drifting continents. Jeffreys' language reveals that Wegener's proposed forces merely provided a convenient weakness through which to attack the larger theory. It is not clear that Jeffreys made a serious attempt to appreciate Wegener's many geological arguments. He even attacked Wegener's geophysical argument against land bridges, a subject in which he should have been expert, but in which his arguments were inconsistent with the well-known observational basis of isostasy, as you will see in the next section. Thus we see again the process of alternating over-reactions generating a heated scientific debate, just as in the nineteenth century arguments over the age of the earth, and in many subsequent topics in many areas.

A primary source of opposition to Wegener's theory was the well-established view amongst geologists and geophysicists that continents are fixed in a strong outer shell of the earth. Compressive mountain building forces were supposed to derive from cooling and contraction of the earth, which generated compressive stresses and occasional failure in this shell, or lithosphere. This theory was attracting its own opponents, because it was far from clear that it could provide for sufficient crustal shortening to account for the major compressional mountain ranges.

Superficially the model of a cooling, contracting earth seems attractive, and very compatible with Kelvin's concept, upon which his cooling age of the earth was based (Section 2.4). However, when the earth is assumed to cool from the outside by conduction, the result is that the deep interior would not have had time to cool at all. The cooling is restricted to a gradually thickening layer at the surface. Initially the surface would be put in tension, as it contracted relative to the constant-volume interior. One has to assume that the resulting tension is relieved by failure, and then this layer would be compressed as the cooling penetrated deeper. The resulting amount of compression is much less than if the whole interior were cooling.

Anyway, Wegener's theory attracted substantial opposition, and for the next several decades it was not very respectable to advocate anything related to continental drift. It is interesting that Gutenberg, in an article written initially in the 1930s [10], reviewed an extensive literature of speculative tectonic theories, with continental drift prominent among them. Gutenberg had been an associate of Wegener's in Germany, before Gutenberg's move to the California Institute of Technology in the late 1920s. However Gutenberg gave much less space to continental drift in his later book, published in 1959 [11].

The only prominent advocates of continental drift in this period were Alex du Toit [12], in South Africa, and Sam Carey [13] at the University of Tasmania. These two geologists enjoyed two advantages. One was that some of the clearest geological evidence for past continental connections exists in the southern continents. The other was that, being located in the further reaches of the civilised western world, they could perhaps be safely ignored in the important centres of learning. du Toit, especially, contributed a great deal of evidence and elaborated Wegener's ideas significantly, and his work attracted a significant minority of followers. Carey also contributed important evidence and arguments, and is otherwise most noted for the radical idea that the earth has expanded substantially since the Palaeozoic, this being his preferred mechanism for continental drift. Carey's ideas were an important stimulus for Wilson (Section 3.4).

3.2 Creeping mantle

The idea of a deformable mantle began to have an empirical basis when it was discovered that the gravitational attraction of mountain ranges is less than would be expected from their topography. An account is given by Daly [14]. The deficit in gravitational attraction was first recorded for the Andes mountains by Bouguer, on an expedition between 1735 and 1745, and later, in 1849, by Petit near the Pyrenees. It was analogous observations arising from Everest's surveying in India that led to a quantified hypothesis.

This came from a discrepancy between two different surveying methods in India, near the Himalayan range, one based on triangulation and the other on astronomical sighting. The astronomical sightings were done relative to the local vertical, as determined by a plumb line, that is a weight hanging on a thread. In 1855 Pratt [15] proposed that the discrepancy arose because the vertical was deflected slightly by the gravitational attraction of the Himalayas. However his calculations revealed that the deflection was only about a third of what would be expected from the visible mountain range.

Explanations were offered later by Pratt [16] and by Airy [17]. Pratt noted that the discrepancy could be accounted for by a hidden excess mass to the south or by a hidden mass deficit to the north. He suggested that such differences in density might have arisen since the earth was young and liquid, but without changing the mass in any vertical column extending down from the surface, for example by differential thermal expansion. In this case different vertical columns of equal surface area would all still contain equal amounts of mass. As an illustration, he calculated that a small density deficit extending to a depth of 100 miles (160 km), and such as to make the mass column including the mountains the same as under the adjacent plains, could reduce the vertical deflection to zero.

Every geology student learns about isostasy, and about Pratt's and Airy's variations on how to distribute the density deficit under mountain ranges. What I had never appreciated until I read Daly's extensive quotation from Airy's short paper was how penetrating and far-reaching was Airy's thinking. He is famous for hypothesising what Dutton later called the condition of isostasy [18], but his thinking goes to the core of the subject of tectonic mechanism.

Rather than assuming that mass columns had remained constant through earth history, as had Pratt, Airy thought it was necessary to consider that the earth was subject to 'disturbing causes' through its history which would change both the topography and the mass within columns. He noted that the shape of the solid part of the earth closely approximates the shape of the liquid ocean surface, that there is not a concentration of land or water near the equator, and that both of these observations had been taken by physicists to indicate 'either that the interior of the earth is now fluid or that it was fluid when the mountains took their present forms'. He goes on

> This fluidity may be very imperfect; it may be mere viscidity; *it may even be little more than that degree of yielding which (as is well known to miners) shows itself by changes in the floors of subterraneous chambers at a great depth* when their width exceeds 20 or 30 feet [7 or 10 metres]; and this degree of yielding may be sufficient for my present explanation. [Emphasis added.]

Here, very clearly, is an empirically based concept of a solid, rocky, but deformable interior.

Airy therefore assumed an outer, non-deforming 'crust' and a denser, fluid interior. He argued first that a broad plateau could not be supported alone by the strength of the crust, demonstrating that the leverage required at its edges required a tensile strength that was very implausible, given that the crust is known to be riven with fractures, even if the crust is 100 miles (160 km) thick. He then asked how else such plateaus might be supported, and answered himself

> I conceive there can be no other support than that arising from the downward projection of a portion of the earth's light crust into the dense [substratum]; ... the depth of its projection downwards being such that the increased power of floatation thus gained is roughly equal to the increase of weight above from the prominence of the [plateau].

He compared the crust to a raft of timber floating on water, wherein a log whose top is higher than the others will be correctly inferred to be larger and thus to project deeper into the water than the others.

Airy then showed how the downward projection of the lower-density crust (the root, as it has become known) will reduce the net gravitational attraction, and that at a distance great compared with the depth of the projection the net gravitational perturbation will approach zero. He noted that one would not expect that there would everywhere be a perfect isostatic balance, but that the strength of the crust would allow some mountains to project higher or some roots to project deeper than in the isostatic condition. Finally he noted that this would be especially true of mountains of small horizontal extent, since the leverage required to hold them up is smaller.

In 1859 Hall [19] presented evidence of slow, continuous adjustment of the earth's surface to changing loads, by demonstrating that sediments now buried deep in thick sedimentary sequences were deposited in shallow water. This observation, and many others of its kind since, went far towards justifying Airy's assumption that the interior of the earth is fluid at present, and that the isostatic condition was not just a relic from early in earth's history.

By 1889 there was accumulating evidence that the crust on broad scales is close to isostatic equilibrium, and Dutton [18] formalised the idea and proposed the name isostasy (Greek: *isos*, equal; *statikos*, stable). (Dutton actually preferred the term isobary, or equal pressure, but this was already in use in another context.)

Helmert [20] conceived in 1909 that the depth of the compensating mass deficit could be constrained by the form and magnitude of the gravity anomaly at the edge of a broad structure, and he and others used observations near continental margins to deduce that the density anomalies extended to depths of the order of 100 kilometres. This implied that the non-deformable crust must extend to such depths, in order for the density anomalies to persist.

In 1914 Barrell [21] proposed the term 'asthenosphere' (weak layer) for the deformable region below the region of strength. By this time the term 'lithosphere' was in use to describe the non-deforming layer near the surface. This had been distinguished from the low-density compositional layer, the 'crust' in modern usage, by the discovery of the Mohorovičić discontinuity in 1909 [22], which was inferred to mark the base of the crust. Barrell was willing to assume that the thickness of the asthenosphere is as great as 600 km, in order to reduce the amount of deformation required to accommodate surface uplifts. This allowed him to argue that a deformable asthenosphere was not incompatible with the solid state, as shown by its ability to propagate seismic shear waves [23].

Thus by 1914 there was a clear picture, well based on observations, of a lithosphere about 100 km thick and strong enough, on geological time scales, to support topography up to a width of the order of 100 km. Topography on broader scales was known to be approximately in isostatic balance, including the earth's first-order topography, the continent–ocean dichotomy. It was inferred that this is because the asthenosphere, below the lithosphere, behaves like a fluid on geological time scales, in spite of being in the solid state.

A different kind of observation was developed through this period which strongly supported this picture, but there were two other kinds of observation that complicated it. The supporting observation was of a protracted 'rebound' of the earth's surface in the Fennoscandian region following melting of the glaciation from the last ice age. It was argued by Jamieson in 1865 [24] that this could be explained by a viscous outflow from under the icecap, with a return flow after the icecap melted. The delayed response, by more than 10 000 years, required more than just an elastic yielding, which would rebound immediately the ice load was removed. This hypothesis was debated for a long time, but by the 1930s well-founded estimates of the viscosity of the asthenosphere had been derived by several workers. The result obtained depends substantially on the assumed thickness of the asthenosphere. For example, van Bemmelen and Berlage [25] assumed a thickness of 100 km and derived a viscosity of 1.3×10^{19} Pa s, whereas Haskell [26] in 1937 assumed an essentially unlimited thickness and obtained a viscosity of 3×10^{20} Pa s.

The first of the complicating observations was the discovery that some earthquakes occur down to depths of nearly 700 km [27]. The second was that even on the largest scale the earth is not quite in hydrostatic equilibrium. A completely hydrostatic earth should have an equatorial bulge due to rotation, but it was found that the equator bulges by about 20 m more than this, and that the equator itself is not uniform, bulging more in some longitudes than in others. Significant stress is required to support these bulges.

Jeffreys [9], whose classic work demonstrated the existence of the bulges, argued that these and the deep earthquakes required the interior to have substantial strength, by which he meant that it could not be deformable over geological time scales. An alternative explanation of the excess bulges is that they are supported by stresses in a fluid mantle, which implies that the mantle would be in sustained internal motion. However this possibility does not seem to have been seriously advocated until it was taken up by Runcorn in 1962 [28]. In 1969 Goldreich and Toomre [29] argued further that the variations around the equator were not consistent with the previously preferred explanation that the 'equatorial bulge' was frozen in from times when the earth's rotation was faster, and they demonstrated that bulges generated by internal fluid motions would cause the earth to tilt so as to bring the largest bulges to the equator.

Jeffreys also argued that the approximate isostatic balance of mountain ranges was due to the fracturing of the crust by the tectonic forces, and subsequently by secondary gravitational (buoyancy) forces induced by the (supposed) resulting topography. He drew attention to the distinction between the strength of unfractured rock and the much lower strength of fractured rock. He supposed that it was the tectonic forces that first fractured the rock, and that the strength implied by remaining isostatic imbalance is a measure of the strength of fractured rock.

Daly ([14] p. 400) disputed Jeffreys on several grounds. He pointed out that Jeffreys' hypothesis could not account for slow isostatic adjustment away from mountain belts in response to erosion and sedimentation, nor for observed continuing adjustment to deglaciation. Daly also noted experiments by Bridgman that had shown that fractures healed quickly at high pressures. As well, we can note the internal contradiction in Jeffreys' argument that the remaining isostatic imbalance should still have reflected the strength of unfractured rock: any unfractured parts could still be out of equilibrium, and it would have been necessary to overcome the unfractured strength in order to bring them closer to balance.

Daly's ideas deserve more recognition. His thinking was wide-ranging and adventurous, and he came remarkably close to some modern concepts. The evidence at the time appeared contradictory, but I see Jeffreys' attempts to resolve the contradiction as limited and superficial in comparison with Daly's. Not all of Daly's ideas were well-based. For example he regarded the asthenosphere as being in a vitreous (glassy) rather than a crystalline state, despite his evident awareness of Airy's point that crystalline rocks were known to deform in deep mines, and despite his colleague Griggs' experiments on rock deformation [30]. He proposed to explain the large-scale bulges of the earth by supposing that below the asthenosphere is a 'mesosphere' of greater strength, though this neglects to explain how stresses maintained in a strong mesosphere would be transmitted through the asthenosphere to the surface.

Admitting that he was indulging in conjecture, Daly offered several suggestions to explain the occurrence of deep earthquakes. He proposed that the asthenosphere is heterogeneous, being strong enough to bear brittle fracture in some places. He proposed ways that this might come about, the most interesting being that blocks of lithosphere might founder and sink through the asthenosphere. Furthermore, noting that suddenly imposed stresses might induce fracture even in the deformable asthenosphere, he suggested that pressure-induced phase transformations, of the kind recently observed by Bridgman, in such sinking blocks might be a suitable trigger. This is an idea still very seriously entertained.

Daly proposed that the foundered or 'stoped' lithospheric blocks could plausibly originate during compressional mountain building:

> mountain making of the Alpine type seems necessarily accompanied by the diving of enormous masses of simatic, lithospheric rock into the asthenosphere. Thus the belt under the growing mountain chain is chilled by huge, downwardly-directed prongs of the lithosphere, as well as by down-stoped blocks. (p. 406.)

He noted that this would explain the occurrence of deep earthquakes 'under broad belts of recent, energetic orogeny'. This picture of the lithosphere, including 'prongs' projecting down under zones of compression, is remarkably close to the modern picture of a subduction zone, which we will get to later.

To summarise the evidence for a creeping mantle, gravity measurements established that mountains are close to an isostatic balance, and observations of associated sedimentary sequences showed that there are slow and continuous adjustments of the earth's surface to changing loads. Observations of post-glacial rebound of the earth's surface supported this inference and yielded quantitative estimates of the viscosity of the mantle. Observation of non-hydrostatic bulges were at first taken as evidence for a rigid interior, but were later reinterpreted as indicating a fluid interior with a viscosity comparable to that inferred from post-glacial rebound. Deep earthquakes remained a puzzle, but Daly conjectured that the asthenosphere in which they occur is abnormal, and that the abnormalities might be associated with the active mountain belts that overlie them.

3.3 A mobile surface – re-emergence of the concept

Having set the scene for mobility in the earth's interior, I will now turn to the surface again, to describe how the surface came to be viewed as moving, the conception of moving rigid plates, and the strong evidence supporting this idea.

Although continental drift was not entirely ignored after about 1930 [10], it was certainly very unfashionable and was dismissed by many geologists, often with some passion. Against this, it was widely recognised that a really satisfactory theory of mountain building did not exist. The old idea of a contracting earth did not seem to provide for sufficient contraction to explain the observed crustal shortening, nor for zones of extension, without *ad hoc* elaborations of the theory. Expansion of the earth was proposed by a few people, and occasionally mantle convection was appealed to in contexts other than continental drift. Most geologists worked on narrower problems, and little progress was made on the question of fundamental mechanism, despite much conjecture [6].

This situation prevailed until about the mid-1950s, at which time two new kinds of evidence began to emerge that raised questions so serious they were harder to ignore. One kind of evidence was from palaeomagnetism, the other from exploration of the sea floor.

When a rock forms, it can record the direction of the local magnetic field, because any grains of magnetic minerals incorporated into the rock tend to align with the field like a compass needle. Collectively these grains then produce a small magnetic field that may be measurable in the laboratory. If a sufficiently large body of rock is magnetised in this way, the effect may be measurable in the (geological) field as a detectable perturbation of the earth's magnetic field.

Three distinct questions have been addressed through measurements of rock magnetism. First, have the rocks moved around on the earth's surface? Second, has the magnetic field changed through time? Third, can rock magnetism be used to map the sequence of formation of rocks, or to date their formation? The second and third questions will be discussed in Section 3.5.

The first question was pursued by British geophysicists in the 1950s, with a view to testing for continental drift. There were many complications to be dealt with, such as being sure that the original orientation of the rock could be reliably established and separating magnetisations acquired by the rock at different times through different microscopic mechanisms. There were also the possibilities that the magnetic field had not always been approximately aligned with the earth's spin axis, that it had not always been approximately dipolar, as at present, and that the earth had tilted relative to the spin axis.

By the late 1950s, these difficulties had been substantially overcome and strong evidence was emerging that North America and Europe had been closer together in the past [31], and that Australia had moved northward from near the south pole [32]. For those with knowledge of and confidence in the palaeomagnetic data, this was strong evidence that continental drift had occurred. However, the difficulties of the method were well-known, and it was hard for all but the minority involved in the measurements to know how much confidence to put in them. Nevertheless, these data were very influential in reinstating continental drift as a respectable scientific topic.

The second important kind of evidence came from exploration of the sea floor, which increased greatly during and after World War II. An intimate and insightful account of this work was given by Menard [6]. One of the early and most startling discoveries was the absence of thick sediment on the sea floor. If the continents and oceans were permanent features, there should have been a continuous sedimentary record of most of earth history, but few rocks older than the Mesozoic were found on the sea floor, and those had affinities suggesting they are fragments of continents. Through the decade of the 1950s, the global extent of the 'midocean ridge' system was revealed, along with great 'fracture zones' on the sea floor. Fracture zones are narrow scars having the appearance of great faults thousands of kilometres long. Vast areas of the sea floor, where it was not covered with thin sediment, comprised monotonously rough 'abyssal hills' whose origin was unknown.

'Guyots' were found over a broad area of the central Pacific. Guyots are submarine mountains with flat tops. They were presumed to be of volcanic origin, and were and still are interpreted as former islands whose tops were eroded to sea level and which subsequently subsided below sea level. Both Hess [33, 34] and Menard [6, 35] inferred the former existence, about 100 Ma ago, of a midocean rise that has now subsided. Menard called it the

'Darwin Rise', in honour of Charles Darwin's correct explanation for the formation of coral atolls upon such drowned islands.

It was found that the heat flux conducted through the sea floor is as large or larger than on continents, despite the continental crust having a considerably higher content of radioactive heat sources.

Many of these discoveries were quite unexpected and difficult to make sense of. We must realise that the area being explored was vast, and that the picture was at first very patchy and incomplete. Nevertheless it was clear that old ideas had to be revised in major ways. The fracture zones are uniquely long and linear features, and it is hard to interpret them as anything other than strike-slip faults with large displacements, but they seem to disappear at continental slopes and have no obvious extension into the continents. The thin sediment covering on the sea floor required either that the rate of sedimentation had been very much less in the past than at present or that the sea floor is no more than about 200 Ma old. The abyssal hills topography looks chaotic, suggesting widespread tectonic disruption but the sediments overlying them on the older sea floor are largely flat-lying and undisturbed.

The relationship of fracture zones to midocean rises, if any, was unclear. In the north-east Pacific several major fracture zones connect to nothing obvious at either end. In the east they run up to the edge of the continent and appear to stop, while in the west they peter out. In the Atlantic, the rough topography and mostly east–west surveys left the picture confused, with Heezen [36] inferring that east–west troughs were part of a continuous graben on the ridge crest. Only later were they interpreted as fracture zones offsetting the ridge crest.

The origin of the midocean rise system was obscure. Where it was traced onto land in Iceland and East Africa, it was undergoing extension. This was consistent with the presence of an axial trough along much of the crest of the Mid-Atlantic Ridge, and Heezen inferred that the entire system of rises was extensional [36]. However, for some time seismic reflection data seemed to show a covering of sediment over the East Pacific Rise, and Menard inferred that it might be young and had not yet begun active rifting. Menard and Hess inferred that rises are ephemeral, and Menard proposed that the East Pacific Rise is young, that the Mid-Atlantic Ridge is mature, with active rifting, and the Darwin Rise is extinct.

Menard and Hess proposed variations on ephemeral convective upwellings to explain the existence of the rises. Heezen had traced the Mid-Atlantic Ridge around Africa and into the Indian Ocean and had inferred that it is all extensional. He reasoned from this that the earth had to be expanding, otherwise Africa would be undergoing active compression because of being squeezed from both sides by the extending ridges. The idea that the sea floor is or has been mobile was implicit in the interpretation of ridges and fracture zones. The uniformity of the sea floor and the absence of widespread evidence of deformation of sediments suggested that large areas of it were moving coherently. For example, Menard thought that the pieces between fracture zones moved independently, driven by separate convection 'cells'.

I recount these things to give some flavour of the ferment of ideas that was induced by the new kinds of observations. These were so puzzling, especially while they were incomplete, and sometimes misleading, that people were willing to appeal even to such disreputable ideas as mantle convection or earth expansion. Menard ([6] p. 132) makes the points, however, that most geologists at the time were busy with other things and unaware of or unconcerned with the sea floor, and that the oceanographers' research also was 'narrow, mostly marine geomorphology, but the areas were hemispheric and the conclusions correspondingly grand'.

That is the context in which two people proposed a third type of explanation for the midocean rises. Not earth expansion and not ephemeral convection cells, but continuous convection, coming right to the earth's surface at ridge crests and descending again at deep sea trenches. Hess wrote his paper in 1960, but it was not published until 1962 [34], while Dietz's paper was written and published in 1961 [37]. Menard argues persuasively that their work was independent ([6] Chapter 13).

Hess and Dietz accepted Heezen's arguments that the midocean ridges are extensional rifts, but they did not accept his conclusion that the earth expands. Hess had a long-standing interest in ocean trenches. Vening Meinesz [38] had measured gravity at sea in submarines, and found large negative gravity anomalies over trenches that he attributed to a down-buckling of the crust where two mantle convection currents converged. He developed a 'tectogene' theory that trenches were the early stages of geosynclines where thick sediments accumulated, later to be thrust upward in association with volcanic activity. Dietz also had an interest in geosynclines, arguing in later papers that they represent former passive continental margins that are activated by subduction. Thus both Hess and Dietz were disposed to the idea of crustal convergence and descending convection at trenches.

The central ideas that have survived from these papers are that convective upwelling of the mantle reaches the surface in a narrow rift at the crest of midocean rises and forms new sea floor. This then drifts away on both sides of the rift, ultimately to descend again into the earth at an ocean trench. A continent can be carried passively by the horizontal part of the convection flow, rather than having to plough through the sea floor, as supposed by Wegener. The youth of the sea floor and the thinness of sediments would be accounted for. A uniformly thick crust might be formed, if it is all formed by the same process at a ridge crest. Dietz recognised that the abyssal hills topography might also be a residue of rifting at the ridge. The high heat flow on ridges would be explained by the close approach of hot mantle to the surface. Dietz coined the concise term 'seafloor spreading'.

Not all of the ideas from Hess's paper have survived. For example, the composition of the oceanic crust was not definitely known at the time, and he supposed it to be serpentine (hydrated mantle peridotite), whereas Dietz more correctly assumed it to be basalt produced by melting the mantle under the ridge. Hess still thought ridges were ephemeral, being misled by the assumption that the Darwin Rise was of the same type as the modern midocean ridges. The Darwin Rise loomed large in Hess's thinking, because of his discovery of guyots. Some have viewed guyots as a key link to the idea of seafloor spreading, e.g. Cox [3], but I think they distracted him into thinking more about vertical motions than horizontal, and his thinking was still a bit confused in this paper. Hess did not think fracture zones were related to ridges. Dietz did, and he proposed that the convection proceeded at different rates on either side of a fracture zone, so the sea floor is displaced by different amounts.

Hess made another important, though somewhat separate point in his paper: that continents would be piled up by convection and also eroded down towards sea level. The consequence would be that the level of the continents would be near sea level, and this would be the result of a dynamic equilibrium between the piling up and the erosion. Thus he correctly recognised the explanation for the bimodal distribution of the elevation of the earth's surface that had been an important argument of Wegener's.

It may seem curious that Hess's and Dietz's papers became famous for proposing seafloor spreading, but not for the complementary removal of sea floor at trenches, which was an integral part of their concept. The reason is probably that the understanding of trenches and their associated mountains (island arcs or active continental margins) was in a state of confusion at the time, and as a result neither of them put much stress on what we now call subduction. Although there was a widespread concept that trenches were the sites of compression and some downward buckling (Vening Meinesz [38]) or faulting (Benioff [39]) the amount of crustal motion envisaged was usually limited. As well, attempts to determine the direction of slip in earthquakes from seismic waves were yielding confusing and inconsistent results. It was not until after a world-wide network of standardised seismographs was in place in about 1963 (to monitor underground nuclear explosions) that clear results of this type emerged. However the confusion did not hinder Wilson, as you will see in the next section.

This account of seafloor spreading has been expressed very much in terms of mantle convection, because that is how Hess and Dietz conceived it. You will see in the next section that there are advantages in looking just at the surface of the earth, without worrying about what is happening underneath. However, the question of how mantle convection relates to the surface becomes more acute as the surface picture is clarified. Already with seafloor spreading there is the novel idea that mantle convection rises right to the earth's surface, but only in a very narrow rift zone at the crest of a midocean ridge. This is a novel form of convection. As you will see later, Holmes had proposed a picture rather similar to that of Hess and Dietz, even to the point of having a regenerating basaltic oceanic crust, but his concept was conditioned by conventional ideas about convection, and he supposed that seafloor extension occurred over a broad region.

3.4 Wilson's plates

J. Tuzo Wilson was a physicist turned geologist. He is best known for recognising a new class of faults, and for naming them 'transform faults', in a paper published in 1965 [40]. This paper is widely recognised as a key step towards the formulation of plate tectonics. It is more than that. It is in this paper that the concept of plate tectonics first appears in its complete form.

Wilson's paper is called 'A new class of faults and their bearing on continental drift'. It is worth quoting the opening of the paper.

> Many geologists [41] have maintained that movements of the earth's crust are concentrated in mobile belts, which may take the form of mountains, midocean ridges or major faults with large horizontal movements. These features and the seismic activity along them often appear to end abruptly, which is puzzling. The problem has been difficult to investigate because most terminations lie in ocean basins.
>
> This article suggests that these features are not isolated, that few come to dead ends, but that they are connected into a continuous network of mobile belts about the Earth which divide the surface into several large rigid plates [(Figure 3.1)]....

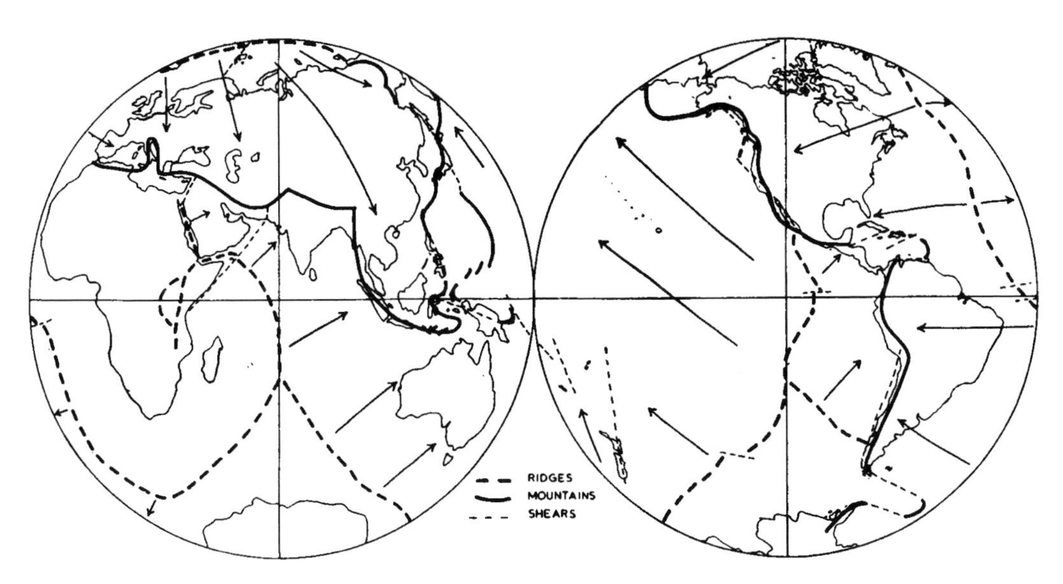

Figure 3.1. Wilson's sketch map from 1965 [40]. The original caption is as follows. 'Sketch map illustrating the present network of mobile belts around the globe. Such belts comprise the active primary mountains and island arcs in compression (solid lines), active transform faults in horizontal shear (light dashed lines), and active midocean ridges in tension (heavy dashed lines).' Reprinted from *Nature* with permission. Copyright Macmillan Magazines Ltd.

Others might have got tantalisingly close, but here, in four of the most pregnant sentences in all of geology, Wilson has defined the problem and presented its solution with simple clarity. His sketch map (Figure 3.1) gave the world its first view of the tectonic plates.

Wilson had very broad interests in geology, but he had been studying in particular large transcurrent faults. It was his recognition of the North American equivalent (the 'Cabot fault') of Scotland's Great Glen fault that first aroused his interest in continental drift [42]. He was also puzzled by the great fracture zones that were being discovered on the ocean floor, because they seemed to be transcurrent faults of large displacement, but they stopped at the continental margin, with no equivalent expression on the adjacent continent. He actually had not believed in continental drift until about 1960, but the publication of Dietz's seafloor spreading paper in 1961 convinced him that it must be right and he set about finding more evidence from the ages of oceanic islands (see Section 3.7).

Wilson's clinching insight was his recognition of the way these great faults can connect consistently with midocean ridges or with 'mountains' (meaning island arcs or subduction zones) if pieces of the crust are *moving relative to each other as rigid blocks* without having to conserve crust locally. Continuing the above quotation,

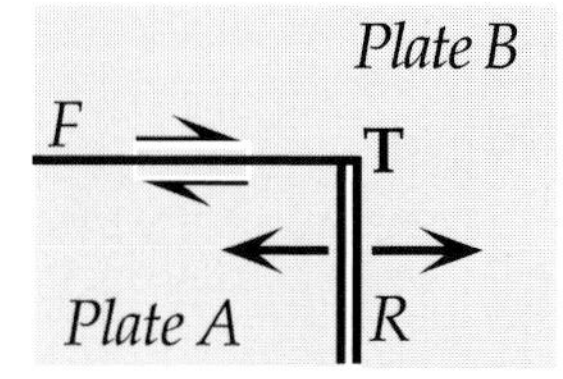

Figure 3.2. Illustration of Wilson's idea of how a transform fault (*F*) is transformed at the point **T** into a midocean ridge spreading centre (*R*). The concept depends on the pieces of crust on either side of *F* and *R* moving independently as rigid blocks or plates without requiring the area of each plate to be conserved locally.

Any feature at its apparent termination may be transformed into another feature of one of the other two types. For example, a fault may be transformed into a midocean ridge as illustrated in [Figure 3.2]. At the point of transformation the horizontal shear motion along the fault ends abruptly by being changed into an expanding tensional motion across the ridge or rift with a change in seismicity.

'... with a change in seismicity'? I'll return to that.

Wilson explains how his 'transform' faults may connect a ridge to a trench, or to another ridge segment, or may connect two trenches. He points out the crucial properties that transform faults may grow or shrink in length as a simple consequence of symmetric ridge spreading and asymmetric subduction, that the sense of motion on a transform fault joining two ridge segments is the reverse of the superficial appearance (Figure 3.3), and that the traces left by such faults beyond the ridge segments they connect are inactive. He does a fast tour of the world, explaining relationships between major structures, explicating what we now know as plate boundaries.

The language of the paper is terse. One senses the excitement of the rush of insights as pieces of a puzzle (literally) fall into place, and the desire to pack as much as possible into a short, crucial paper. Key information is almost lost. He forgets to spell out that it was known that the only seismically active parts of the great fracture zones cutting across the equatorial Atlantic sea floor are the parts between the ridge segments [43] (Figure 3.3), and that this was a major puzzle. That information appears only in the caption of his

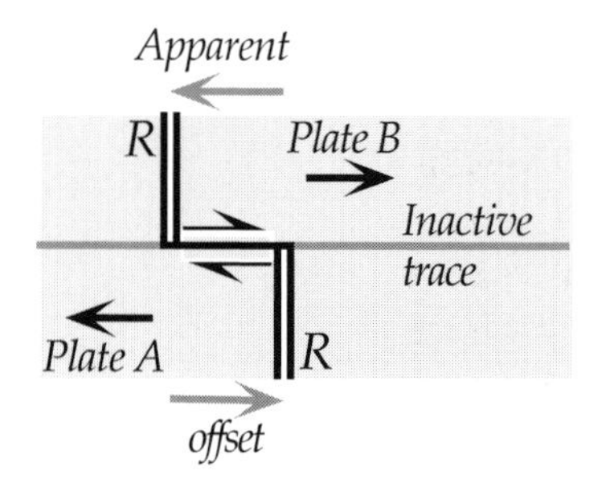

Figure 3.3. Illustration of the distinction between Wilson's 'transform fault' interpretation of ridge segments offset by a fracture zone and the 'transcurrent fault' interpretation. In Wilson's interpretation, the sense of motion across the active transform fault joining two ridge segments is right lateral in this example (looking across the fault, the other side moves to the right). In the transcurrent interpretation, the apparent offset is left lateral. Also in Wilson's interpretation the extensions of the fracture zone beyond the ridges are inactive scars within the plates, whereas in the transcurrent interpretation the extensions would also be active. Wilson noted the crucial observation that earthquakes occur mostly on the segment connecting the ridges, and only infrequently on the extensions.

sketch map, and ambiguously, where he distinguishes active faults as solid lines and 'inactive traces' as dashed lines, without making clear that this had already been observed, and was not just a prediction of his theory. The cryptic 'with a change in seismicity' noted above means that the type of earthquake changes from strike-slip to normal faulting where a transform fault joins a ridge segment.

Wilson was thinking as a structural geologist, and that was crucial. He envisaged rigid blocks bounded by three types of boundary that correspond to the three standard fault types: strike-slip (transform fault), normal (ridge) and reverse (subduction zone). Conceptually he narrowed the old notion of mobile belts down to sharp boundaries, and he explicitly adopted the long-standing implication of that old term, that there is little deformation outside the mobile belts, taking it conceptually to the limit of proposing that there is *no* deformation. He was explicit in the fourth sentence, quoted above, that the plates are 'rigid'. The point is explicit also within the paper: 'These proposals owe much to the ideas of S. W. Carey, but differ in that I suggest that the plates between the mobile belts are not readily deformed except at their edges.'

This was why Wilson was able to see the plates in all their simplicity. A unique and crucial feature of mantle convection, as distinct from other forms of convection, is that part of the medium behaves as a viscous fluid and part as a brittle solid, as I will explain in later chapters. This had been a source of confusion in attempts to formulate and relate ideas about continental drift, seafloor spreading and mantle convection. This can be seen for example by contrasting Holmes's concept of a new ocean in which there is broad deformation across the sea floor, reflecting the behaviour of a viscous fluid, with Hess's and Dietz's narrow spreading centres and the angular, segmented geometry of midocean ridges. By focussing on the motions that can be discerned at the surface, Wilson recognised the behaviour of a brittle solid, and successfully defined plate tectonics in those terms.

There can be no doubt that Wilson was aware of the implications of his new structural concepts for continental drift, justifying the second part of his title '... and their bearing on continental drift'. That is explicit in his explanation of the transform concept and in the last sentence of the paper where, referring to transform faults, he says 'proof of their existence would go far towards establishing the reality of continental drift and showing the nature of the displacements involved.' Perhaps too modestly, he implies here that he has not already pointed out compelling evidence, in the form distribution of earthquakes on fracture zones and implicitly in the wealth of geological and seismological evidence that had given rise to the concept of mobile belts and the complementary idea of internally stable blocks.

Contrast Wilson's paper with a little-known paper by Coode [44], also published in 1965. In this very brief note, Coode elegantly presents the conception of a ridge-ridge transform fault, along with a diagram explaining how both the ridge crest and magnetic anomalies (next section) are offset. That is all Coode does. The further implications are not developed. The paper was almost unknown until it was pointed out by Menard [6] (though this was also because it was in a journal where oceanographers were unlikely to see it).

There can be no doubt also that Wilson appreciated that he had taken a major step towards a unifying dynamic theory of the earth that would probably involve mantle convection. Two years earlier he had published several papers containing the fruits of another remarkable burst of creativity, including the seminal insight that led to the idea of mantle plumes [45, 46], and a wide-ranging article in *Scientific American* on continental drift [42]. In the latter it is clear that he has a comprehensive grasp not only of a large number of geological observations but also of the arguments from isostasy, post-glacial rebound, materials science and gravity observations over ocean trenches that the mantle is deformable and undergoing convection. His map of convection currents bears a strong resemblance to his 1965 map of the plates, and he writes of moving crustal blocks.

Reading the 1965 paper, we may see a structural geologist presenting a brilliant and novel synthesis. Reading it in conjunction with the 1963 papers, we see more: a scientist in the full pursuit of the secrets of the earth, chasing whatever kind of evidence will serve. Reading them all, I see a man move, in little more than five years, from first conversion to mobilism through to clarity of understanding of geology's major unifying concept.

In frankly championing Wilson, I do not wish to detract from the contributions of many others. I just think that his grasp of what he was doing has not been fully appreciated, perhaps because his 1965 paper is so terse, concentrates necessarily on the novel technicalities of transforms, and its title does not fully portray the unity and simplicity of his concept. I think he deserves a special place in the pantheon of geology for being the first to see the plates in complete and simple form.

I will explain further what I mean, so as to avoid unnecessary confusion. I take the essence of plate tectonics to be the concept of rigid, moving pieces of the earth's surface meeting at three kinds of boundary. I distinguish this two-dimensional concept, which can be displayed on a map, from the three-dimensional concepts of thick lithosphere and of mantle convection; these have continued to be debated and refined without detracting from the plate concept. Wilson's concept was not confined to planar geometry. Although Wilson sketched the transform concept in planar maps, that idea transfers completely to a sphere, because it involves the relationships of boundaries meeting at a point (Figure 3.2). There is no doubt that Wilson was thinking of rigid plates on a sphere. I also distinguish the *concept* from its quantitative, mathematical *description*. The idea of using Euler's theorem of rotation to describe the motions of plates on a sphere (Section 3.6) was powerful and productive, but it was a quantification of the *pre-existing* idea of moving, rigid, spherical plates.

Comparing the plate-tectonic revolution to the Copernican revolution in his preface to a collection of *Scientific American* articles [47], Wilson made the following observation.

> That the earth is the centre of the universe and that it rests on a fixed support was the obvious and early interpretation. To realise that the earth is spinning freely in space and that the sun, and not the earth, is at the focus of the solar system required a prodigious feat of imagination.... Changing the basic point of view created a new form of science with a different frame of reference. It was this change in the manner of interpreting the observations that constituted the scientific revolution.

Though others were close, both before and after, I think Wilson was the first to complete the change in point of view. Once Wilson had stood upon the far shore, it was easier for others, knowing it was there, to follow.

3.5 Strong evidence for plates in motion

3.5.1 Magnetism

About 1960 studies of palaeomagnetism began to focus strongly on the second question posed earlier (Section 3.3): has the earth's magnetic field changed through time? Specifically, has it reversed polarity? Matuyama, in 1929 [48], had studied the magnetisation through a sequence of lava flows erupted by a Japanese volcano. He found that the younger flows near the top were magnetised parallel to the present earth's magnetic field lines, but that the older flows near the bottom of the sequence were magnetised in the opposite direction. During the 1950s the question of whether this was due to reversal of the earth's field or to a peculiar response of some rocks was vigorously debated [3]. It seemed that there may have been many reversals of the earth's field, but this was difficult to demonstrate convincingly. From 1963 two groups in particular used a combination of magnetisation measurements and potassium–argon dating to try to resolve the question and to establish a chronology of reversals. These groups were at the U.S. Geological Survey in Menlo Park, California [49], and at the Australian National University in Canberra [50]. They found that the ages of normally and reversely magnetised rocks correlated around the world, which supports the idea that the earth's field had indeed reversed. By about 1969, the time sequence of reversals was established with some detail to an age of about 4.5 Ma, beyond which the K–Ar dating method did not have sufficient accuracy [51].

Meanwhile Ron Mason, of Imperial College, London and the Scripps Institute of Oceanography in California, was trying to identify magnetic reversals in oceanic sedimentary sequences. Because of this work, but still almost by chance, a magnetometer was towed behind a ship doing a detailed bathymetric survey off the west coast of the U.S. ([6], p. 72). From this magnetic survey there emerged a striking and puzzling pattern of variations in magnetic intensity: alternating strips of the sea floor had stronger and weaker magnetic field strengths [52]. The pattern was parallel to the local fabric of seafloor topography, and later was found to be offset by fracture zones, by about 1000 km in the case of the Mendocino fracture zone. It was presumed that the pattern might be explained by strips of sea floor with differing magnetisations, but its origin was obscure. In retrospect it was unfortunate that in this area the ocean spreading centre at which the sea floor formed no longer exists, and so there was no obvious association with midocean ridges.

In subsequent surveys, elsewhere, it was found that ridge crests have a positive magnetic anomaly (meaning merely that the field strength is greater than average), which some people presumed to indicate 'normal' (i.e. not reversed) magnetisation. Beyond this, on either side of the ridge crest, there was a negative (i.e. weaker than average) anomaly. In 1963 Fred Vine, then a graduate student at Cambridge University, was analysing the results of one such survey over the Carlsberg Ridge in the Indian Ocean. He noticed that the seamounts near the ridge crest were reversely magnetised. This is easier to infer for seamounts, because they are more like point sources and produce a more distinctive three-dimensional pattern of anomalies, whereas a long strip of sea floor produces a two-dimensional pattern that is more ambiguous. While his supervisor

Drummond Matthews, who had collected the data, was away, Vine conceived an explanation for the magnetic stripes ([6], p. 219).

Lawrence Morley in Canada was involved in aeromagnetic surveys over Canada, and was familiar with many aspects of geomagnetism. In his seafloor spreading paper, Dietz had commented on how the magnetic stripes off the western U.S. seemed to run under the continental slope, and had suggested that they were carried under the continent by subduction and destroyed by subsequent heating. It is well known that magnetisation does not survive if rocks are heated. Conversely, it is reacquired by magnetic materials upon cooling. Morley realised that the oceanic crust could be magnetised as it formed and cooled at a spreading ridge ([6], p. 217).

What has become known as the Vine–Matthews–Morley hypothesis combines the hypotheses of seafloor spreading and magnetic field reversals. The idea is that oceanic crust becomes magnetised as it forms at a spreading centre, and a strip of sea floor accumulates that records the current magnetic field direction (Figure 3.4a). If the magnetic field then reverses and the seafloor spreading continues, a new strip will form in the middle of the old strip (Figure 3.4b), the two parts of the old strip being carried away from the ridge crest on either side. Subsequent reversals would build up a pattern of normal and reverse strips, and the pattern would be symmetric about the ridge crest (Figure 3.4c).

Vine later commented that the hypothesis required three assumptions each of which was, at the time, highly controversial: seafloor spreading, magnetic field reversals, and that the oceanic crust (the seismic 'second layer') was basalt and not consolidated sediment ([6], p. 220). Morley submitted a paper about the beginning of 1963, which was rejected by two journals in succession, the second with unflattering comments. Vine and Matthews submitted

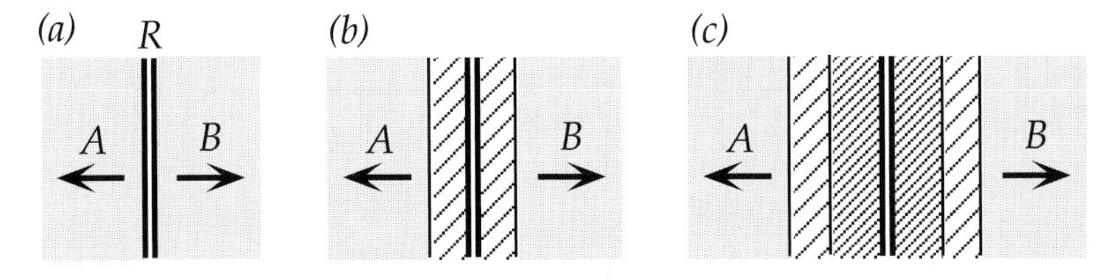

Figure 3.4. Illustration in map view of how seafloor spreading and magnetic field reversals combine to yield strips of sea floor that are alternately normally and reversely magnetised. The resulting pattern is symmetric about the crest of the ridge if the spreading itself is symmetric (meaning that equal amounts of new sea floor are added to each plate).

a paper in about July 1963, which was published in September [53]. Morley's story emerged later [3, 5, 6].

Subsequent exploration revealed extensive patterns of magnetic stripes on the sea floor, with an astonishing degree of symmetry about ridge crests (Figure 3.5), and which correlated with the field reversal chronology established on land [54, 55]. These magnetic stripes provided strong and startling evidence in favour of seafloor spreading. They also opened the prospect of assigning ages to vast areas of the sea floor on the basis of the reversal sequence, which was rapidly correlated from ocean to ocean [56]. Thus was the third question addressed through rock magnetism (Section 3.3) answered with a resounding yes: rocks can be dated using rock magnetism.

We should reflect on the magnitude of that last paragraph. Assigning ages to rocks always has been and still is a central occupation of geologists. It is painstaking work, whether the method is correlation of fossils or measurement of radioactive decay. It has taken much of this century to develop the ability to get reliable ages accurate to within a small percentage or less for many kinds of rocks. As Menard remarked ([6], p. 212)

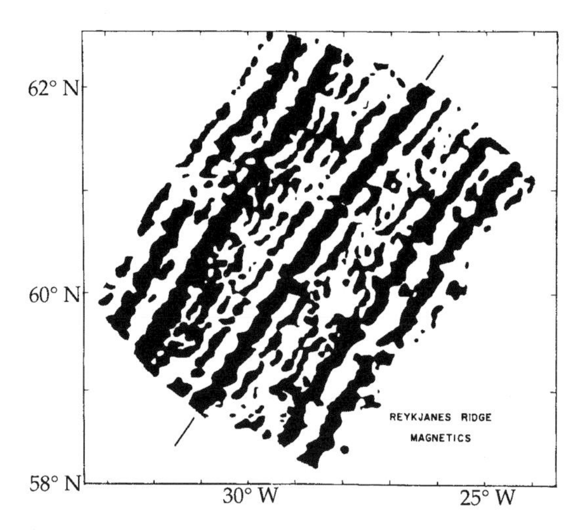

Figure 3.5. The pattern of magnetic anomalies across the Mid-Atlantic Ridge south of Iceland, where it is known as the Reykjanes Ridge. Black indicates a positive anomaly, inferred to be due to normally magnetised crust, and white indicates a negative anomaly, inferred to be due to reversely magnetised crust. The short lines mark the location of the Ridge crest, along which there is a positive anomaly. Despite the irregularities, the pattern shows a striking symmetry about the Ridge crest. From Heirzler *et al.* [57]. Copyright by Elsevier Science. Reprinted with permission.

To general astonishment, magnetic reversals provide the long-sought global stratigraphic markers that are revolutionising most of geology. At sea, as though by a miracle, magnetic anomalies give the age of the sea floor without even collecting a sample of rock.

3.5.2 Seismology

Seismology had already provided a key piece of evidence even before Wilson and Coode conceived of transform faults, as will be reiterated shortly. The Lamont (now Lamont–Doherty) Geological Observatory of Columbia University in New York state, directed by Maurice Ewing, had pioneered the exploration of the Atlantic sea floor, and then of other oceans. After Dietz's paper on seafloor spreading, Ewing turned much of the effort to testing the hypothesis. Part of this programme was to study the earthquakes in oceanic regions, and it was already known that these occur mainly on midocean ridges. By 1963 there was a better distribution of modern seismographs around the world, including the World-Wide Standardised Seismograph Network already mentioned. This permitted earthquakes in remote regions to be located with an accuracy about ten times better than previously.

Lynn Sykes, working at Lamont, found that the earthquakes are located within a very narrow zone along the crests of midocean ridges, and along the joining segments of fracture zones, where these were known or could be inferred [43, 58]. He made the explicit point that earthquakes on fracture zones occur predominantly on the segments that connect segments of ridge crest, and hardly at all on segments beyond ridge crests (Figure 3.6). This had been very puzzling when it was thought that fracture zones had offset ridges by motion along the length of the fracture zone. However it was explicitly predicted by the transform fault concept, and was noted (barely) by Wilson as evidence in its favour (Figure 3.3).

When Sykes saw the evidence of his colleagues for symmetric magnetic anomalies, he was convinced of seafloor spreading, but realised that he could make another decisive test through seismology. The elastic waves emitted by an earthquake have a distinctive four-lobed pattern. In two opposite lobes, the waves that arrive first are compressional. In the intervening two lobes the 'first arrivals' are dilatational. These waves spread through the earth's interior in all directions. With a global distribution of seismographs, it is possible to sample these waves with sufficient density to reconstruct the orientation of the lobed 'radiation pattern' and the orientation of its two 'nodal planes'. One of these planes corresponds to the fault plane, and the other is perpendicular, though you can not tell

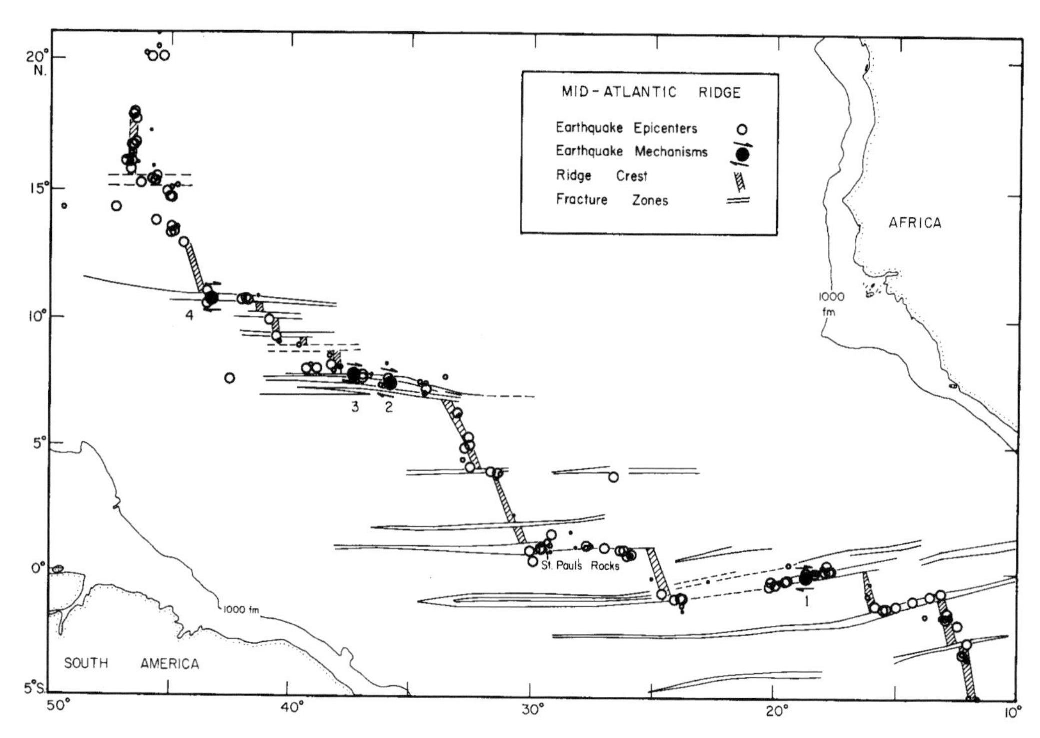

Figure 3.6. Earthquakes along the Mid-Atlantic Ridge. Open symbols show locations (epicentres) of earthquakes, and solid symbols with arrows show the sense of slip inferred from fault plane solutions. The fracture zones (oriented east–west) have earthquakes mainly between segments of the ridge crest and only rarely on the extensions beyond ridge crests. These locations and the sense of slip on the active segments are consistent with Wilson's transform fault hypothesis. From Sykes [59]. Copyright by the American Geophysical Union.

which is which just from the seismic waves. It is also possible to infer directly the orientation of stresses at the earthquake source. The result of this determination was called a 'fault plane solution'.

Sykes knew that some previous fault plane solutions on ridges were suggestive, but that he could get much more reliable results from the new global seismographic network. This he did [59]. He found that for earthquakes located on segments of ridge crest, the solutions indicated normal faulting, consistent with the ridge crest being extensional. Earthquakes located on active segments of fracture zones had one nodal plane approximately parallel to the fracture zone, consistent with strike-slip faulting (Figure 3.6). Most importantly, the sense of strike-slip motion was consistent with that predicted by the transform fault hypothesis, and opposite to that predicted by the simple transcurrent offset interpretation. This was another kind of observation strongly supportive of seafloor spreading.

3.5.3 Sediments

Ewing, during the same period, had used seismic refraction to determine the thickness of sediments in the Atlantic. If seafloor spreading were occurring, the thickness of sediments should increase with distance from the ridge crest. The results were confusing [60], partly because of the rough seafloor topography of the Atlantic, and Ewing was reluctant to come out in support of seafloor spreading.

Later a different approach became possible through a deep-sea drilling programme, which allowed the recovery of long sediment cores. An early cruise in the South Atlantic Ocean was aimed specifically at testing seafloor spreading. The results were spectacular [61]. It was found that the age of the oldest sediment, just above the basaltic basement, determined from micro fossils, increased in simple proportion to distance from the ridge crest, exactly as predicted by assuming seafloor spreading at a nearly constant rate (Figure 3.7). The results also provided an important calibration of the magnetic reversal chronology, which until then was well-calibrated only for the first few million years.

Menard and others have remarked that most scientists are converted to a new idea by observations from within their own speciality. Thus palaeomagnetic polar wandering converted a small minority of geophysicists to continental drift. Later the dramatic evidence of seafloor magnetic stripes, earthquake distributions and fault plane solutions converted a majority of geophysicists to seafloor spreading. To many of the more traditional geologists, however, such geophysical observations were still unfamiliar, and they were unsure how to regard them. However, fossil ages are a long-standing concept in geology, and something most geologists can readily relate to. Thus, although the deep-sea sediment ages were not published until 1970, they were important for spreading the word to the great majority of geologists who work on continental geology.

This completes my short survey of some of the most direct and compelling evidence that led to the acceptance of plate tectonics by a majority of geologists. There is much other evidence and there were many more players, but a knowledge of these observations suffices to place plate tectonic motions on a firm empirical footing.

3.6 Completing the picture – poles and trenches

With compelling evidence for seafloor spreading and strong evidence for the rigidity of plates from the lack of deformation of

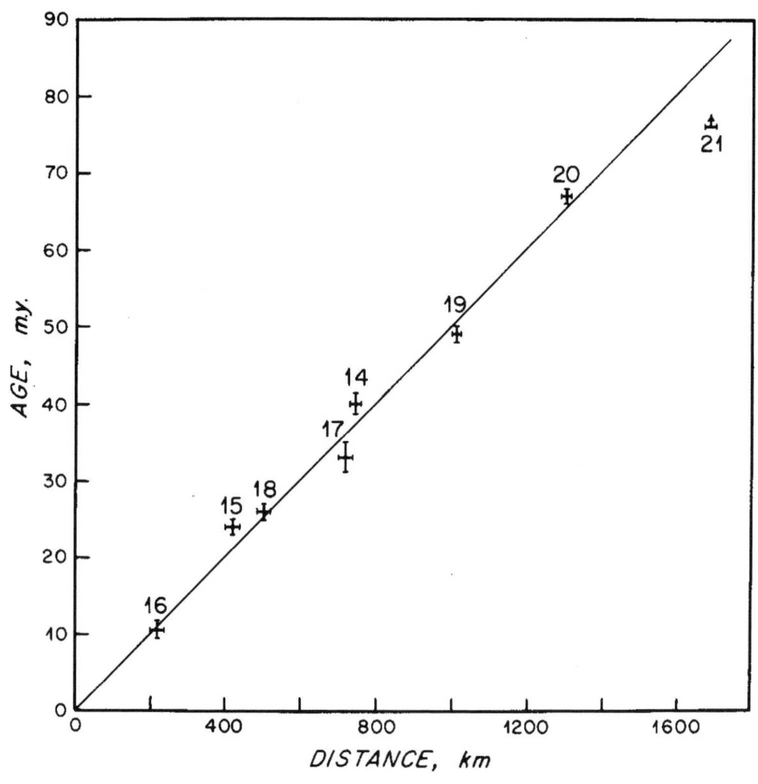

Figure 3.7. Ages of sediments immediately above the basaltic basement of the sea floor of the South Atlantic, plotted against distance from the crest of the Mid-Atlantic Ridge. The ages are inferred from micro fossils. From Maxwell *et al.* [61]. Copyright American Association for the Advancement of Science. Reprinted with permission.

sediments and magnetic anomalies over large areas of sea floor, Wilson's concept of moving rigid plates was strongly supported. However the focus had been mainly on ridges and seafloor spreading. There remained the untidiness at the other end of the conveyor belt: the ocean trenches. Also, would Wilson's tightly integrated global concept withstand further close quantitative examination? Were such quantitative tests possible? Was there, perhaps, some possibility that earth expansion had not been excluded?

3.6.1 Euler rotations

Two people independently conceived that there is a simple way to describe quantitatively the motions of rigid plates on a sphere: Jason Morgan and Dan McKenzie. The idea had been used in 1965 by E. C. (Teddy) Bullard and others at Cambridge to test the idea that continental outlines on either side of the Atlantic matched [62], but it was originally due to the mathematician Euler. Euler's theorem is that any motion of a rigid piece of a sphere over the surface of the sphere can be described as a rotation about some axis through the centre of the sphere. The intersections of the axis with the surface are called poles of rotation. Menard coined the terms Euler pole and Euler latitude ([6], p. 324). An Euler latitude is analogous to geographic lines of latitude relative to the earth's geographic north pole. Lines of latitude are 'small circles' on a sphere (that is, the intersection of a plane with the surface of the sphere), except for the equator, which is a 'great circle' (that is, the intersection of a plane passing through the centre of the sphere with the surface of the sphere). Morgan and McKenzie each realised that in Wilson's theory plate motions should be described by rotations about Euler poles.

We must stop, at this point, and think about what plates are moving relative to. It is natural at first to think that there is some part of the earth's interior that is not moving, and to think of the plates as moving relative to that region. The problem is that we can't see into the earth very clearly, to identify such a region. In fact there is no evidence for such a region. Everything – crust, plates, mantle and core – seems to be in relative motion. It is true that the volcanic hotspots (discussed in the next section) seem to be moving only slowly relative to each other, and it has been useful to assume them to be 'fixed' for some purposes, but there is no reason to think they are not moving at all.

The better way to approach this problem is to realise that it is not necessary to think about the interior at all. The plate motions can be described entirely in terms of *relative* motions, that is of motions relative to each other, without reference to some internal, external or 'absolute' frame. The most convenient approach is to describe the relative motion of plates in adjoining pairs. This relative motion determines the nature of their interaction – whether they are pulling apart (at a ridge or 'spreading centre'), pushing together (at a trench or 'subduction zone') or sliding past each other (along a 'transform fault').

Extending this point, it is best to stop thinking about convection, or driving forces, and to realise that we are doing geometry. Strictly speaking, it is kinematics, the description of motion, without reference to causes or forces. If you like, it is the moving geometry of the earth's surface. Wilson succeeded, in his formulation of plates, also by ignoring convection and looking only at the surface.

In 1967, Menard published a short paper describing further mapping of the great fracture zones of the north-east Pacific [63]. In one illustration in that paper he used a great circle projection in order to show that the fracture zones are remarkably straight (a great circle being the nearest thing to a straight line on a sphere – it is the shortest distance between two points). Morgan was struck that the fracture zones could be more accurately represented as a series of small circles and that the radii of the small-circle segments increased from north to south in passing from one fracture zone to the next ([6], p. 285). In fact they could be thought of as sets of Euler latitudes. Combining Euler's theorem with Wilson's transform fault concept, he realised that the sets of small-circle fracture zones represented traces on a sphere of transform faults offsetting a former ridge that must have existed there. (Morgan may not have realised at this stage that a third plate, the Farallon plate, must have existed between the Pacific and North American plates.) Morgan then analysed the fracture zones along several presently active spreading centres and showed that they were consistent with Wilson's concept of transform faults between rigid plates moving on the spherical surface of the earth [64].

Meanwhile McKenzie also was struck, while reading the Bullard *et al.* paper [62], that relative plate motions could be represented in terms of rotations. His approach was to test present motions, as determined from earthquake fault plane solutions in the manner of Sykes. His colleague and co-author Robert Parker introduced the idea of using a Mercator projection oriented relative to the Euler pole. In this projection, earthquake slip directions (and fracture zones) would plot as horizontal lines (like lines of latitude). In this manner, they demonstrated that motions over a huge area, from the Gulf of California across the North Pacific and to Japan, were consistent with the motion of a single rigid Pacific plate relative to adjacent areas [65]. (In fact Japan is part of the Eurasian plate, but its motion is slow relative the North American plate, so this didn't affect the results much.)

McKenzie and Morgan, in these papers and in a subsequent joint publication [66], developed the ideas of rotation vectors, local velocity vectors and triple junctions. Triple junctions are points where three plates meet, and their evolution can be deduced from the behaviour of the types of plate boundary involved. This will be taken up in Chapter 9. The important point here is that the sizes and shapes of plates can evolve just by the way they are created and removed at ridges and trenches, without any changes in the motion of plates. With an understanding of this, it is possible to reconstruct the past evolution of plates.

In this way, McKenzie and Morgan were able to develop the suggestion by McKenzie and Parker that there had been another plate (the Farallon plate) between the Pacific and North American plates that has been consumed by subduction. The great fracture zones of the north-east Pacific discovered by Menard were formed at a ridge between the Pacific and Farallon plates. That ridge was subsequently replaced by the San Andreas transform fault, which represents motion between the Pacific and North American plates. Finally the mystery of these 'dangling' fracture zones was solved.

With the concept of Euler rotations, the quantitative descriptions of plate motions became possible, and it was demonstrated that indeed much of the earth's surface comprises rigid moving plates. There are some areas where large-scale rigidity does not apply, mainly where plate boundaries enter continents (for example, central Asia and western North America).

3.6.2 Subduction zones

As I mentioned in Section 3.3, the deep ocean trenches and their associated deep earthquake zones were the subjects of controversy until late in the development of plate tectonics. The trenches were found to have large negative gravity anomalies by Vening Meinesz [38], and he and others developed the idea that they are the sites of crustal compression and possibly of convective downwelling in the mantle (see Menard [6]; Chapter 10). Wadati [67] and Benioff [39] mapped the deep seismic zones. Benioff interpreted them as tracing great reverse (thrust) faults, but with limited displacements. Attempts to determine the sense of motion from fault plane solutions generated confusion because of the difficulty of obtaining consistent high-quality readings of seismograms from a sufficient global distribution of instruments. Japanese seismologists [68] detected a zone of relatively high attenuation (suggesting higher temperatures) above the deep seismic zone. This indicated some kind of spatial heterogeneity, as had been conjectured by Daly to reconcile the occurrence of deep earthquakes with strong independent evidence for an asthenosphere.

Clarity only began to emerge with the advent of a world-wide network of seismographs in the early 1960s, mentioned earlier, and with the installation by Lamont seismologists of several seismographs in the Tonga–Kermadec region north of New Zealand, this being the site of the most active deep earthquake zone. What emerged first was a higher resolution map of the Tonga–Kermadec zone, showing among other things that the seismic zone is quite thin, defining a surface dipping to the west away from the trench, and that at the northern end the trench, the volcanic line and the deep earthquakes all curve sharply to the west, suggesting an intimate association between them [69]. Second, it was found that the attenuation of seismic waves was greater both above and below the deep seismic zone [70]. This defined a tongue of mantle of anomalously low attenuation, suggesting low temperatures, about 100 km thick and including the seismic zone, which was continuous from the surface to the deepest earthquakes.

This was a spectacular confirmation of Daly's 'prongs' of lithosphere sticking down from the surface. In fact it is revealing to quote Oliver and Isacks [70] in the light of Daly's hypothesis that the asthenosphere may be laterally heterogeneous (Section 3.2).

> In retrospect, it appears quite reasonable that the zones of deep shocks should be anomalous if only because earthquakes occur there and not elsewhere. Yet most models of the earth's interior include a mantle without lateral variation, and, except in one or two cases, the models that do take into account lateral variation have not associated such variation with the entire zone of deep earthquakes. In general, in hypotheses relating to the mechanism of deep earthquakes the emphasis has been on process alone, whereas it should be on both process and the nature of the material.

Subsequent work showed that shallow earthquakes have predominantly thrust mechanisms [71], confirming the conjectures of Holmes, Vening Meinesz, Hess, Deitz, Wilson and many others that these are zones of horizontal convergence. All the findings together painted a clear picture of surface lithosphere turning and descending under the trench and island arc [71, 72].

I indulge here in a personal footnote that carries a small lesson and an indication of my own perspective. In September of 1968 I travelled from Australia to begin graduate work at the California Institute of Technology. My first publication in geophysics, in 1971 with seismologist Jim Brune, was an estimate of the rate of convergence in subduction zones using a catalogue of this century's earthquakes and magnitudes [73]. The results were of the right order (centimetres per year) predicted by plate tectonics. At the time it was clear that there were uncertainties of up to a factor of two, and subsequent work revealed that the largest earthquakes were not adequately measured by the old magnitude determinations. Nevertheless it was for some a satisfying closing of the circle that subduction was occurring at about the same rate as seafloor spreading, and no expansion of the earth was called for. To me the conclusion was no big deal: I was new to the field and it seemed quite obvious to me that plate tectonics was correct.

Had that study been done ten years earlier, it might have accelerated the formulation of plate tectonics, balancing somewhat the concentration on ridges and seafloor spreading. Had it been done twenty or thirty years earlier, it might have crystallised some of the ideas of the time into a rudimentary form of plate tectonics (see Section 3.9). I didn't know until recently that Charles Darwin had pioneered the approach 135 years earlier.

A great deal else has also been deciphered from the record of the seafloor magnetic anomalies and the rules of plate motion. It is not necessary to recount them all here. Chapter 9 explains the principles and principal consequences of plate evolution, with some representative examples. The plates and their motion are a principal boundary condition on mantle convection, and this aspect will be taken up in Part 3.

3.7 Plumes

Perhaps Charles Darwin's best-known contribution to geology is his theory of the formation of coral atolls. He proposed that the various forms of coral reefs could be arranged in a sequence (island and fringing reef, island and barrier reef, atoll), and that the sequence made sense if islands were progressively eroded and submerged, with the coral reef growing as the island shrank away. Darwin did not actually observe his proposed sequence in a contiguous set of islands. Dana, on the U.S. Exploring expedition in 1838–1842, did find Darwin's proposed sequence in place in several linear island chains, and extended the sequence to include the initial active volcano that formed the island ([6], p. 195). Dana observed the sequence in the Society, Samoan and Hawaiian islands, and correctly inferred that the Hawaiian and Society island chains age to the north-west.

When Wilson read Dietz's seafloor spreading paper, he thought of using ocean islands as probes of the sea floor, reasoning that islands should be progressively older at greater distances from spreading centres [46]. The idea was good, but the data were scattered and somewhat misleading, since some islands include fragments of continental crust, and other ages were not representative of the main phase of island formation. Even an accurate age for the main phase of building an island gives only a lower bound on the age of the sea floor upon which it is built. The sea floor will have the same age as the island if the island formed at a spreading centre, but it will be older if the island formed away from the spreading centre.

Despite the limited data available at the time, two clear ideas emerged from Wilson's work. One was that some 'lateral ridges' could be explained if they represent the traces of extra volcanism at a spreading centre. In fact, such volcanism might produce a complementary pair of ridges, one on each plate moving away from the spreading centre. Wilson cited the Rio Grande Ridge and the Walvis Ridge in the South Atlantic as an example of such a pair, the active volcanism of Tristan da Cunha being the current site of generation. A closely related idea had been proposed by Carey in 1958 [13], and Wilson has acknowledged his debt to Carey [40].

The second idea was a mechanism to explain age progressions in island chains. Wilson recognised that there is active volcanism on some islands that are located well away from spreading centres, so that some islands clearly had not formed at a spreading centre, the Hawaiian islands being an outstanding example. With the idea of seafloor moving sideways, he realised that Dana's inferred age sequence for the Hawaiian islands could be produced if there was a (relatively) stationary source of volcanism deep in the mantle that had generated the islands successively as the seafloor passed over [45]. He conjectured that this 'hotspot' source might be located near the slowly moving centre of a convection 'cell'.

In 1971 Morgan [74, 75] developed this idea by proposing that there are plumes of hot material rising from the lower mantle. His proposal actually had three parts: that the island volcanism is produced by a plume rising through the mantle, that the plume comes from the lower mantle, and that plume flow drives the plates. He also presented reconstructions of plates to argue that the volcanic centres are relatively fixed, meaning that they have low horizontal velocities relative to each other. It was a common assumption at the time that the lower mantle, or 'mesosphere', has a very high viscosity and does not partake in convection (e.g. [71]), and so presumably the assumption that plumes come from the lower mantle was a way of accounting for their slow motions. Morgan devoted most space to demonstrating hotspot 'fixity' and to arguing that plate motions are driven by plumes. Hotspot fixity has been a useful approximation that has helped to refine details of plate motions. Plumes as the primary means of driving plates received little support, and I think it is not viable, as will become clear in Part 3.

I abandon here Wilson's original meaning of the term hotspot, since his hypothesis has been superseded by Morgan's plume hypothesis. Wilson's 'hotspot' was a small, hot volume in the mantle of unspecified origin. I think it is more useful to have a term for the surface manifestation from which plumes are inferred, and that a suitable term is 'volcanic hotspot', or 'hotspot' for short. I will therefore use these terms to refer to a volcanic centre *on the earth's surface* that has the characteristics that have come to be associated with (Wilson's) hotspots and plumes: persistent volcanism in a location that is relatively independent of plate motions and moves only slowly relative to other hotspots, often with an associated topographic swell.

The reality of plumes as a source of island volcanism became commonly accepted, though not without debate. Morgan noted that volcanic hotspots often have a topographic swell associated with them, and this observation was documented more thoroughly by Crough [76]. The details of how these swells are generated has been the subject of a confused debate. This will be taken up in Chapters 11 and 12.

The concept of plumes developed only slowly as a physical and quantitative theory, with the unfortunate result that plumes came to be invoked often in very *ad hoc* ways to explain a wide range of geological observations throughout earth history and on other planets. Even when important developments in the understanding of plumes occurred, the implications were frequently overlooked.

In 1975 Whitehead and Luther [77] reported laboratory experiments that showed that the viscosity of the plume fluid has a strong effect on the form of a newly forming plume. If the plume is of higher viscosity than the surroundings, it rises as a finger. If it is of lower viscosity, it rises in a 'mushroom' or 'head and tail' form: a large spherical 'head' preceeding a narrower conduit or 'tail' up which fluid continues to flow from the source.

Morgan, in 1981 [78], pointed out that a number of 'hotspot tracks' (the volcanic chain produced on a plate as it passes over a plume) originate in flood basalt provinces. Flood basalts are the largest known volcanic eruptions in the geological record, and typically comprise basalts of the order of 1 km thick over an area up to 2000 km across. Morgan proposed that this association could be explained if the flood basalt was produced from a plume head arriving at the base of the lithosphere and the hotspot track was produced by the following plume tail.

Loper and Stacey in 1983 [79, 80] developed the quantitative theory of flow in a thermal plume tail for the case when the viscosity of the material is strongly temperature-dependent. In this case, the plume material has a low viscosity because it is hot, and the plume tail can be quite narrow, of the order of 100 km in diameter. Loper and Stacey developed the analogous theory for a hot thermal boundary layer, from which the plume was assumed to grow. Olson and Singer [81] quantified the growth and ascent of plume heads in the case where they are compositionally distinct, and some aspects of plume tail behaviour in the presence of horizontal shear flow in the surrounding fluid.

Griffiths and Campbell in 1990 [82] presented a physical theory of thermal plume heads and tails, confirmed and calibrated by laboratory experiments. They demonstrated an important distinction between compositional plume heads and thermal plume heads. In the latter, a boundary layer of adjacent material is heated by conduction, becomes buoyant, and then rises with and is entrained into the plume head. The result in the mantle can be that the plume head reaches a diameter of about 1000 km, two to three times larger than a compositional, non-entraining, plume head.

Morgan's idea that flood basalts are produced by plume heads was revived by Richards, Duncan and Courtillot in 1989 [83], with more information on hotspot track ages. Campbell and Griffiths [84] developed this hypothesis further in 1990, arguing that first-order features of flood basalts (size, temperature, composition) could be accounted for by thermal plume heads rising from the base of the mantle.

By this stage plumes were well quantified and their physics quite well understood, both for compositional and thermal plumes, and quantitative predictions were being made and tested. Many details are still debated, but a basic theory is in place and there is much observational support for the broad concept. The physical theory of plumes will be developed in some detail in Part 3.

3.8 Mantle convection

We have looked at evidence for drifting continents and for moving plates, evidence for a deformable mantle, evidence for mantle plumes, and at the development of these concepts over the past century or more. The idea of mantle convection, which arises from the convergence of these other concepts, also goes well back into the nineteenth century. Here I briefly recount the development of ideas of mantle convection that precede the conception that will be developed in Part 3.

As I have indicated, ideas about mantle convection were at times intimately linked with ideas of continental drift and the emerging idea of plate tectonics. So they should be, but the source of the convection and its relationship to surface tectonics were for a very long time unclear and puzzling. The convection was usually assumed to have a particular form, like that of the classical Benard convection [85], with steady flow, 'cells', hot upwellings and cold downwellings. It was often assumed also to occur below the crust or lithosphere, which was assumed to be dragged around by the underlying convection. These ideas are distinct from the concept that will be presented in Part 3, which has resulted from some relatively recent conceptual shifts.

An early mention of mantle convection is by Hopkins in 1839 [86]. Fisher, in his 1881 book *Physics of the Earth's Crust* [87], proposed mantle convection as a tectonic agent, with flow rising under the oceans and descending under continents. He assumed the mantle to be relatively fluid, drawing on the concepts of isostasy being developed at that time. He envisaged that this flow would expand the oceans and compress the continents at their edges, generating mountains.

According to Hallam ([1], p. 140), the idea of a fluid mantle was more widespread in continental Europe, particularly in Germany, than in Britain and America. He cites a number of instances of this, noting that this implies a more sympathetic climate around the turn of the century within which Wegener's ideas of continental drift could develop. However, Wegener himself did not appeal to mantle convection, and his concept that continents plough through oceanic crust seems to owe little to any idea of a deformable mantle.

It was Arthur Holmes who most seriously advocated mantle convection, and he proposed it explicitly as a mechanism for continental drift, first in a talk and brief note in 1928 [88], then in a paper in 1931 [89], and finally in his book *Principles of Physical Geology*, the first edition of which appeared in 1944 [90]. Holmes's basic proposal was that convection occurs under the lithosphere and drags the continents around. His proposed flow was different from Fisher's, in that Holmes, in his initial version, reasoned that convection might rise under a continent because of the thermal blanketing effect of continental radioactivity, a subject that he was very familiar with.

Holmes then envisaged that the rising and diverging convection might rift a continent and carry the pieces apart. In his earlier version, he supposed that a piece of continent might be left over the upwelling site, because the horizontal flow would be relatively stagnant there. In his later version, he proposed instead that the crust between the diverging continental fragments might be broadly stretched and the extension accommodated by the intrusion or eruption of basaltic melts generated in the (presumed) warmer upwelling mantle.

Holmes also envisaged that a basaltic oceanic crust would be returned to the mantle. He presented the case with admirable simplicity ([90] ; see Cox [3], p. 21):

The obstruction that stands in the way of continental advance is the basaltic layer, and obviously for advance to be possible the basaltic rocks must be continuously moved out of the way. In other words, they must founder into the depths, since there can be nowhere else for them to go.

Holmes, in this later version, proposed a different driving force for his convecting system. He contrasted sialic rocks, whose density is not much affected by pressure, with basaltic compositions, which are converted by pressure first to granulites and then to eclogite, undergoing in the process a density increase from about 2.9 Mg/m^3 to 3.4 Mg/m^3. Given that it was not known then that the oceanic basaltic crust is quite thin (about 7 km), this was quite a plausible suggestion. He continues

Since this change is known to have happened to certain masses of basaltic rocks that have been involved in the stresses of mountain building, it may be safely inferred that basaltic roots would undergo a similar metamorphism into eclogite. Such roots could not, of course, exert any [positive] buoyancy, and for this reason it is impossible that tectonic mountains could ever arise from the ocean floor. On the contrary, a heavy root formed of eclogite would continue to develop downwards until it merged into and became part of the descending current, so gradually sinking out of the way, and providing room for the crust on either side to be drawn inwards by the horizontal currents beneath them.

Thus Holmes, in this later version, proposed the generation of a basaltic crust over mantle upwellings and its removal into downwellings, concluding

To sum up: during large-scale convective circulation the basaltic layer becomes a kind of endless travelling belt on the top of which a continent can be carried along, until it comes to rest (relative to the belt) when its advancing front reaches the place where the belt turns downwards and disappears into the earth.

Menard ([6], p. 157) has commented on how closely this anticipates Dietz's version of seafloor spreading, the only essential difference being that Dietz proposed that the basaltic oceanic crust is produced in the narrow rift zone at the crest of the midocean rise system, whereas Holmes assumed it would emerge over a broad extensional area.

There is another brief passage worth quoting from this section of Holmes:

The eclogite that founders into the depths will gradually be heated up as it shares in the convective circulation. By the time it reaches the bottom of the substratum it will have begun to fuse, so forming pockets of magma

> which, being of low density, must sooner or later rise to the top. Thus an adequate source is provided for the unprecedented plateau basalt that broke through the continents during Jurassic and Tertiary times. Most of the basaltic magma, however, would naturally rise with the ascending currents of the main convecting systems ...

Here Holmes has proposed that the subducted eclogite might rise in two distinct ways: most of it carried up by the main circulation to form new oceanic crust, but some of it forcing its way up independently and breaking out on the surface as flood basalt. Aside from his assumption of melting at great depth, rather than as the material approaches the surface, this is broadly similar to current ideas that subducted oceanic crust is concentrated near the base of the mantle and recycled to the surface through plumes to form flood basalts and hotspot tracks [84, 91, 92].

Holmes's ideas were not entirely ignored, although they did not become part of mainstream thinking. During the 1930s, Pekeris [93] showed that convection driven by the differential thermal blanketing of continents and oceans could result in velocities of millimetres per year and stresses sufficient to maintain observed long-wavelength gravity anomalies. Hales [94] showed that plausible convection could be maintained by a mean vertical temperature gradient (above the adiabatic gradient) of as little as 0.1 K/km. Haskell's estimate of mantle viscosity from post-glacial rebound, assuming flow to penetrate deep into the mantle, appeared during this period [26]. Griggs [30] developed a number of ideas, a central one being that experimentally observed non-linearities in rock rheology could result in episodic convection. He also presented a simple laboratory realisation of the way crust might be piled into mountains over a convective downwelling. This experiment probably had a positive influence on concepts of subduction and the interpretation of the Wadati–Benioff deep earthquake zones.

I described earlier some of the ways that concepts of mantle convection entered the thinking of those who developed the ideas of seafloor spreading and plate tectonics. After the general acceptance of plate tectonics, there was a great deal of discussion of 'the driving mechanism'. I defer a detailed discussion of the ideas from this time until Chapter 12, after I have presented what I think is the most useful way to think of the relationship between plate tectonics and mantle convection. It will then be easier to discuss the limitations of some of the early plate-tectonic views. However I summarise here the nature of the required shift in thinking.

Many of the earlier discussions of mantle convection conceived it as something that happens *under* the lithosphere. One key shift in perspective is to regard the lithosphere as *part* of mantle convection. Another is to realise that the (negative) thermal buoyancy of the cold lithosphere can provide the driving force. A third is to realise that convection need not have active upwellings, but can comprise cold, negatively buoyant, *active* downwellings and complementary *passive* upwellings. A fourth is that the flow pattern is likely to be *unsteady*, especially if it is strongly affected by the changing plate configuration (Chapter 9), and in that case it is not useful to think of a 'cell' of convection.

Holmes's early idea of thermal blanketing was plausible, though the quantitative effect is rather smaller than is required. His later idea of invoking the basalt-eclogite transformation was also plausible within the uncertainties of the time about the thickness of the oceanic crust. It is a possibility still worth entertaining for some earlier stage of earth history (Chapter 14), though as an adjunct to thermal convection, not as a substitute. Daly in 1940 [14] got close to the modern concept of mobile lithosphere, with his idea that the lithosphere would be thrust downwards in compressional zones, but the mobility he envisaged was limited. He also got close by invoking gravity sliding off topographic highs, which is a form of driving by thermal buoyancy, but evidently he did not also think of the weight of his downward-projecting 'prongs' as a possible driving component.

With the advent of plate tectonics, the idea of the lithosphere being an active component was soon advanced, though there was a debate with those who thought it must still be carried passively by convection (necessarily of uncertain origin) underneath. There were also those who evidently did not think of the motions induced by an active lithosphere as convection and for whom the term mantle convection still referred to something happening under the plates [95].

The almost universal assumption that upwellings under mid-ocean ridges would necessarily be hot and active, and usually fixed as well, was a hindrance to the emergence of the concepts of sea-floor spreading and plate tectonics, because in that case neither the offsets of spreading centres along transform faults nor the relative motion of different spreading centres made sense. With passive upwelling, both problems go away. Clear and direct evidence that passive upwelling under spreading centres is the norm comes from seafloor topography, as I will argue in Chapters 8, 10 and 12.

Plumes are also a form of mantle convection, as I will argue in Part 3. Furthermore, they are active upwellings. However, they do not occur universally at spreading centres but are to a substantial degree independent of the plates, and the proportion of the total length of spreading centres affected by plumes is small. Therefore they do not contradict the point just made that upwelling under spreading centres is normally passive.

I will conclude here with a quite different kind of argument. In 1965, Tozer [96] argued that mantle convection was inevitable under very general assumptions. He observed that the viscosity of rocks is very strongly temperature-dependent, decreasing by roughly one order of magnitude for each 100 °C increase in temperature, for temperatures near that of the mantle (about 1300 °C). This has the effect of feeding back on thermal convection, so that large changes in heat transport can be accomplished by small changes in mantle temperature.

Tozer supposed that there is a certain amount of radioactive heat generation at present in the mantle. There will then be a certain mantle temperature at which the buoyancies and the viscosity are such that the convective heat loss just balances the radioactive heating. This is a stable equilibrium temperature. If the mantle were ever hotter than this, the viscosity would have been substantially less and convection very much faster, and the mantle would have rapidly cooled towards the equilibrium value. If the mantle were cooler, the viscosity would have been much higher and convective heat loss much less, and the mantle would have been warmed by the radioactivity until it approach the equilibrium temperature.

These arguments were quantified and confirmed fifteen years later [97]. The time scale of approach to the equilibrium from higher temperatures is a few hundred million years. Only if the earth started very cold does the argument fail, because it would not yet be hot enough to convect, but there is abundant evidence for a tectonically active and therefore hot earth through most of its history. An analogous argument will be given in Part 3 for the existence of thermal plumes in the mantle.

3.9 Afterthoughts

I have tried to trace the emergence of key ideas that comprise our current understanding of mantle convection. I am struck by the fact that most of them were in place well before plate tectonics was invented and mantle convection was accepted by implication. It is of course easy to be wise in retrospect, but consider that the following ideas were well established, if not widely appreciated, by about 1945.

There is a lithosphere about 100 km thick underlain by an asthenosphere with a viscosity of the order of 10^{21} Pa s [14, 26].
The major tectonic events of the Phanerozoic are recorded in long, narrow mobile belts [41].
Current tectonic activity, comprising most of the earthquakes and volcanoes, occurs in a nearly continuous network of belts [98].
By implication, there is relatively little deformation of the crust outside current and past mobile belts.
Large earthquakes involve metres of fault displacement, and recur about once a century in the most active regions. Repetition of such earthquakes would yield slip rates of centimetres per year.
Circumstantial evidence for continental drift would require drift rates of centimetres per year [7, 12].
Oceanic trenches are far from isostatic balance (having large negative gravity anomalies as well as large negative topographic anomalies), which requires a force to pull them down. It was suggested that descending convective flow would provide this force [38].
The lithosphere would be forced or required to move downward in compressional mountain belts [14].
The oceanic crust might descend under its own weight because of the transformation of basalt to dense eclogite [90].

I have not found a suggestion prior to plate tectonics that the lithosphere would sink because it is *colder* and denser, but Holmes' compositional density idea could have served to stimulate thinking about an active lithosphere.

Another point worth mentioning is that most of the key insights in the development of plate tectonics were by people once-removed from observational work. Hess, Dietz and Morley were involved in science administration. Wilson and Morgan were at institutions without big field or oceanographic programmes. Vine was a new graduate student interpreting data much of which had been obtained by his supervisor, Matthews. McKenzie had done a theoretical thesis on mantle viscosity, and his plate tectonics work was with theory and archived data.

This is not meant to disparage observational science, which is essential, and theories are only useful if their proponents attempt to connect them with observations. It does suggest, though, that often observational programmes become too dominating, and that people could take more time to think more broadly. It also clearly demonstrates that the full significance and meaning of an observation may not be evident for some time after it is made. Seafloor magnetic stripes and fracture zones are obvious examples, but there are many in my experience.

Being a theoretician, I have often encountered the attitude that theoreticians only play games, and that observational and experimental work is the 'real' science. Too many of my theoretical peers do not pay enough attention to observational constraints, so there is a grain of truth in the perception. However, the theoretical versus observational debate is sterile. Science requires both, and in my experience the most stimulating scientists are those who straddle both to some degree.

3.10 References

1. A. Hallam, *Great Geological Controversies*, 244 pp., Oxford University Press, Oxford, 1989.
2. A. Hallam, *A Revolution in the Earth Sciences*, 127 pp., Clarendon Press, Oxford, 1973.
3. A. Cox, ed., *Plate Tectonics and Geomagnetic Reversals*, 702 pp., W.H. Freeman and Company, San Francisco, 1973.
4. W. Glen, *Continental Drift and Plate Tectonics*, 188 pp., Charles E. Merrill Publishing Company, Columbus, 1975.
5. W. Glen, *The Road to Jaramillo*, Stanford University Press, Stanford, 1982.
6. H. W. Menard, *The Ocean of Truth*, 353 pp., Princeton University Press, Princeton, New Jersey, 1986.
7. A. Wegener, *Die Entstehung der Kontinente und Ozeane*, Vieweg and Son, Brunswick, 1915.
8. A. Wegener, *The Origin of Continents and Oceans*, Methuen, London, 1966.
9. H. Jeffreys, *The Earth, its Origin, History and Physical Constitution*, Cambridge University Press, 1926.
10. B. Gutenberg, Hypotheses on the development of the Earth, in: *Internal Constitution of the Earth*, B. Gutenberg, ed., Dover, New York, 178–226, 1951.
11. B. Gutenberg, *Physics of the Earth's Interior*, Academic Press, New York, 1959.
12. A. L. du Toit, *Our Wandering Continents*, Oliver and Boyd, Edinburgh, 1937.
13. S. W. Carey, The tectonic approach to continental drift, in: *Continental Drift; a Symposium*, S. W. Carey, ed., University of Tasmania, Geol. Dept., Hobart, 177–358, 1958.
14. R. A. Daly, *Strength and Structure of the Earth*, 434 pp., Prentice-Hall, New York, 1940.
15. J. H. Pratt, *Philos. Trans. R. Soc. London* **145**, 53–5, 1855.

16. J. H. Pratt, *Philos. Trans. R. Soc. London* **149**, 779, 1859.
17. G. B. Airy, *Philos. Trans. R. Soc. London* **145**, 101–4, 1855.
18. C. E. Dutton, *Bull. Phil. Soc. Washington* **11**, 51, 1889.
19. J. Hall, *Geology of New York State*, **3**, 69, 1859.
20. F. R. Helmert, *Sitzungsber. Preuss. Akad. Wiss.* 1192, 1909.
21. J. Barrell, The strength of the earth's crust, *J. Geol.* **22**, 655–83, 1914.
22. A. Mohorovičić, Das Beben vom 8.x.1909, *Jahrb. Met. Obs. Zagreb (Agram.)* **9**, 1–63, 1909.
23. R. D. Oldham, Constitution of the interior of the earth as revealed by earthquakes, *Quart. J. Geol. Soc. London* **62**, 456–75, 1906.
24. T. F. Jamieson, *Quart. J. Geol. Soc. London* **21**, 178, 1865.
25. R. van Bemmelen and P. Berlage, *Gerlands Beitr. zur. Geophysik* **43**, 19, 1934.
26. N. A. Haskell, The viscosity of the asthenoshpere, *Am. J. Sci.*, ser. 5 **33**, 22–8, 1937.
27. K. Wadati, Shallow and deep earthquakes, *Geophys. Mag. (Tokyo)* **1**, 162–202, 1928.
28. S. K. Runcorn, Paleomagnetic evidence for continental drift and its geophysical cause, in: *Continental Drift*, S. K. Runcorn, ed., Academic Press, New York and London, 1–40, 1962.
29. P. Goldreich and A. Toomre, Some remarks on polar wandering, *J. Geophys. Res.* **74**, 2555–67, 1969.
30. D. T. Griggs, A theory of mountain building, *Am. J. Sci.* **237**, 611–50, 1939.
31. S. K. Runcorn, Rock magnetism, *Science* **129**, 1002–11, 1959.
32. E. Irving, Paleomagnetic pole positions, *Geophys. J. R. Astron. Soc.* **2**, 51–77, 1959.
33. H. H. Hess, Drowned ancient islands of the Pacific Basin, *Am. J. Sci.* **244**, 772–91, 1946.
34. H. H. Hess, History of ocean basins, in: *Petrologic Studies: a Volume in Honor of A. F. Buddington*, A. E. J. Engel, H. L. James and B. F. Leonard, eds., Geol. Soc. Am., Boulder, CO, 599–620, 1962.
35. H. W. Menard, *Marine Geology of the Pacific*, McGraw-Hill, New York, 1964.
36. B. C. Heezen, The rift in the ocean floor, *Sci. Am.* **203**, 98–110, 1960.
37. R. S. Dietz, Continent and ocean evolution by spreading of the sea floor, *Nature* **190**, 854–7, 1961.
38. F. A. Vening Meinesz, *Gravity Expeditions at Sea, 1923–32. Vol.2: Interpretation of the Results*, Publ. Neth. Geod. Comm., Waltman, Delft, 1934.
39. H. Benioff, Seismic evidence for crustal structure and tectonic activity, *Geol. Soc. Amer. Spec. Paper 62*, 61–74, 1955.
40. J. T. Wilson, A new class of faults and their bearing on continental drift, *Nature* **207**, 343–7, 1965.
41. W. H. Bucher, *The Deformation of the Earth's Crust*, 518 pp., Princeton University Press, Princeton, 1933.
42. J. T. Wilson, Continental drift, *Sci. Am.*, April, 1963.

43. L. R. Sykes, Seismicity of the South Pacific Ocean, *J. Geophys. Res.* **68**, 5999–6006, 1963.
44. A. M. Coode, A note on oceanic transcurrent faults, *Can. J. Earth Sci.* **2**, 400–1, 1965.
45. J. T. Wilson, A possible origin of the Hawaiian islands, *Can. J. Phys.* **41**, 863–70, 1963.
46. J. T. Wilson, Evidence from islands on the spreading of the ocean floor, *Nature* **197**, 536–8, 1963.
47. J. T. Wilson, *Continents Adrift and Continents Aground,* 230 pp., W. H. Freeman and Company, San Francisco, 1976.
48. M. Matuyama, On the direction of magnetization of basalt in Japan, Tyosen, and Manchuria, *Proc. Japan Acad.* **5**, 203–5, 1929.
49. A. Cox, R. R. Doell and G. B. Dalrymple, Geomagnetic polarity epochs and Pleistocene geochronometry, *Nature* **198**, 1049–51, 1963.
50. I. McDougall and D. H. Tarling, Dating of polarity zones in the Hawaiian islands, *Nature* **200**, 54–6, 1963.
51. A. Cox, Geomagnetic reversals, *Science* **163**, 237–45, 1969.
52. R. G. Mason, A magnetic survey off the west coast of the United States, *Geophys. J. R. Astron. Soc.* **1**, 320–9, 1958.
53. F. J. Vine and D. H. Matthews, Magnetic anomalies over oceanic ridges, *Nature* **199**, 947–9, 1963.
54. F. J. Vine, Spreading of the ocean floor: new evidence, *Science* **154**, 1405–15, 1966.
55. W. C. Pitman III, Magnetic anomalies over the Pacific-Antarctic ridge, *Science* **154**, 1154-1171, 1966.
56. J. R. Heirtzler, G. O. Dickson, E. M. Herron, W. C. Pitman III and X. le Pichon, Marine magnetic anomalies, geomagnetic field reversals, and motions of the ocean floor and continents., *J. Geophys. Res.* **73**, 2119–36, 1968.
57. J. R. Heirzler, X. Le Pichon and J. G. Baron, Magnetic anomalies over the Reykjanes ridge, *Deep-Sea Res.* **13**, 427–43, 1966.
58. L. R. Sykes, The seismicity of the Arctic, *Bull. Seismol. Soc. Am.* **55**, 501–18, 1965.
59. L. R. Sykes, Mechanism of earthquakes and nature of faulting on the midocean ridges, *J. Geophys. Res.* **72**, 2131–53, 1967.
60. M. Ewing, J. Ewing and M. Talwani, Sediment distribution in the oceans: the Mid-Atlantic Ridge, *Bull. Geol. Soc. Am.* **75**, 17–36, 1964.
61. A. E. Maxwell, R. P. Von Herzen, K. J. Hsu, J. E. Andrews, T. Saito, S. F. Percival, E. D. Milow and R. E. Boyce, Deep sea drilling in the South Atlantic, *Science* **168**, 1047–59, 1970.
62. E. C. Bullard, J. E. Everett and A. G. Smith, The fit of the continents around the Atlantic, *Philos. Trans. R. Soc. London* **258**, 41–51, 1965.
63. H. W. Menard, Extension of northeast Pacific fracture zones, *Science* **155**, 72–4, 1967.
64. W. J. Morgan, Rises, trenches, great faults and crustal blocks, *J. Geophys. Res.* **73**, 1959–82, 1968.

65. D. P. McKenzie and R. L. Parker, The north Pacific: an example of tectonics on a sphere, *Nature* **216**, 1276–80, 1967.
66. D. P. McKenzie and W. J. Morgan, Evolution of triple junctions, *Nature* **224**, 125–33, 1969.
67. K. Wadati, Shallow and deep earthquakes, *Geophys. Mag. (Tokyo)* **4**, 231–85, 1931.
68. T. Utsu, Regional differences in absorption of seismic waves in the upper mantle as inferred from abnormal differences in seismic intensities, *J. Fac. Sci. Hokkaido Univ. Japan, Ser. VII* **2**, 359–74, 1966.
69. L. R. Sykes, The seismicity and deep structure of island arcs, *J. Geophys. Res.* **71**, 2981–3006, 1966.
70. J. Oliver and B. Isacks, Deep earthquake zones, anomalous structures in the upper mantle, and the lithosphere., *J. Geophys. Res.* **72**, 4259–75, 1967.
71. B. Isacks, J. Oliver and L. R. Sykes, Seismology and the new global tectonics, *J. Geophys. Res.* **73**, 5855–99, 1968.
72. B. Isacks and P. Molnar, Distribution of stresses in the descending lithosphere from a global survey of focal-mechanism solutions of mantle earthquakes, *Rev. Geophys. Space Phys.* **9**, 103–74, 1971.
73. G. F. Davies and J. N. Brune, Regional and global fault slip rates from seismicity, *Nature* **229**, 101–7, 1971.
74. W. J. Morgan, Convection plumes in the lower mantle, *Nature* **230**, 42–3, 1971.
75. W. J. Morgan, Plate motions and deep mantle convection, *Mem. Geol. Soc. Am.* **132**, 7–22, 1972.
76. T. S. Crough, Hotspot swells, *Annu. Rev. Earth Planet. Sci.* **11**, 165–93, 1983.
77. J. A. Whitehead and D. S. Luther, Dynamics of laboratory diapir and plume models, *J. Geophys. Res.* **80**, 705–17, 1975.
78. W. J. Morgan, Hotspot tracks and the opening of the Atlantic and Indian Oceans, in: *The Sea*, C. Emiliani, ed., Wiley, New York, 443–87, 1981.
79. D. E. Loper and F. D. Stacey, The dynamical and thermal structure of deep mantle plumes, *Phys. Earth Planet. Inter.* **33**, 304–17, 1983.
80. F. W. Stacey and D. E. Loper, The thermal boundary layer interpretation of D$''$ and its role as a plume source, *Phys. Earth Planet. Inter.* **33**, 45–55, 1983.
81. P. Olson and H. A. Singer, Creeping plumes, *J. Fluid Mech.* **158**, 511–31, 1985.
82. R. W. Griffiths and I. H. Campbell, Stirring and structure in mantle plumes, *Earth Planet. Sci. Lett.* **99**, 66–78, 1990.
83. M. A. Richards, R. A. Duncan and V. E. Courtillot, Flood basalts and hot-spot tracks: plume heads and tails, *Science* **246**, 103–7, 1989.
84. I. H. Campbell and R. W. Griffiths, Implications of mantle plume structure for the evolution of flood basalts, *Earth Planet. Sci. Lett.* **99**, 79–83, 1990.

85. H. Benard, Les tourbillons cellulaires dans une nappe liquide transportant de la chaleur par convection en regime permanent, *Ann. Chim. Phys.* **23**, 62–144, 1901.
86. W. Hopkins, Researches in physical geology, *Philos. Trans. R. Soc. London* **129**, 381–5, 1839.
87. O. Fisher, *Physics of the Earth's Crust*, Murray, London, 1881.
88. A. Holmes, Continental drift: a review, *Nature* **122**, 431–3, 1928.
89. A. Holmes, Radioactivity and earth movements, *Geol. Soc. Glasgow, Trans.* **18**, 559–606, 1931.
90. A. Holmes, *Principles of Physical Geology*, Thomas Nelson and Sons, 1944.
91. A. W. Hofmann and W. M. White, Mantle plumes from ancient oceanic crust, *Earth Planet. Sci. Lett.* **57**, 421–36, 1982.
92. G. F. Davies, Mantle plumes, mantle stirring and hotspot chemistry, *Earth Planet. Sci. Lett.* **99**, 94–109, 1990.
93. C. L. Pekeris, Thermal convection in the interior of the earth, *Mon. Not. R. Astron. Soc., Geophys. Suppl.* **3**, 346–67, 1935.
94. A. L. Hales, Convection currents in the earth, *Mon. Not. R. Astron. Soc., Geophys. Suppl.* **3**, 372-79, 1936.
95. D. P. McKenzie, A. B. Watts, B. Parsons and M. Roufosse, Planform of mantle convection beneath the Pacific Ocean, *Nature* **288**, 442–6, 1980.
96. D. C. Tozer, Heat transfer and convection currents, *Philos. Trans. R. Soc. London, Ser. A* **258**, 252–71, 1965.
97. G. F. Davies, Thermal histories of convective earth models and constraints on radiogenic heat production in the earth, *J. Geophys. Res.* **85**, 2517–30, 1980.
98. B. Gutenberg and C. F. Richter, *Seismicity of the Earth*, Geol. Soc. Amer. Spec. Paper 34, 1941.

PART 2

FOUNDATIONS

Part 2 assembles the observations, concepts and tools that are required for a quantitative discussion of mantle convection. Mantle convection itself is the subject of Part 3.

Although our knowledge of the interior of the earth is incomplete because it is accessible only to indirect observations, there are observations from the earth's surface that impose important constraints on mantle convection. As well, quite a lot is known about mantle structure and properties. This information is summarised in Chapters 4 and 5. It is collected in these chapters so that the discussion of mantle convection will not be cluttered by descriptions of the observations, and so that it is in a readily identifiable place. These two chapters could be read quickly and then used as a reference.

Convection involves two basic physical processes, fluid flow and heat conduction, and the fundamentals of these topics are developed in Chapters 6 and 7. Since fluid flow in particular may not be very familiar to many geological scientists, it is developed at some length. In these chapters, the key ideas and results are developed in as simple a way as possible, so as to make them accessible to as wide an audience as possible. Although there are sections that include more mathematical treatments for those who are interested, these are identified as being more advanced and are not essential to understanding later chapters. Therefore if you are less mathematically inclined, you need not be unduly concerned by glimpses of elaborate equations.

CHAPTER 4

Surface

Observations of the earth's surface provide important constraints on mantle dynamics. As well, our knowledge of the earth's interior comes entirely from observations made at or close to the surface. The deepest borehole penetrates only about 10 km, compared with the nearly 3000 km depth of the mantle.

Seismology provides by far the most detailed and accurate information on the structure of the earth's interior. Observed variations in the strength of gravity at the earth's surface yield complementary constraints on variations in density in the interior. The topography of the earth's surface and geographical variations in the rate at which heat conducts through the surface provide important constraints on internal dynamical processes.

This chapter focusses on observations that, for the moment, are best left in the form of their surface geography, namely topography, heat flow and gravity. The geography of the tectonic plates is also summarised here. The full interpretation and implications of these observations are intimately related to mantle dynamics, and they will be considered as part of that topic in Part 3. However, some of the significance of the observations of topography and heat flow can be appreciated more directly, and this will be demonstrated in Chapter 7.

On the other hand, although seismological observations are made at the surface, they are usually converted into graphs or images of internal structure. Therefore the presentation of seismological constraints is deferred to Chapter 5, where they will be presented in conjunction with our knowledge of the material properties of the interior.

4.1 Plates

The existence of the tectonic plates, moving pieces of the lithosphere, is an observation central to the topic of mantle convection.

The evidence for their existence was described in Chapter 3. The way they move is described in Chapter 9. Here I will briefly emphasise key features of the plates.

A map of the plates is shown in Figure 4.1. They have a considerable range of sizes. The large Pacific plate is about 14 000 km across at its widest. The other large plates are 5000–10 000 km across. The Nazca plate, to the west of South America, is about 4000 km across. There are several smaller plates in the range 1000–2000 km across, and various fragments that are smaller still. A notable comparison is between three plates in the Pacific basin: the large Pacific plate, the middle-sized Nazca plate and the small Cocos plate (off Central America), all of which move rapidly (50–100 mm/a relative to each other and to the hotspots).

The shapes of the plates are varied and irregular. The shapes of the main plates are better compared in Figure 4.2, which shows each at the same scale in a Lambert equal area projection centred on each plate. The Indo-Australia plate is boomerang-shaped (though there is evidence that it is deforming in the middle and might be better considered to be two plates in slow relative motion). The Eurasian plate also has an odd shape, with Southeast Asia projecting. There is considerable deformation within China and Southeast Asia. Although there is no clear evidence of deformation between the North American and South American plates, the join between them is so narrow that they are often regarded as separate plates.

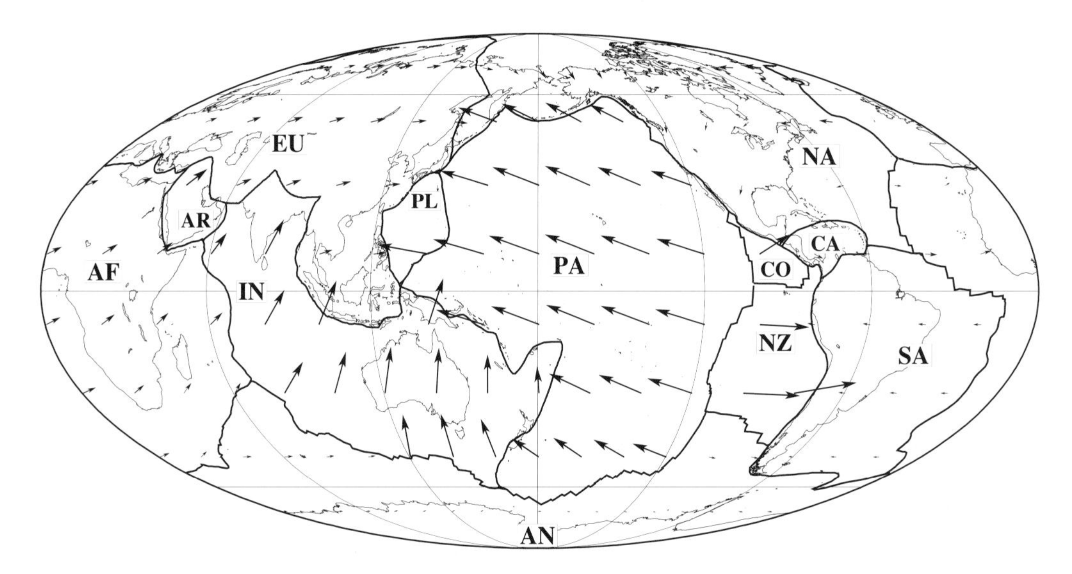

Figure 4.1. Map of the tectonic plates and their velocities. From Lithgow-Bertelloni and Richards [1] using the velocities of Gordon and Jurdy [2]. Copyright by the American Geophysical Union.

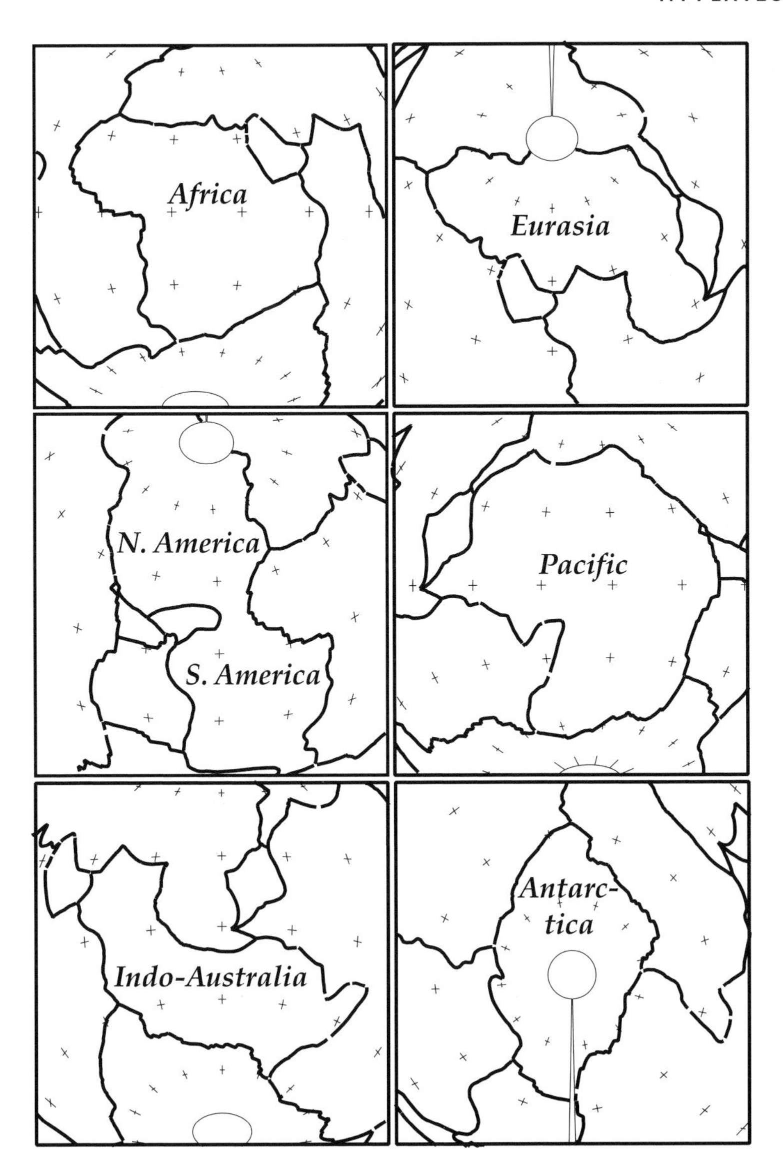

Figure 4.2. Outlines of the large plates at the same scale and in a Lambert equal area projection, centred on the plate to minimise distortion. Ticks are at 30° intervals of latitude and longitude and the polar circles are at 80° latitude.

Many plate margins are angular, mainly as a result of orthogonal relationships between spreading centres and transform faults. Subduction zone margins tend to be curved where they are not constrained to follow a continental margin, but this is not universal. The curvature tends to be convex towards the subducting plate – this is the well-known pattern of island arcs, most clearly expressed by the Aleutian Island arc in the North Pacific.

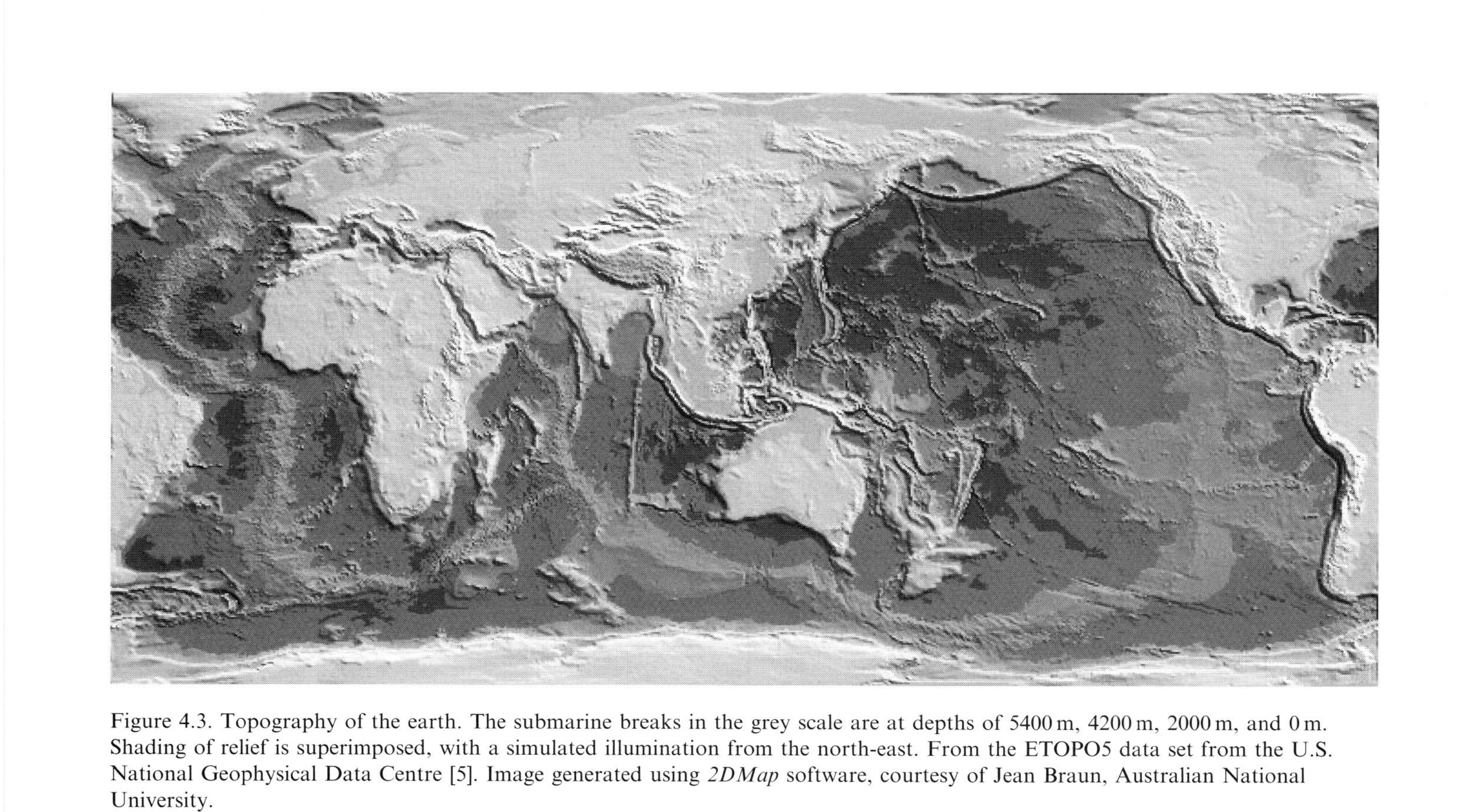

Figure 4.3. Topography of the earth. The submarine breaks in the grey scale are at depths of 5400 m, 4200 m, 2000 m, and 0 m. Shading of relief is superimposed, with a simulated illumination from the north-east. From the ETOPO5 data set from the U.S. National Geophysical Data Centre [5]. Image generated using *2DMap* software, courtesy of Jean Braun, Australian National University.

Plates change with time, even when no new plate margins form. There are actually three kinds of change recorded in the record of seafloor magnetic stripes: steady growth or shrinkage of plates, sudden changes in plate velocity, and the formation of new plate margins by plate breakup. The first kind of change is a consequence of the difference in behaviour between spreading margins and converging margins. This will be explored in detail in Chapter 9. Thus plates may grow and shrink, and some plates may disappear, through the normal evolution of their margins. Examples of sudden changes in plate velocity and of plate breakup will also be noted in Chapter 9.

These characteristics of plates, namely their range of sizes, their odd shapes and angularity, and the particular ways in which they change, are not typical of a convecting fluid. This will emerge in Part 3 as a distinctive and illuminating feature of the mantle dynamical system.

4.2 Topography

4.2.1 Continents

The division of the earth's surface into continents and ocean basins is so familiar that it is easy to overlook its significance. With the ocean water removed, it is obvious that the continents are the primary topographic feature of the surface of the solid earth (Figure 4.3). Furthermore, the continents are not just the parts of the solid earth that happen to protrude above the ocean surface. They are plateaux whose tops are remarkably flat, and very close to sea level, apart from restricted areas of mountain ranges. Since much of the sea floor is also very flat, this gives the earth's topography a bimodal distribution of area versus elevation, with peaks near −4 km and 0 km (Figure 4.4).

The continental crust is known from seismology to be 35–40 km thick and less dense (about 2700 kg/m^3) than the mantle (3300 kg/m^3) or the oceanic crust (2900 kg/m^3) . The oceanic crust is only about 6 km thick. The differences in thickness and density between continental and oceanic crust have been long recognised as the explanation for the higher elevation of the continental surfaces: the continents are relatively buoyant and float higher. This was emphasised particularly by Wegener, who used it to argue against the idea of former land bridges between the present continents.

However, buoyancy alone does not explain the bimodality of the earth's topography: why is the continental material piled up usually to near sea level, rather than to a wide range of heights

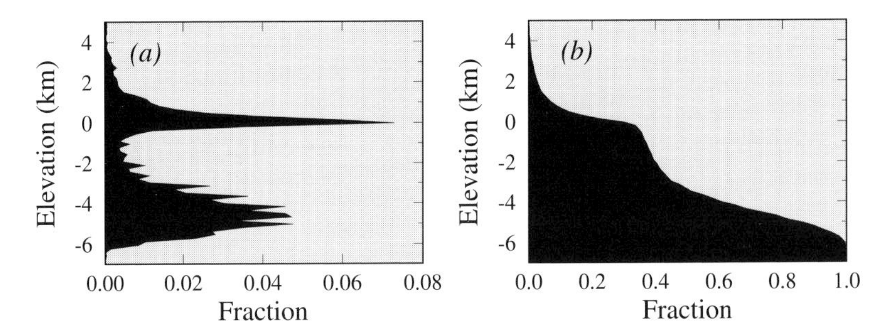

Figure 4.4. Distribution of elevations of the earth's solid surface (from the ETOPO5 data set [5]). (a) A histogram of elevations, relative to sea level. (b) Cumulative fraction of the earth's surface above a given elevation.

above the deep sea floor? Why aren't the continental margins eroded into broad slopes and fans, like the margins of mountain ranges? The explanation evidently lies in the combined workings of subaerial erosion, submarine erosion and plate tectonics. This was only clearly recognised with the formulation of plate tectonics, but the recognition was early. Both Dietz [3] and Hess [4] saw that seafloor spreading (and subduction) provided a sweeping mechanism whereby continental material on the sea floor could be carried to continental margins and piled up there. This would restrict the areal extent of the continental material. The vertical distribution evidently is controlled by rapid subaerial erosion, which reduces the surface to near sea level, and very slow submarine erosion, which allows the relatively steep continental slopes to survive.

4.2.2 Sea floor

Turning now to the sea floor, the dominant feature is the system of 'midocean' ridges or rises. The rises reach elevations of 2–3 km above the surrounding deep sea floor, or 2–3 km below sea level, averaging about 2.6 km depth, apart from some relatively shallow regions that we will distinguish shortly. They are thousands of kilometres wide, but they do not have clear edges: the slopes of their flanks simply decrease with distance away from the crest. A typical profile is shown in Figure 4.5.

The next most prominent seafloor features are plateaux, ridges and chains of seamounts. Some of the plateaux are known to be composed of continental-type crust, such as the Campbell Plateau south-east of New Zealand and the Seychelles plateau north-east of Madagascar. These are interpreted to be fragments of continental crust that have been isolated by the vagaries of plate break-ups.

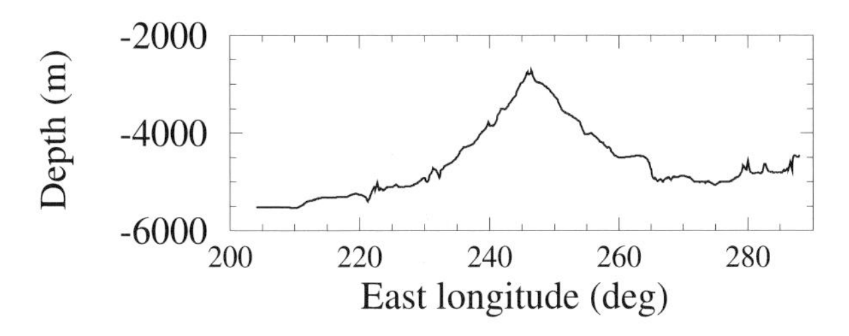

Figure 4.5. Profile across the East Pacific Rise in the south-east Pacific. From the ETOPO5 data set [5].

Other plateaux, such as the Ontong–Java Plateau north-east of New Guinea and the Kerguelen Plateau in the southern Indian Ocean, are believed to be more closely related to oceanic crust on the basis of dredge and drill samples and some seismic profiles.

There are narrow, often linear, features that are in some cases continuous ridges, such as the Ninetyeast Ridge in the eastern Indian Ocean (at 90° E longitude), but more commonly are less continuous chains of seamounts, such as the Hawaii–Emperor chain extending north-west from Hawaii. Many of these have been sampled and found to be basaltic and closely related to the oceanic crust.

These classes of topographic features (plateaux, ridges and seamount chains) are due to greater than normal thicknesses of crust (whether oceanic or continental in character). In the case of the more extensive plateaux, this crust may be floating in isostatic balance. In the case of seamounts that are not more than 100–200 km in width, the strength of the lithosphere may be supporting the extra weight out of isostatic balance.

A number of these ridges have currently active volcanic centres, usually at one end. These volcanic centres are *volcanic hotspots*, of which there are about 40 scattered around the earth's surface, not all of them in ocean basins. (I use the term 'hotspot' to refer to the surface volcanic feature, as I discussed in Section 3.7, rather than to the putative hot volume in the mantle that Wilson conceived of, but for which there is currently little support.) The number of such hotspots depends on how they are defined. Geologists have recognised that there are substantial volcanic centres that are either isolated from plate boundaries (like Hawaii) or show a substantial excess of volcanism on a spreading centre (like Iceland). It is in the interpretation of the word 'substantial' that subjective judgement enters – probably not every isolated volcano belongs to this class. This uncertainty of definition is particularly pertinent in Africa, which has many volcanic centres scattered across it.

Another class of seafloor topographic feature that has been clearly identified is broad, low swells, typically 0.5–2 km in height and 1000–2000 km in width. The clearest example is the Hawaiian swell (Figure 4.3), which extends for about 500 km around the most prominent part of the Hawaiian seamount chain. These were characterised by Crough [6, 7], who recognised that they are associated with volcanic hotspots and coined the term 'hotspot swell'.

4.2.3 Seafloor depth versus age

The depth of the sea floor increases with distance from the crest of a midocean rise. However the East Pacific Rise is broader than the Mid-Atlantic Rise, as can be seen in Figure 4.3. It is also true that seafloor spreading is faster at the East Pacific Rise than in the Atlantic Ocean. In 1971 it was found by Sclater and others [8] that the depth of the sea floor depends primarily on the *age* of the sea floor. Qualitatively, this explains why the East Pacific Rise is broader, because sea floor of a given age has travelled further from the rise crest than in the Atlantic, but has about the same depth.

It turns out that this kind of behaviour can be explained quite well simply in terms of conductive cooling of the hot crust and mantle as it drifts away from a rise crest. In the course of a mathematical analysis of this process, which I will present in Chapters 7 and 10, Davis and Lister [10] showed that seafloor depth is approximately proportional to the square-root of age, and that this is most readily demonstrated by plotting depth versus the square-root of age. Examples of such plots for various regions of the sea floor are shown in Figure 4.6.

4.3 Heat flow

4.3.1 Sea floor

The heat being conducted out of the sea floor is determined by measuring the temperature gradient with depth, $\partial T/\partial z$, and the conductivity, K, of the rocks through which this temperature gradient is measured. The product of these quantities then gives the heat flux, q:

$$q = -K\partial T/\partial z$$

where the minus is because the heat flows from the hotter to the cooler rocks. During the 1950s and 1960s, techniques were devel-

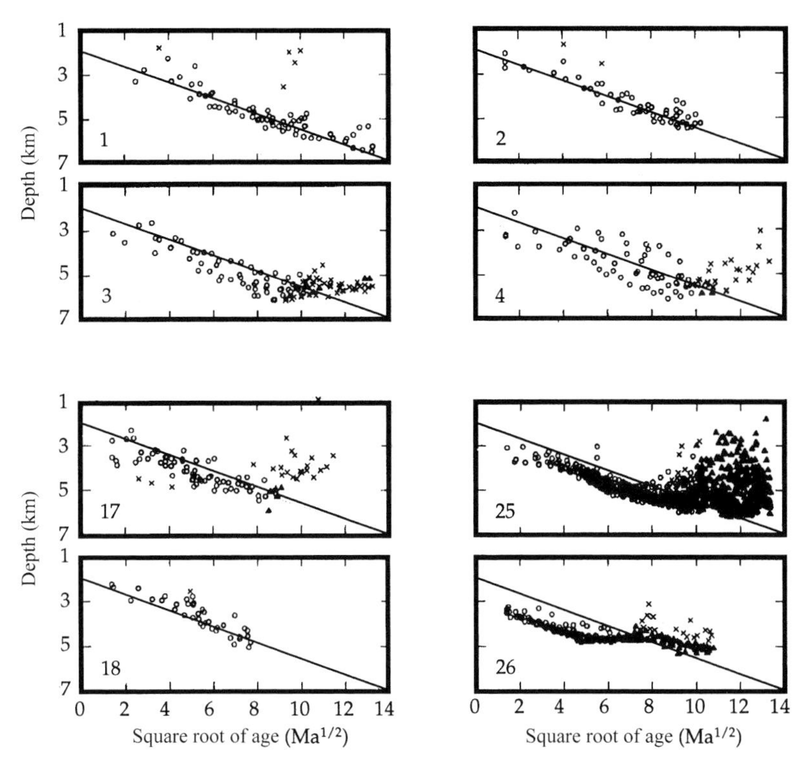

Figure 4.6. Seafloor depth versus square-root of age for a selection of regions of the sea floor. The same reference line is shown in each plot. After Marty and Cazenave [9]. Copyright by Elsevier Science. Reprinted with permission.

oped to measure these quantities in the deep sea floor by dropping probes from ships.

With the advent of the theory of seafloor spreading, a systematic programme was pursued to test the theory by determining whether heat flow was high near midocean rises. It emerged that it is, but the results were quite scattered. Away from rise crests, the heat flow seemed to settle to an approximately constant value. This bit of history is important, because it led to a model in which the lithosphere is assumed to have a constant thickness, consistent with the heat conducted through it being asymptotically constant [11, 12, 13].

Subsequently, it was found that measurements on sea floor younger than about 40 Ma were affected by the circulation of water through the rocks [14, 15]. This reduced the temperature gradient, so that the measured heat flux was only part of the total heat flux, the balance being transported by the hydrothermal circulation. It was also found that more reliable measurements could be obtained by carefully choosing sites where sediments are undisturbed and of a kind that restricted the circulation of water. Heat flux values obtained in this way were in many cases higher and less scattered than previous measurements. The net result was that heat flux was found to decrease steadily with seafloor age (Figure 4.7A).

The same theory of conductive cooling of the oceanic crust and mantle as predicts the subsidence of the sea floor (Figure 4.6) also predicts that the heat flux should be inversely proportional to the square-root of the seafloor age (Chapters 7 and 10). On a logarithmic plot, this would be represented by a straight line of slope $-1/2$. Figure 4.7B shows the observations on such a plot, with a line of slope $-1/2$ ('simple cooling model', dashed) for reference. (The corresponding curve is included in Figure 4.7A). It is clear that the observations are consistent with this theory.

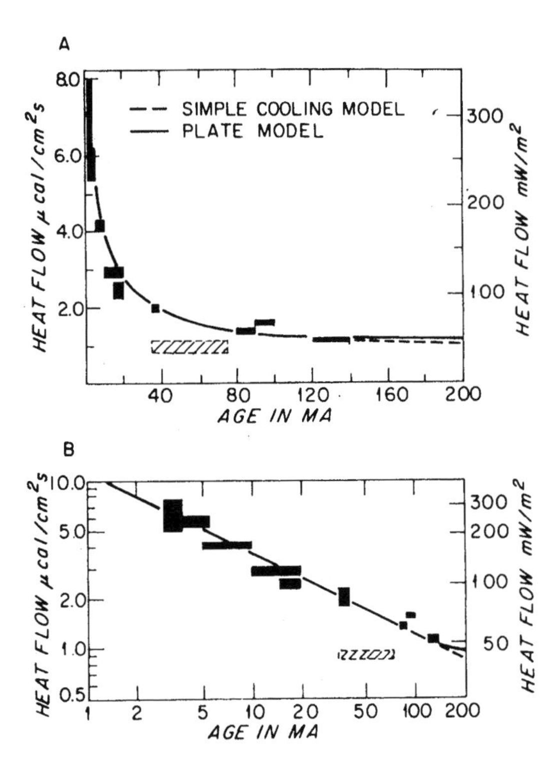

Figure 4.7. (A) Oceanic heat flow versus age of sea floor. The two curves are explained in the text. The shaded box represents observations that were superseded by the data at 40 Ma. (B) The same data and curves on a logarithmic plot, in which the 'simple cooling model' predicts a straight line with a slope of $-1/2$. From Sclater *et al.* [16]. Copyright by the American Geophysical Union.

The predictions of the model based on the assumption of a lithosphere of constant thickness are included in Figure 4.7 as solid lines ('plate model'). The predictions of the two models diverge only at ages greater than the age of the oldest measured sea floor. In fact no asymptotically constant value is evident in the newer data. Thus the original observational basis for the constant-thickness lithosphere model has been removed.

4.3.2 Continents

The increase in temperature with depth in mines had long been known, even in 1855 when Airy remarked on its effect on rock deformation (Chapter 3). Reasonable estimates were also available by then for the flux of heat being conducted out of the earth's continental crust (see Fisher [17] and Chapter 2). A modern average heat flux for continental crust is about 55 mW/m^2, compared with about 100 mW/m^2 in the oceans [16].

Heat flux tends to be higher in areas that have been tectonically active in the more recent past. It has been proposed that there is a relationship between heat flux and 'tectonic age' for continents that is analogous to the relationship found for the sea floor (Figure 4.7), though with a longer time scale of about 500 Ma (Figure 4.8). However the data are quite scattered, and this has not yielded any great insights. This may be because tectonic age is not a well-defined quantity, and also because there is a complex and not well-understood relationship between erosion, the amount of

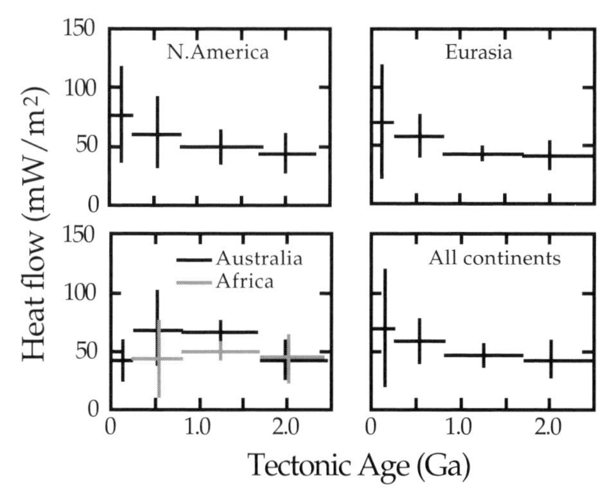

Figure 4.8. Continental heat flow versus tectonic age. From Sclater *et al.* [16]. Copyright by the American Geophysical Union.

(a)

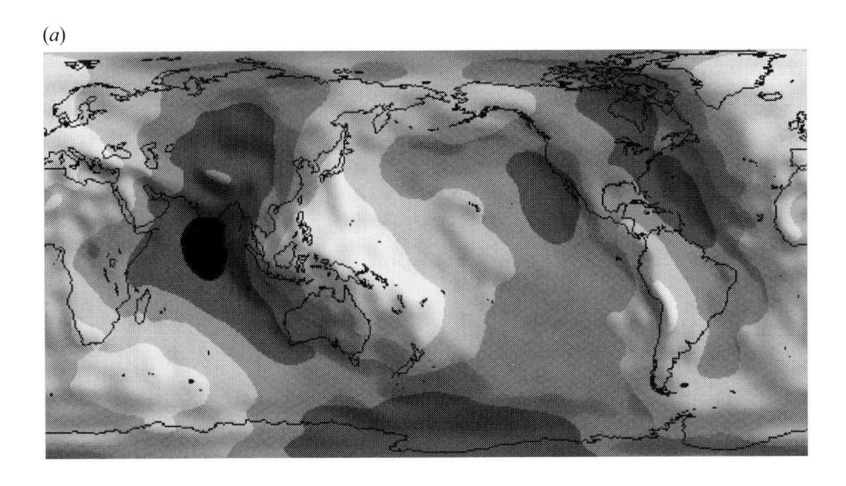

(b)

(c)

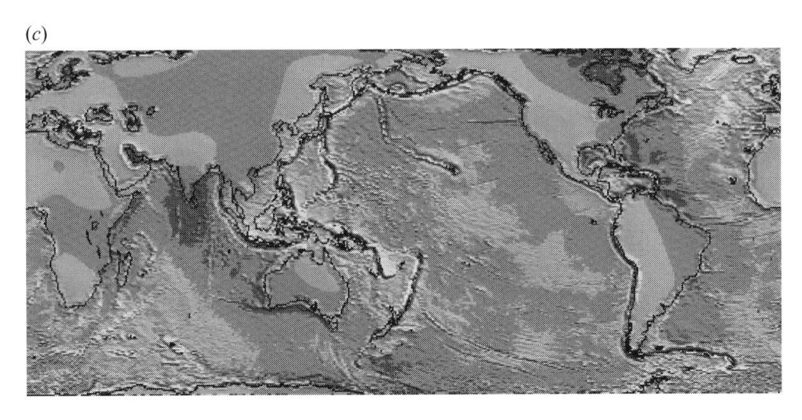

Figure 4.9. Three representations of the earth's gravity field. Deviations of the geoid (a, top) and gravity (b, middle) from that expected for a rotating, hydrostatic earth, to a resolution of about 2000 km. Data from model PGS 3520, similar to model GEM-T3 [18, 19]; (c, bottom) Variations of gravity at high resolution derived from radar measurements of the height of the sea surface (Haxby [20] and NGDC [5]). Images produced by *2DMap* software, courtesy of Jean Braun, Australian National University.

radioactivity removed by erosion, and the extra heat flow due to the excavation of deeper, hotter rocks by erosion. The vertical distribution of radioactivity within the continental crust is not well-understood either, and may be quite variable. This will be discussed further in Chapter 7.

It was expected, when the first measurements of seafloor heat flux were made, that it would be less than that for continents, because continental rocks have a much higher rate of radioactive heat generation (Chapter 8). The presumption was that similar amounts of heat would be emerging from the earth's interior in the continents and oceans, and that the continental radiogenic contribution would add to this. However, the early seafloor values were quite comparable to continental values. In fact, more recent estimates of the average seafloor heat flux are about twice the continental value [16]. The resolution of this puzzle has been that in the oceans heat is transported from the deep interior by the mass motions involved with seafloor spreading, not just by conduction, as we will see in Chapters 7 and 10.

4.4 Gravity

The strength of the gravity field at the earth's surface varies with location by as much as 0.05%. This is due to variations in the density of the earth's interior. Since convection is driven by density differences, we might expect that there is important information about mantle convection in the gravity variations. This is undoubtedly the case. However, the effect of convection on the surface gravity is more subtle than might at first appear. This is because an internal density anomaly causes a vertical deflection of the earth's surface (i.e. it causes topography) which also contributes to the net gravity. However, the contribution from the induced topography is of the opposite sign to that from the internal density anomaly. Thus the two gravity contributions tend to cancel. The sign of the resulting net gravity perturbation depends on details of the mechanical properties of the interior (the elasticity or viscosity). A consequence is that the interpretation of the gravity field is not straightforward, and is still the subject of considerable debate. Since I want to present here the most robust and accessible arguments, I will not discuss the interpretation of the gravity field in great detail. I will return to it briefly in Part 3 to summarise the current situation.

Nevertheless it will be useful to present here the main features of the gravity field. Maps of variations in the gravity field are shown in Figure 4.9 in three different ways. Figure 4.9a shows the geoid, while Figures 4.9b,c show variations in the strength of gravity at different spatial resolutions. The geoid is defined as the equipotential surface of the gravity field that is coincident with the surface of the oceans. It is useful because it is more sensitive than gravity to deeper or larger-scale density variations. The information in the geoid map is therefore somewhat complementary to that in the gravity maps.

We can see this relationship between geoid and gravity by recalling that gravitational attraction is proportional to $1/r^2$, where r is the distance from an attracting mass. On the other hand, gravitational potential is proportional to $1/r$. As a result, the geoid, which is a measure of gravitational potential, is sensitive to mass over a greater distance range. For our purposes here, you can think of the geoid as 'feeling' to greater depth in the earth than gravity. Correspondingly, the large-scale or long-wavelength components of the geoid are enhanced relative to those of gravity.

The long-wavelength components of the gravity field (or, more specifically, the low spherical-harmonic components) have been most accurately measured through the gravitational potential, rather than the strength of gravity. Thus the early determinations were from perturbations to the orbits of artificial earth satellites. On the other hand, short-wavelength components were most accurately determined from measurements of the strength of gravity at or near the earth's surface. Detailed measurements were later obtained by measuring the topography of the sea surface using radar measurements of distance from satellites. The latter measurements contain a great deal of short-wavelength information.

Turning now to the maps, the largest-amplitude features in the geoid (Figure 4.9a) are at the largest scale. There are broad highs over the western Pacific Ocean and Africa, separated by a band of lows that follow approximately a great circle through the western Atlantic Ocean, the poles, Central Asia, India and the Indian Ocean. There is a smaller but distinct high over the Andes mountains in South America, and less-distinct highs over the Alpine–Himalayan and North American mountain belts. The deepest low is to the south of India. The midocean rises have little expression in the geoid.

The largest gravity anomalies at intermediate wavelength (Figure 4.9b) are along subduction zones and convergence zones. The gravity and geoid expressions of these regions correspond fairly closely, but the geoid contains stronger long-wavelength components. Thus there are gravity highs over the circum-Pacific subduction zones and along the Alpine–Himalayan belt. The main band of lows evident in the geoid map is also discernible in this gravity map, though it is less pronounced. There is a suggestion of gravity highs over some hotspots, such as Hawaii, Iceland, and Kerguelen in the southern Indian Ocean.

The more detailed gravity map derived from satellite radar altimetry (Figure 4.9c) reveals many of the same intermediate-scale features in oceanic regions. Comparison with Figure 4.9b shows that the broad features, at near-continental scale, correspond quite well. However, many of the smaller-scale details of Figure 4.9b are not present. This indicates that the wavelength truncation inherent in the spherical-harmonic representation of Figures 4.9a,b has introduced an artificial 'ripple' (known to the initiated as the Gibbs effect).

There is much detail in Figure 4.9c. There is a characteristic gravity signature of subduction zones, though it is not so easy to see at the scale of Figure 4.9c. Typically there is a gravity low of several hundred milligals over the trench, and a high of lesser amplitude on the landward side of the trench. There is a low-amplitude but clearly discernible gravity high over the Mid-Atlantic Ridge, and less obvious highs over the Pacific and Indian Ridges. There are clear highs over some hotspot swells. The Hawaiian–Emperor chain of seamounts, extending to the north-west from Hawaii, has highs over the volcanic chain and flanking lows due to the local depression of the sea floor from the weight of the seamounts. These are superimposed on the broader high over the hotspot swell. At yet smaller scale, many fracture zones and individual seamounts are clearly discernible through their gravity signature.

4.5 References

1. C. Lithgow-Bertelloni and M. A. Richards, The dynamics of cenozoic and mesozoic plate motions, *Rev. Geophys.* **36**, 27–78, 1998.
2. R. G. Gordon and D. M. Jurdy, Cenozoic global plate motions, *J. Geophys. Res.* **91**, 12 389–406, 1986.
3. R. S. Dietz, Continent and ocean evolution by spreading of the sea floor, *Nature* **190**, 854–7, 1961.
4. H. H. Hess, History of ocean basins, in: *Petrologic Studies: a Volume in Honor of A. F. Buddington*, A. E. J. Engel, H. L. James and B. F. Leonard, eds., Geol. Soc. Am., Boulder, CO, 599–620, 1962.
5. NGDC, National Geophysical Data Center, U.S. National Oceanic and Atmospheric Administration, 325 Broadway, Boulder, CO 80303–3328.
6. S. T. Crough, Thermal origin of mid-plate hotspot swells, *Geophys. J. R. Astron. Soc.* **55**, 451–69, 1978.

7. S. T. Crough, Hotspot swells, *Annu. Rev. Earth Planet. Sci.* **11**, 165–93, 1983.
8. J. G. Sclater, R. N. Anderson and M. L. Bell, Elevation of ridges and evolution of the central eastern Pacific, *J. Geophys. Res.* **76**, 7888–915, 1971.
9. J. C. Marty and A. Cazenave, Regional variations in subsidence rate of oceanic plates: a global analysis, *Earth Planet. Sci. Lett.* **94**, 301–15, 1989.
10. E. E. Davis and C. R. B. Lister, Fundamentals of ridge crest topography, *Earth Planet. Sci. Lett.* **21**, 405–13, 1974.
11. M. G. Langseth, X. LePichon and M. Ewing, Crustal structure of midocean ridges, 5, Heat flow through the Atlantic Ocean floor and convection currents, *J. Geophys. Res.* **71**, 5321–55, 1966.
12. D. P. McKenzie, Some remarks on heat flow and gravity anomalies, *J. Geophys. Res.* **72**, 6261–73, 1967.
13. J. G. Sclater and J. Francheteau, The implications of terrestrial heat flow observations on current tectonic and geochemical models of the crust and upper mantle of the Earth, *Geophys. J. R. Astron. Soc.* **20**, 509–42, 1970.
14. C. R. B. Lister, On the thermal balance of a midocean ridge, *Geophys. J. R. Astron. Soc.* **26**, 515–35, 1972.
15. J. G. Sclater, J. Crowe and R. N. Anderson, On the reliability of oceanic heat flow averages, *J. Geophys. Res.* **81**, 2997–3006, 1976.
16. J. G. Sclater, C. Jaupart and D. Galson, The heat flow through the oceanic and continental crust and the heat loss of the earth, *Rev. Geophys.* **18**, 269–312, 1980.
17. O. Fisher, *Physics of the Earth's Crust*, Murray, London, 1881.
18. J. G. Marsh, C. J. Koblinsky, F. J. Lerch, S. M. Klosko, J. W. Robbins, R. G. Williamson and G. B. Patel, Dynamic sea surface topography, gravity, and improved orbit accuracies from the direct evaluation of Seasat altimeter data, *J. Geophys. Res.* **95**, 13 129–50, 1990.
19. F. J. Lerch *et al.*, A geopotential model for the earth from satellite tracking, altimeter, and surface gravity observations: GEM-T3, *J. Geophys. Res.* **99**, 2815–39, 1994.
20. W. F. Haxby, *Gravity Field of the World's Oceans*, Lamont-Doherty Geophysical Observatory, Palisades, NY, 1987.

CHAPTER 5

Interior

In this chapter we look at the structure, chemical composition and physical state of the earth's interior, and especially of the mantle. The internal structure of the earth defines the mantle, and hence the arena of mantle convection. There is also important structure within the mantle. The physical state and chemical composition of the mantle determine the properties of the mantle that permit mantle convection to occur and that control its form.

The primary internal structure of the earth is defined by its main layers, and most of this chapter is about the layers and their inferred composition. We finish with a summary of three-dimensional structure, which has been increasingly well-resolved over the past decade or so and which has recently become clear enough to begin to relate directly to mantle convection.

The internal structure is determined mainly from seismology (e.g. [1]). With the main layers so determined, composition is inferred from three main sources: field occurrences of rocks inferred to come from the shallower parts of the earth; igneous rocks thought to be formed from magma derived from melting the mantle; and laboratory measurements of the physical properties of minerals and rocks.

The term physical state here includes not just, for example, whether the material is solid or liquid, but also the particular assemblage of mineral phases in which each part of the mantle exists. This mineralogy depends on chemical composition, temperature and pressure. Pressure-induced changes are of particular importance. They define the main internal mantle structure and they have the potential to substantially modify mantle convection. These potential dynamical effects arise from the interaction of temperature and composition with the pressure-induced phase transformations, so we will need to look at those aspects in some detail.

5.1 Primary structure

5.1.1 Main layers

The main seismological divisions of the earth's interior are the crust, the mantle and the core. A jump in seismic velocities at a depth of about 60 km was inferred by Mohorovičić in 1909 [2], based on his identification of a seismic wave that was refracted through the interface and travelled, at higher velocity, just below the interface. Such a discontinuity has been identified in many continental areas, usually at a depth of 35–40 km.

Until this time, the term 'crust' was often used loosely to refer either to an outer layer of different composition or an outer layer of greater strength. After this time, the term became restricted to the seismically defined layer, which has been inferred to have a different composition from the substratum. The stronger layer is usually thicker, includes the crust, and is referred to as the 'lithosphere' [3].

According to Bullen [4], evidence for the existence of a core with different seismic velocity was first presented by Oldham in 1906 [5]. More detailed evidence was presented by Gutenberg in 1913, who determined the depth of the interface with the core to be about 2900 km. A recent seismological model gives the depth of the interface as 2889 km [6].

The core is inferred to be liquid because it transmits compressional seismic waves, but not shear waves. This cannot be observed directly, because the waves that travel through the core still have to travel through the intervening mantle before they can be observed by seismologists. What is observed is that only one kind of wave travels through the core. It is characteristic of solids that two kinds of elastic waves propagate through them with different velocities: compressional waves, like familiar sound waves in air, and shear waves that involve shearing deformations which change the shape of the material at constant volume. Liquids, on the other hand, have no shear strength and propagate only compressional waves. It is inferred from this that the core must be liquid and the waves that pass through it must be compressional waves.

The main layers of the earth's interior are evident in Figure 5.1, which shows profiles of seismic velocities and density with depth, as inferred from seismology [6]. The largest discontinuity in properties occurs at the core–mantle boundary. The crust is so thin on this scale that it is barely visible. It is simplified as a two-layer structure extending to 35 km depth. Other structure is evident with the mantle, which will be discussed in the next section. An inner core is defined by the jumps near 5150 km depth. The non-zero shear

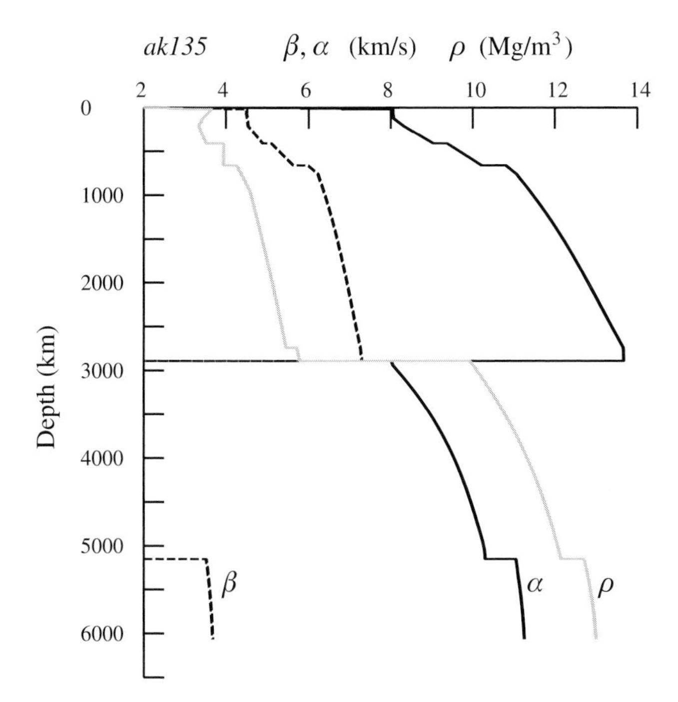

Figure 5.1. Profiles of seismic velocities and density with depth in the earth. Compressional velocity: α; shear velocity: β; density: ρ. From the model *ak135* [6, 7].

velocity (β) below this depth indicates that the inner core is solid. The inner core was discovered by Inge Lehmann in 1936 [8].

Since the core is not the main focus of this book, we will discuss it only briefly here. It is inferred to be composed mainly of iron. The main clues to this come from meteorites, in which metallic iron is a not uncommon constituent, and from the sun, in which iron is the most abundant heavy element. Comparison with the properties of iron at appropriate pressures and temperatures, determined from shock-wave experiments, confirms that iron is a plausible primary component of the core, but also indicates that a significant proportion of the core material (perhaps 10–20%) comprises elements of lower atomic weight than iron, such as oxygen, sulphur or silicon [9].

The principal relevance of the core to mantle convection is as a heat reservoir. The fact that it is liquid and the inference that it convects vigorously (to generate the magnetic field) also implies that it maintains the base of the mantle at a spatially uniform temperature, like a heat bath. Since the core is hotter than the mantle, there is a flow of heat from the core into the mantle.

5.1.2 Internal structure of the mantle

The mantle comprises the region between the core and the crust. There is considerable structure within the mantle. First there is a steady increase in the density and seismic velocities with depth (except in the uppermost mantle). This is due to the increase of pressure with depth, and will be discussed in Section 5.1.4. There are also three jumps in these properties evident in Figure 5.1, and these will be discussed here.

A steepening of seismic velocity gradients at a few hundred kilometres depth was first indicated by a change in the slope of travel-time versus distance curves at a distance of about 20° found by Byerly [10] and Lehmann [11]. This was modelled by Jeffreys and Bullen [12] as a 'transition region' extending between depths of about 400 and 1000 km within which seismic velocities rise more steeply with depth than above or below this region. From about 1965, D. L. Anderson and his students developed evidence, using arrays of seismographs, that the seismic velocity increases are concentrated near depths of 400 km and 650 km [13, 14]. Other structure has been proposed within this depth range, but these two rapid jumps in seismic velocity have been reliably established. In the model shown in Figure 5.1 they are at depths of 410 km and 660 km.

From 1939, Jeffreys, Gutenberg and Richter were finding indications of an anomalous zone near the base of the mantle [15, 16]. Bullen in 1949 named this region D″ (breaking his region D into two parts: D′ and D″) [17]. By 1974 it was clear that seismic velocity gradients with depth are lower in the lowest 200 km or so of the mantle [18]. There were suggestions that the gradients become negative, but these have not been strongly supported. Currently there is good evidence that the zone is heterogeneous on both small and large scales [19]. The small-scale heterogeneity is deduced from scattering of high-frequency body waves, while large-scale variations are deduced from differences between studies of different geographic regions. There is some evidence for a sharp discontinuity of seismic velocities whose depth may be different in different regions, though this conclusion is debated [20]. It is possible that this region or layer is very thin or absent in some regions.

Recently a thin layer of much lower velocity has been resolved at the base of the mantle [21, 22]. It has been most clearly resolved beneath part of the Pacific, where there are also lower than normal seismic velocities through a large volume of the lower mantle, and it is not resolvable in surrounding regions, indicating that it is quite variable laterally. Reductions of up to 10% in compressional velocity are inferred in a layer up to about 40 km thick. This compares with compressional velocity variations of 1% or less through the rest of the lower mantle (Section 5.4).

There is more detail known about the upper part of the mantle than shows in Figure 5.1, which was designed as a useful global mean reference structure. An important feature is the low velocity zone. Gutenberg, starting in 1939 and over a period of many years [23], argued for the presence of a zone of anomalously low seismic shear velocities between depths of about 80 km and 200 km. This was confirmed when detailed studies of seismic surface waves were developed [24], though it does not exist under older parts of continents.

5.1.3 Layer names

The naming of mantle layers and interfaces is not in a very satisfactory state. In current terminology, the upper mantle usually refers to the mantle above the first major seismic discontinuity, currently placed at 410 km. The transition zone usually refers to the region between the two major seismic discontinuities (410 to 660 km depth) and the lower mantle usually refers to the mantle below the second discontinuity. However these were not the original usages.

Other variations on this usage are often encountered, depending on the context. It is common among mantle dynamicists to use 'upper mantle' to refer to everything above the 660 km discontinuity. The transition from the upper mantle assemblage of minerals to the lower mantle assemblage probably extends from about 350 km depth to about 750 km depth, so the term 'transition zone' would more logically apply to this larger depth range, in keeping with its original usage. Other problems are that the '660-km discontinuity' keeps moving (from 650 km to 670 km to 660 km) and it is clumsy terminology. The same applies to the '410-km discontinuity' and as well it may not be a sharp discontinuity. The term 'D″ layer' is unhelpful and dry.

It would be helpful if the different criteria and concerns of mineral physicists, seismologists and dynamicists were distinguished by different terminologies. A step towards this is presented in Figure 5.2. The terminology of Bullen is included in Figure 5.2 for reference; the only term still in common usage is D″.

The dominant rock type in the uppermost mantle is peridotite (Section 5.2.1) and the dominant mineral in the deep mantle is a magnesium silicate in a distorted perovskite structure, so the relevant mineralogical zones are so named. The original usage of 'tran-

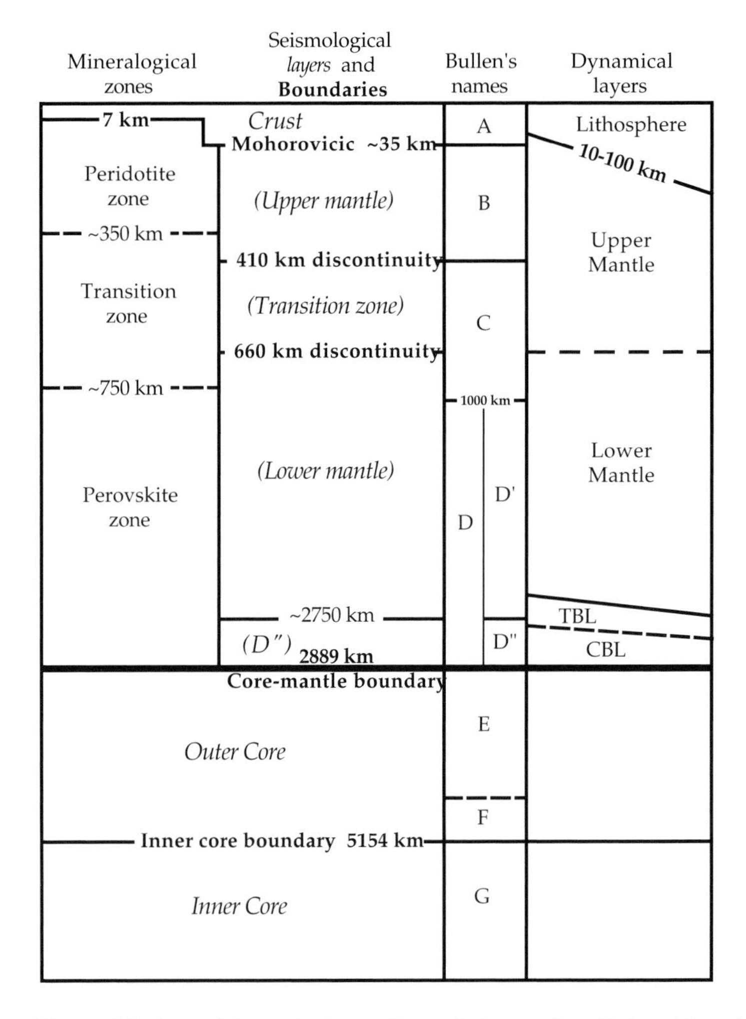

Figure 5.2. A partial terminology of mantle layers that distinguishes the different concerns and usages of mineral physics, seismology and dynamics. TBL: thermal boundary layer; CBL: chemical boundary layer.

sition zone' referred to the whole region through which the mineralogical transformations occurred, so it is appropriate to return to this usage.

The currently common dynamical usage is retained for the regions above and below the 660-km discontinuity. This is appropriate because this discontinuity is the most likely location of dynamical effects associated with the mantle's internal structure, as we will see in Section 5.3 and Part 3. The other entities that are important in the dynamical context are thermal boundary layers. The upper thermal boundary layer is closely (but not identically) approximated by the lithosphere. The lower thermal boundary layer has not been resolved seismologically, and the D″ layer probably involves a change in composition. Therefore a thermal boundary layer (TBL) and chemical boundary layer (CBL) are conceptually distinguished.

The terminology of the seismological interfaces and layers is the least satisfactory. The current conventional terminology is repeated in Figure 5.2, but with the layer names in parentheses because they are not strictly consistent with original usage nor with common and proposed dynamical usage. It would be useful if the interfaces were given names that do not depend on their estimated depth, perhaps the names of their chief discoverers. The intervening layers might also be named, perhaps for those chiefly responsible for determining their mineralogical composition. Since there is no obvious set of such names that would not do some disservice to some people, I will not attempt to offer names here.

5.1.4 Pressure, gravity, bulk sound speed

The gradients of seismic velocities and density with depth in the lower mantle are due almost entirely to the effect of the increase of pressure with depth. There are some relationships involving their variation with pressure that have been important in the development of models of the density variation, that are useful in deducing constraints on mantle composition, and that are worth noting in relation to mantle convection.

The variation of pressure, P, with radius, r, is governed by

$$\frac{dP}{dr} = -g(r)\rho(r) \tag{5.1}$$

where $g(r)$ is the acceleration due to gravity at the radius r within the earth. This is given by

$$g = \frac{GM_r}{r^2} \tag{5.2}$$

where G is the gravitational constant and M_r is the fractional mass, that is the mass within radius r. A radial density profile in the earth can be integrated to yield g, which can then be used in Equation (5.1) to calculate P. These are shown in Figure 5.3a. The acceleration due to gravity, g, is nearly constant through the mantle, with a value near $10\,m/s^2$.

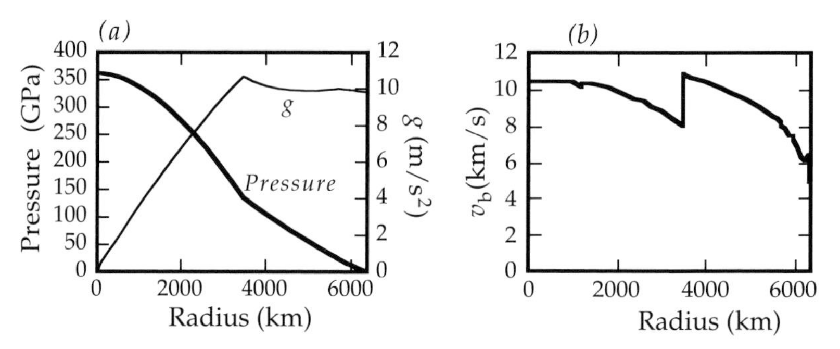

Figure 5.3. Variation of pressure, gravity (g) and bulk sound speed (v_b) with radius in the earth. Data from [25].

Another useful quantity is the 'bulk sound speed', which requires some explanation. The seismic velocities α and β are related to the elastic moduli of the mantle and the density as follows.

$$\alpha = \sqrt{\frac{K + 4\mu/3}{\rho}} \tag{5.3a}$$

$$\beta = \sqrt{\frac{\mu}{\rho}} \tag{5.3b}$$

where K is the bulk modulus, μ is the shear modulus and ρ is the density. In a liquid, μ is zero, so β is zero and $\alpha = \surd(K/\rho)$. The equivalent quantity, $\surd(K/\rho)$, for a solid can be calculated from α and β. Because it depends only on the bulk modulus and because it is the equivalent of the sound speed in a liquid, it is called the bulk sound speed, v_b:

$$v_b = \sqrt{\frac{K}{\rho}} = \sqrt{\alpha^2 - 4\beta^2/3} \tag{5.4}$$

The variation of v_b with radius is shown in Figure 5.3b.

Equations (5.1), (5.2) and (5.4) can be combined with the definition of the bulk modulus to yield an equation from which the variation of density with depth can be calculated from the seismic velocities and a starting density. The bulk modulus is $K = \rho(\mathrm{d}P/\mathrm{d}\rho)$. The resulting equation, called the Adams–Williamson equation, was the means by which the density profile within the earth was determined for much of the twentieth century

[1, 4]. It has been superseded in the past few decades by determinations from the long-period modes of free oscillation of the earth that are excited by large earthquakes, but it is still used to provide starting approximations. The free-oscillation determinations avoid two assumptions that are required for the use of the Adams–Williamson equation. These are that the temperature profile in the mantle is adiabatic (see Section 7.9), and that the long-term compressibility of the mantle is the same as the short-term compressibility sensed by seismic waves: there are relaxation mechanisms of crystal lattices that cause the elasticity to be imperfect and time-dependent. On the other hand, the free oscillations do not yield high resolution. These issues are not of direct concern to us here, but they are worth noting in passing.

5.2 Layer compositions and nature of the transition zone

5.2.1 Peridotite zone

The composition of the upper mantle is inferred from three main sources. First are rocks thought to be parts of the mantle that have been thrust to the surface by tectonic movements. Second are rocks inferred to be pieces of the mantle that are carried to the surface by magmas derived from melting of the mantle. Third are the compositions inferred, with the help of laboratory experiments, to give rise to magmas of the observed range of compositions upon melting of the mantle. A good summary of the evidence and arguments, which is still broadly current, was given by Ringwood [26].

The tectonically emplaced rocks are known as alpine peridotites (because they occur in the European Alps) and ophiolites, and they occur typically in convergence zones where extensive thrust faulting has brought deep material to the surface. Rocks carried to the surface by magmas are known as *xenoliths* (literally, 'strange rocks') because typically their compositions are not directly related to those of the magmas that carry them. There are two main magma types that bear mantle xenoliths: basalts and kimberlites. Kimberlites are notable not only for being the sources of diamonds, but for the extreme velocity of their eruptions, due to relatively high contents of volatiles. In both cases, the erupting magma seems to have broken off pieces of the mantle on its way up and carried them to the surface.

The third type of evidence on mantle composition, from inferred magma source compositions, is less direct. It depends on the fact that the compositions of magmas are generally not the same as the compositions of the rocks they melt from. This is because the different minerals comprising a rock have different melting temperatures and melt to different degrees, so the magma will contain them in proportions different from those in the unmelted material. The process can be complex, and is the subject of a lot of work in experimental petrology.

The conclusion from these lines of evidence is that the upper mantle is predominantly composed of *peridotite*, with a fraction (perhaps 5–10%) of *eclogite*. Peridotite's mineralogy is typically about 60% olivine (gem-quality olivine is known as peridot), 25% orthopyroxene and the balance clinopyroxene and garnet. The eclogites carried up in kimberlite magmas are distinct, being composed of about 60% clinopyroxene, 30% garnet and lesser and variable amounts of other minerals. The predominance of peridotite in the upper mantle leads to the proposed name *peridotite zone* for use in the context of considering its chemical and mineralogical composition (Figure 5.2).

Estimates of the composition of the peridotite zone were encapsulated by Ringwood's definition of *pyrolite* (*pyr*oxene–*ol*ivine rock). By definition, pyrolite has an inferred composition based on all available lines of evidence, so arriving at a particular composition involves inference and judgement. For this reason, the estimated composition of pyrolite has been often revised in detail. However the primary features of pyrolite are well-established. An approximate composition of pyrolite is given in Table 5.1 in terms of principal simple oxide components [27].

5.2.2 Transition zone and perovskite zone

The reason for the more rapid increase of seismic velocities with depth within the transition zone was the subject of controversy at first. The question was whether it was due to a change in composition with depth or to pressure-induced phase transformations of mantle minerals to denser crystal structures. Birch, over a long period, introduced important arguments that weighed strongly in favour of phase transformations, with any change in composition being secondary. I will review these arguments here, since they have formed the basis of all discussions of this question. An excellent account of the current status of this subject is given by Jackson and Rigden [28].

Birch [29] argued first that the gradients of seismic velocities with depth in the transition zone are inconsistent with characteristic rates of change of the elastic properties of solids with increasing pressure. He also argued that the gradients in the perovskite zone

Table 5.1. *Approximate composition of pyrolite.*

Component	Weight %
SiO_2	45.0
MgO	38.0
FeO	7.8
Fe_2O_3	0.3
Al_2O_3	4.4
CaO	3.5

are consistent with the effect of increasing pressure. Compression experiments on a wide range of solids, mainly by Bridgman, had shown that the pressure derivative of the bulk modulus, $K' = \partial K/\partial P$ is typically about 4, decreasing to about 3 at deep mantle pressures. Birch used the seismic velocity profiles of the mantle from Jeffreys and Bullen [12] to calculate K' in the mantle (using Equation 5.4 above). He found that K' is 3–4 in the uppermost mantle and in the perovskite zone, but is scattered and much higher (4–10) in the transition zone. This demonstrated more explicitly the anomalous nature of the transition zone at a time when it had not been resolved into sharp jumps. It also indicated that the variation of properties with depth in the perovskite zone is consistent with it being of uniform chemical and mineralogical composition.

Birch [29] went on to demonstrate that the properties of the perovskite zone could sensibly be extrapolated to zero pressure, and that the density and bulk modulus so obtained are comparable to those of some dense simple oxides, such as MgO, Al_2O_3 and TiO_2, which have an appropriate combination of high density and high bulk modulus. This is in contrast to olivine and pyroxene, which have lower densities and substantially lower bulk moduli.

The steeper increase in density through the transition zone could be due to an increase in the iron content of the mantle minerals with depth. However an increase in iron content has the opposite effect on elastic wave speeds: they decrease. Thus the observation that both density and elastic wave speeds increase with depth indicates that the explanation cannot be a change in iron content.

Birch later quantified this argument and made it more convincing. He demonstrated [30], on the basis of a large number of laboratory measurements, that the compressional velocity of solids depends mainly on their density and mean atomic weight, with crystal structure being of secondary importance. Birch's original plot (Figure 5.4) still makes the point clearly. At a particular mean atomic weight, velocity increases approximately linearly with density. This has been found to be approximately true regardless of whether the change in density is due to a change in composition, a change of crystal structure, a change in pressure or a change in temperature. The effects of changing crystal structures are illustrated in Figure 5.4 by two sequences of minerals that are polymorphs of olivine (Mg_2SiO_4; α–β–γ) and pyroxene ($MgSiO_3$; Px–Ga–Il–Pv) that exist at successively higher pressures.

Birch was then able to demonstrate [33] that the peridotite zone and the perovskite zone fall approximately on the same velocity–

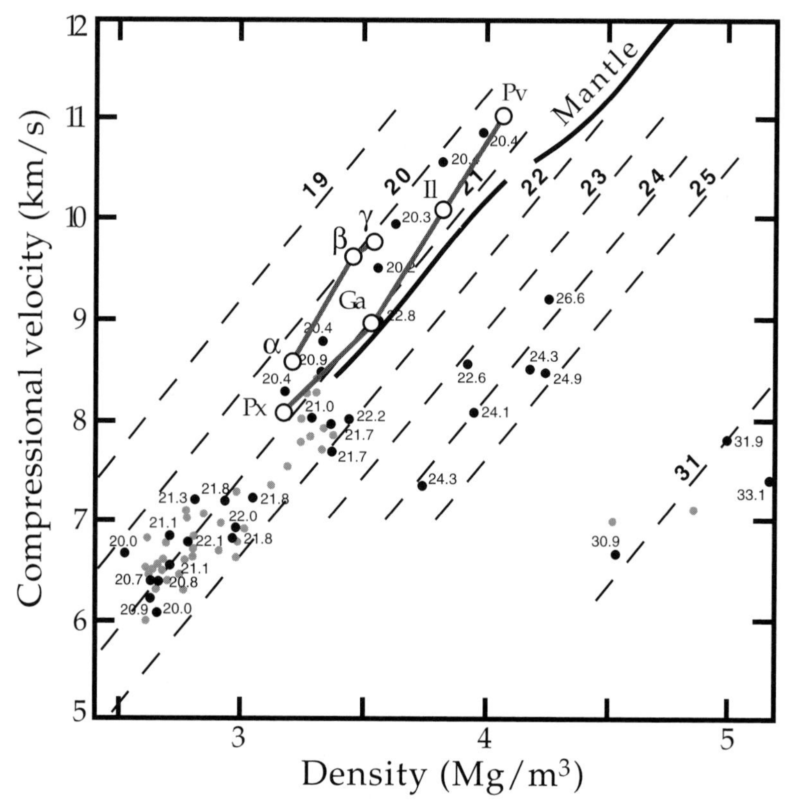

Figure 5.4. Birch's plot [30] of compressional velocity versus density for a collection of oxides and silicates. The dashed lines are approximate contours of mean atomic weight, based on the dark points, which are labelled with specific values of mean atomic weight. More recent data are superimposed for comparison: a mantle model [25] and two sequences of high-pressure polymorphs (Section 5.3). $MgSiO_3$ sequence of structures: Px – pyroxene, Ga – garnet, Il – ilmenite, Pv – perovskite. Mg_2SiO_4 sequence: α – olivine, β – wadsleyite, γ – ringwoodite (data from [31, 32]). Copyright by the American Geophysical Union.

density trend, consistent with them having the same mean atomic weight. A more recent mantle model is included in Figure 5.4. A small increase in iron content through the transition zone is possible but not required by Figure 5.4. (It is a fortunate coincidence that the principal oxide components of the peridotite zone, except for FeO and Fe_2O_3, Table 5.1, have mean atomic weights close to 20.) This strengthened the case that the transition zone is primarily a zone of presssure-induced phase transformations, rather than a zone of compositional change. In particular it discounts the possibility that the increase in density through the transition zone is due entirely or mainly to an increase in iron content, since this would cause the perovskite zone to fall on a trend closer to mean atomic weight 25, with lower seismic velocity.

Conversely, the same relationship allowed Birch [33, 34] to discount the possibility that the core is a high-pressure form of mantle silicates, since the core falls far off the mantle trend of velocity versus density. In Figure 5.5 the mantle and core trends are compared with data for metals, and the core falls closer to the trend for iron. (The data for metals are labelled by atomic number rather than atomic weight.) Where they meet at the core–mantle

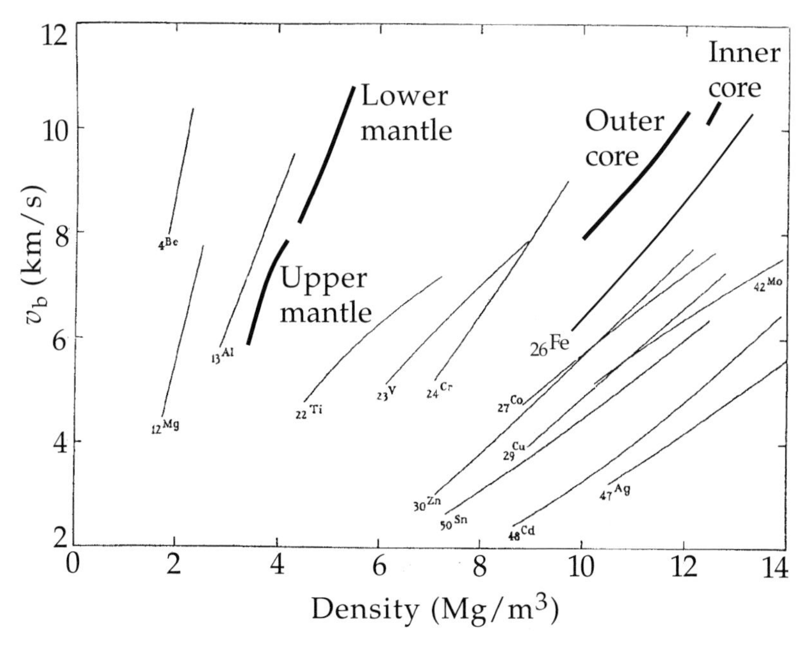

Figure 5.5. Bulk sound velocity versus density for metals, labelled by atomic number, after Birch [33, 34]. Mantle and core curves [25] are superimposed. The core is close to the trend for iron. Copyright by Blackwell Scientific Ltd.

boundary, the core has a much higher density and a lower bulk sound velocity than the mantle (Figures 5.1, 5.3). On the other hand, the core has a slightly lower density and higher bulk sound velocity than iron, and this is the primary evidence that the core contains a complement of lighter elements (Section 5.1.1; [9]).

The conclusion that the transition zone is due mainly to the occurrence of pressure-induced phase transformations has become well accepted, and the likely sequence of such transformations has been demonstrated by laboratory experiments (Section 5.3). There has remained, however, a controversy over whether the transition zone is entirely due to phase transformations, or whether there might be a small change in composition across it. It has been suggested that the perovskite zone might be relatively enriched in iron [35] or silica [36, 37].

This question has been an important one in mantle studies for the past two decades. There have been three main approaches to trying to resolve it: through seismology, dynamics and mineral physics. The main seismological approach has been the detection of subducted lithosphere, discussed below (5.4). The dynamical approach depends on the fact that if the perovskite zone has a higher iron content and thus an intrinsically higher density, this would inhibit convective mixing between the layers, with potentially observable dynamical consequences. This possibility will be discussed in Chapter 12. There could also be dynamical effects from the dependence of phase transformation pressures on temperature and composition that might inhibit convection through the transition zone, and these will be discussed next in Section 5.3.

The third approach, through mineral physics, has been more direct. It is to compare measurements of the physical properties of high-pressure phases with the density and seismic velocity profiles of the mantle. One of Birch's methods, of extrapolating the properties of the perovskite zone to zero pressure and comparing them with measured properties of relevant materials, has been followed often, and an example is shown in Figure 5.6. The laboratory values have been corrected to the estimated temperature at zero pressure (1600 K) for comparison, and there is some uncertainty in both the temperature and the correction.

The results show that the perovskite zone ('Hot decompressed lower mantle') properties are nearly the same, within uncertainties, as those measured for the pyrolite composition in the high-pressure assemblage of phases (Section 5.3). The iron content is constrained by the density to be close to that of pyrolite. However, the perovskite zone has a slightly lower bulk modulus, between pyrolite and pure olivine, which would correspond to being less silica-rich than

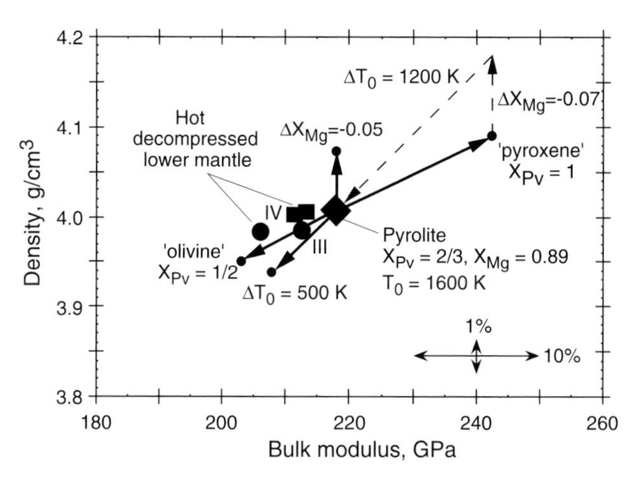

Figure 5.6. Properties of the perovskite zone, extrapolated to zero pressure, compared with properties of proposed mantle composition based on laboratory measurements. Four separate extrapolations are shown, based on a third-order (III) and fourth-order (IV) extrapolation function and for two earth models: PREM (circles) [39] and *ak135* (squares) [6, 7]. Pyrolite at 1600 K estimated from laboratory measurements is the large diamond. Solid arrows show the effects of changes in the mantle temperature (ΔT_0), iron content (ΔX_{Mg}) and silica content (X_{Pv}). Dashed arrow shows a possible trade-off of temperature with composition. From Jackson and Rigden [28].

the pyrolite composition. This is contrary to the suggestions noted above that it might be more silica-rich. These results tend to favour the pyrolite composition rather than a silica-enriched or iron-enriched composition, taking account of the uncertainties. However, we should bear in mind that they depend on the assumption that the perovskite zone is compositionally homogeneous with a temperature profile close to adiabatic, and it turns out that the density profile in model *ak135* is quite strongly subadiabatic in the lower mantle [38].

The alternative approach to this comparison is to extrapolate laboratory measurements of density and elastic properties to high pressures and temperatures. This avoids assumptions about the state of the perovskite zone, but involves uncertainties in the dependence of elastic properties on pressure and temperature, which are not all accurately measured yet. An example of this approach is illustrated in Figure 5.7. This shows that the density and elastic properties (represented by the 'seismic parameter' $\phi = v_{\mathrm{b}}^2 = K/\rho$) of the perovskite zone are closely matched by the extrapolated properties of the pyrolite composition, though again the elastic stiffness is slightly lower, which could be accommodated by lowering the silica content of the perovskite zone.

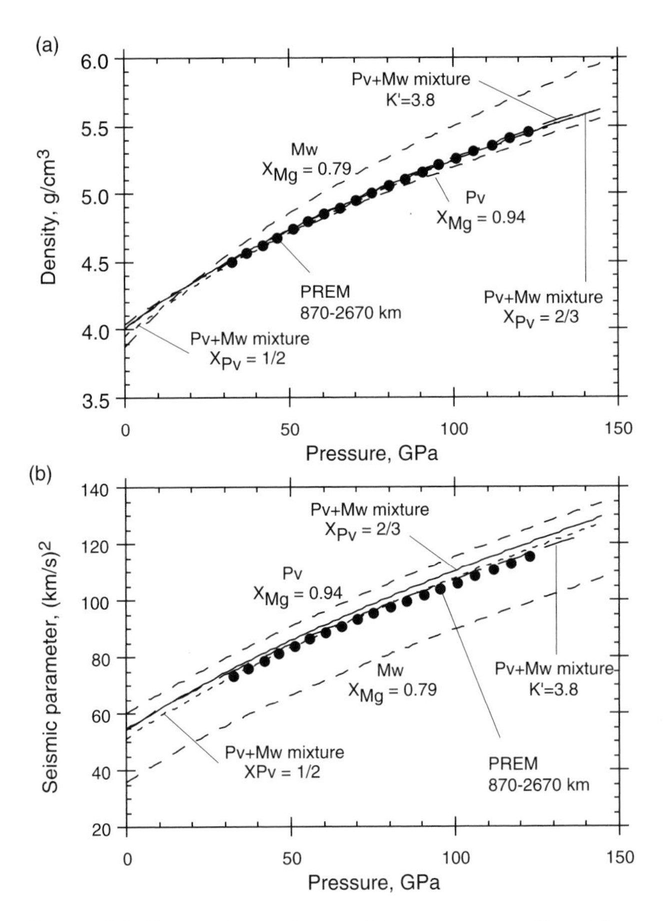

Figure 5.7. Comparison of perovskite zone properties with properties of pyrolite extrapolated to high pressures and temperatures from laboratory measurements. In these conditions pyrolite exists as $(Mg,Fe)SiO_3$ in the perovskite structure (Pv) and (Mg,Fe)O (magnesiowüstite, Mw). Dots are the mantle values from the 'PREM' model [39]. From Jackson and Rigden [28].

The principal uncertainty left by these comparisons of the mantle properties with measured properties of minerals resides in the trade-off possible between temperature and silica content, represented by the proportions of pyroxene and olivine (Figure 5.6). It is possible to match the mantle properties with a purely pyroxene lower mantle, but it must be about 1200 °C hotter than the upper mantle. Such a high temperature would imply a strong, double thermal boundary layer near 660 km depth, with potentially observable consequences in both seismology and dynamics (Chapters 8, 12), but it cannot yet be ruled out on the basis of mineral physics. Aside from this trade-off, there are enough unmeasured properties and innacuracies remaining in this approach that it can still be argued that the perovskite zone has a slightly different iron content than pyrolite and, by inference, the peridotite zone. However the difference must be quite small. What can be said is that the properties of the perovskite zone are consistent, within uncertainties, with its composition being the same as that of the peridotite zone. In fact this has been true ever since Birch's work, but the uncertainties have been steadily decreasing.

5.3 Phase transformations and dynamical implications

5.3.1 Pressure-induced phase transformations

Bernal [40] first proposed that olivine might transform to the spinel structure under the high pressures of the mantle, and Birch showed that such phase transformations were a plausible explanation for the higher densities and elasticities of the transition zone and perovskite zone. However, it was Ringwood who first systematically set about demonstrating this by means of high-pressure experiments, first on analogue compositions and solid solutions closely related to olivine [41, 42].

The subsequent elucidation of the high-pressure phases that might exist in the transition zone was carried out primarily in the laboratories of Ringwood and Akimoto [26]. The existence of a spinel-structure phase was eventually confirmed, but it was also found that there is an intervening phase with a distorted spinel structure. Both of these phases have been found to occur in nature (within the shock zones of meteorite impacts), and bear the mineral names wadsleyite (the intermediate or β-phase) and ringwoodite (the spinel phase). It was also found that the pyroxene component of the upper mantle dissolves progressively into a solid solution with garnet, the fully homogenised form being known as majorite. The mineral proportions as a function of depth are illustrated in Figure 5.8

Experimental testing of Birch's conjecture that the properties of the lower mantle could be explained if it existed as an assemblage of dense simple oxides remained elusive until the advent of laser-heated diamond-anvil technology in the 1960s. Liu, working in association with Ringwood, then discovered that the dominant high-pressure phase of the lower mantle is $(Mg,Fe)SiO_3$ in an orthorhombically distorted perovskite structure [43]. Because of the large volume of the lower mantle it has been justifiably described as the most abundant mineral in the earth. Its proportional dominance is the basis for calling the zone below which the

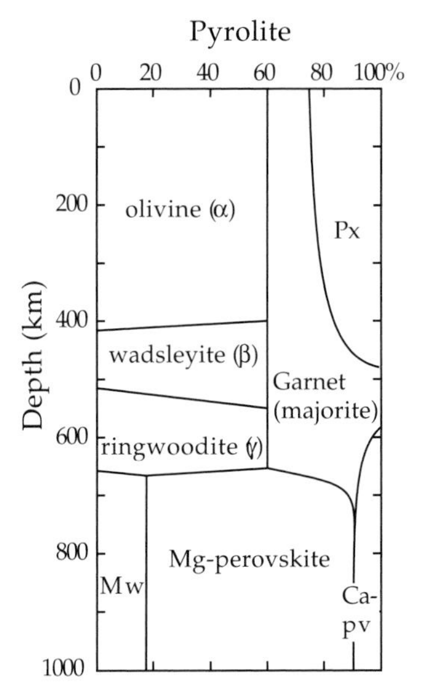

Figure 5.8. Sequence of pressure-induced transformations and reactions as a function of depth in a mantle of pyrolite composition. Px: pyroxene, Mw: magnesiowüstite, pv: perovskite. After Irifune [44].

transformations are essentially complete the *perovskite zone* (Figure 5.2).

The balance of the lower mantle composition is made up of (Mg,Fe)O in the sodium chloride structure, a mineral stable at zero pressure and known as magnesiowüstite, and $CaSiO_3$ in the true perovskite structure [45].

5.3.2 Dynamical implications of phase transformations

The change in density through a phase transformation might at first sight seem to preclude convective flow through the transformation zone, since the rise of the deeper phase would be resisted by its greater density. However, if the flow is slow and the temperature is sufficiently high for the reaction kinetics to proceed, the phase may transform as it rises. Then, to a first approximation, there would be no effect on the convective motion: the material could move up and down, transforming as it did. But this is still not the full story, because the pressure at which transformations occur is usually affected by temperature and by details of composition. This means that the transformation may occur shallower or deeper than in nearby mantle, and this would give rise to a large horizontal density difference that would affect convective motions.

The deflection of phase transformation boundaries is potentially of great importance to mantle convection, since it is in the rising and descending regions that the temperature is different from the average temperature. It is also clear that there is compositional zoning in subducted lithosphere, and it is likely that the composition of mantle plumes is significantly different from that of normal mantle. Thus both thermal and compositional deflections need to be considered.

5.3.3 Thermal deflections of phase boundaries

The principle of thermal deflections and their effect on buoyancy is sketched in Figure 5.9. For most transformations, the transformation pressure increases with increasing temperature. This is usually expressed by saying that the Clausius–Clapeyron slope (or the Clapeyron slope, for short) is positive: $\beta = \partial P_t/\partial T > 0$, where P_t is the transformation pressure. In this case the transformation would occur at a lower pressure, that is at a shallower depth, in subducted lithosphere, where the temperature is lower than in surrounding mantle. As a result, there would be a region in and near the lithosphere (shaded, Figure 5.9a) within which the higher-density phase existed at the same depth as the low-density phase in surrounding mantle. This would cause a negative buoyancy force (broad arrow) that would add to the negative thermal buoyancy of the cold slab, aiding its descent.

There are some transformations, however, for which the Clapeyron slope is negative. In this case (Figure 5.9b) the transformation would be delayed to greater pressure and depth within a

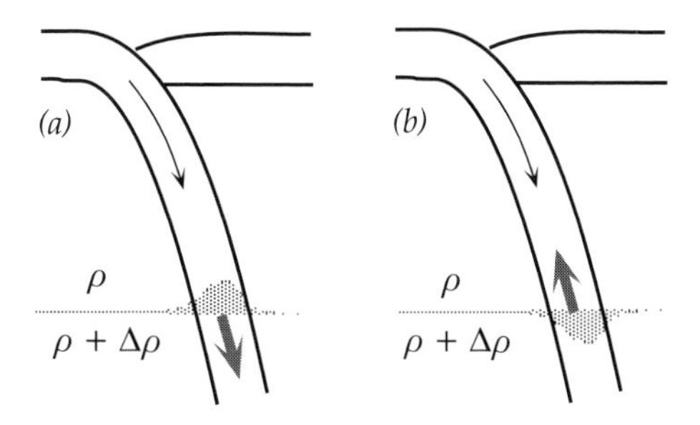

Figure 5.9. Sketch of the deflection of a phase boundary due to the lower temperature in subducting lithosphere. A phase of density ρ transforms to a phase of density $\rho + \Delta\rho$ across the dashed line. The pressure and depth at which the transformation occurs are different in the cooler lithosphere. (a) Positive Clapeyron slope. (b) Negative Clapeyron slope.

descending slab, producing a positive buoyancy that would oppose the slab's descent.

Complementary effects would occur in a hot rising column of mantle. A positive Clapeyron slope would cause the transformation at a greater depth, yielding a positive buoyancy that would enhance the column's rise. A negative Clapeyron slope would inhibit its rise. Thus the effect of a phase transformation on convective motion depends on the sign of the Clapeyron slope of the transformation: for a positive slope, convection is enhanced, while for a negative slope convection is inhibited.

This discussion has assumed that the transformation reactions occur at their equilibrium pressure. It is possible, particularly in cold subducted lithosphere, that the temperature is too low for the thermally activated reactions to occur. In this case the phase would persist metastably, that is outside its equilibrium stability range. In subducting lithosphere this effect would produce a positive buoyancy that would always oppose the descent of the slab. This possibility has been difficult to evaluate. One argument has been that deep earthquakes within subducted lithosphere are triggered by the sudden transformation of low-pressure phases that have been carried metastably into the high-pressure range [46]. If this could be established it would provide useful empirical constraints on metastability, but it remains controversial.

A great deal of attention has been focussed on the possible dynamical implications of the ringwoodite → perovskite + magnesiowüstite transformation (Figure 5.8), which has a negative Clapeyron slope of about −2 MPa/K [47]. Inclusion of the buoyancy effect associated with just this transformation has caused some numerical convection models to become episodically layered (Chapters 12, 14). However, there are significant uncertainties in the thermodynamic parameters, so it is not clear if there is a strong effect in the mantle.

As well, there is another reaction, the garnet → perovskite transformation (Figure 5.8), at nearly the same depth, and this one may have a strongly positive Clapeyron slope (about 4 MPa/K) and a substantial density increase, yielding a negative buoyancy in opposition to the transformation of the ringwoodite component [48]. The net effect of these two transformations is unclear, and should really be evaluated in the multicomponent system in which all components can react to mutual equilibrium.

At depths near 400 km, the situation is similar, but with the roles of the components reversed. The olivine → wadsleyite reaction has a positive Clapeyron slope (about 3 MPa/K), but the pyroxene → majorite transformation may have a strongly negative slope [47, 48]. Again, the net effect is unclear.

As a further caution, the mechanical strength of subducted lithosphere is probably sufficient for stresses from 410 km to be transmitted to 660 km, so the opposing buoyancies from the different depths will also tend to cancel.

5.3.4 Compositional deflections and effects on density

Subducted lithosphere is compositionally stratified, and this causes the density of the lithosphere to differ from that of the adjacent mantle, especially in the vicinity of phase transformations. The stratification of oceanic lithosphere originates at seafloor spreading centres. There, the mantle melts to produce about 7 km thickness of oceanic crust, broadly of basaltic composition. The underlying residual zone from which most of this melt is drawn is about 20–30 km thick, and has a composition depleted in basaltic components relative to average mantle.

Ringwood and associates [27, 49] have demonstrated that the different proportions of mineral components within these zones of the subducted lithosphere compared with surrounding mantle will result in a net difference in density. The compositional stratification also affects the depths at which phase transformations occur. Each of these effects contributes a net buoyancy to subducted lithosphere, the magnitude and sign of which fluctuates with depth. Figure 5.10 summarises the sequence of transformations in the depleted zone and the oceanic crust of subducted lithosphere. The sequence is substantially different in subducted oceanic crust compared with pyrolite (Figure 5.8), the pyroxene–garnet fields being dominant. Conversely, these fields are much reduced in the depleted mantle sequence.

The effect of lithosphere stratification on buoyancy is important right at the earth's surface because, perhaps surprisingly, both parts of the lithosphere have reduced density. The density of the basaltic crust is about 2.9 Mg/m^3, compared with 3.3 Mg/m^3 for normal mantle. The density of the depleted zone is also reduced, by about −0.08 Mg/m^3, mainly because of the preferential partitioning of iron into the basaltic melt that is removed. The oceanic lithosphere at the earth's surface thus has a net positive compositional buoyancy. This means that oceanic lithosphere is gravitationally stable until it has cooled and accumulated enough negative thermal buoyancy to outweigh the compositional buoyancy. This takes about 15 Ma. Earlier in earth history the effect would have been larger, because the mantle was hotter, there was

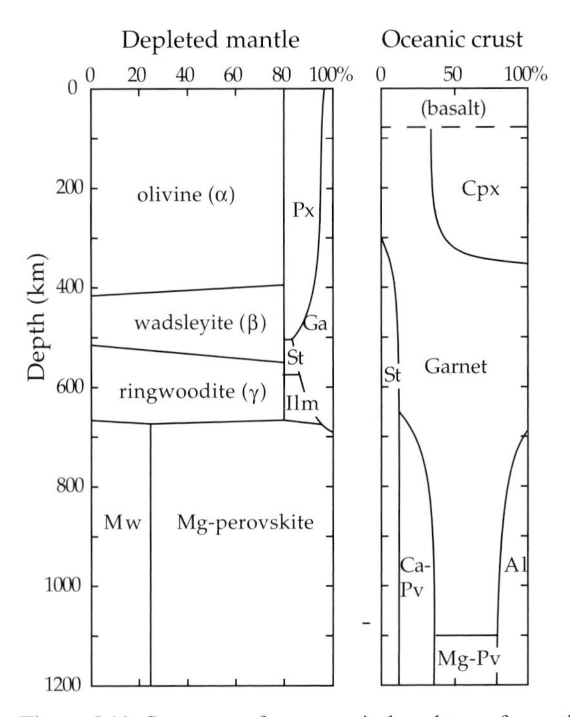

Figure 5.10. Sequence of pressure-induced transformations and reactions as a function of depth in residual mantle depleted of basaltic components, and of the complementary basaltic oceanic crust. Px: pyroxene, Cpx: clinopyroxene, Ga: garnet, St: stishovite, Ilm: ilmenite, Mw: magnesiowüstite, Pv: perovskite, Al: aluminous phase. After Irifune [44] and Kesson *et al.* [49, 50].

more melting at spreading centres and thus thicker oceanic crust. The effect may have been sufficient to prevent or modify subduction of plates (Chapter 14).

As subducted lithosphere sinks, the first major change with depth is the transformation of the oceanic crust to eclogite (garnet plus pyroxene), with a density of about $3.5\,\mathrm{Mg/m^3}$, at about 60 km depth. This makes the net compositional contribution to buoyancy negative, and promotes the subduction of the lithosphere. The net buoyancy is likely to fluctuate with depth as phase transformations occur at slightly different depths in the oceanic crust and depleted parts of the subducted lithosphere relative to the surrounding mantle (Figures 5.8, 5.10). An estimate of the variation in density of the crustal and depleted parts of subducted lithosphere with depth down to 800 km is shown in Figure 5.11. There are substantial uncertainties in these estimates for two reasons. First, there are uncertainties in extrapolating the densities of all the relevant phases to appropriate depths and temperatures. Second, the sequence of

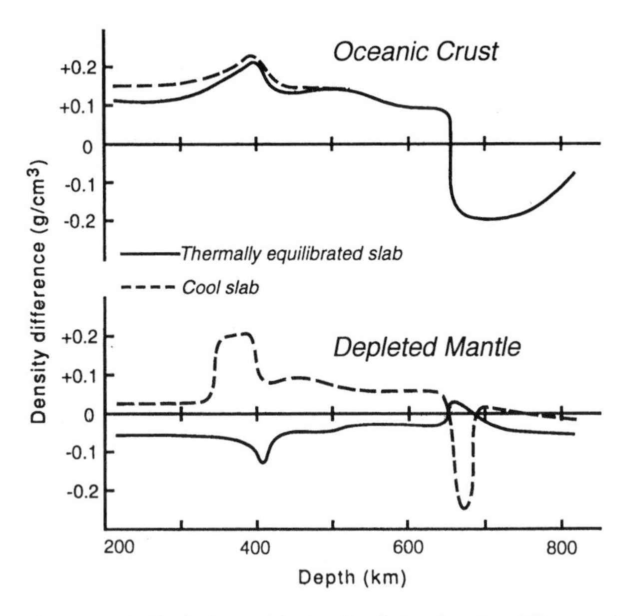

Figure 5.11. Variation with depth of the density difference between parts of the subducted lithosphere and normal mantle. After Irifune [44].

phases within the subducted lithosphere depends on its temperature. The latter effect is potentially large, as is illustrated in Figure 5.11, which compares the cases of a cool 'slab' and a thermally equilibrated slab.

An important feature of the crustal sequence at depths below 660 km is the persistence of the majorite garnet phase in the oceanic crust zone to depths of about 1100 km, well into the lower mantle [49, 50] (Figure 5.10). The density deficit of the crust is significant ($-0.2\ Mg/m^3$), but the relative thinness of the crust zone means that the net buoyancy is not very large, though it is positive. It has been suggested that this buoyancy might be sufficient to prevent subducted lithosphere from sinking into the lower mantle [27, 51]. Quantitative evaluations of the dynamics have not favoured this for the earth in its present state (Chapter 12), but it may have been more likely in the past (Chapter 14).

Within the perovskite zone, the density excess of subducted oceanic crust may change from about $0.04\ Mg/m^3$ at about 1100 km to about $-0.03\ Mg/m^3$ at the base of the mantle [49, 50]. This implies that the net chemical buoyancy of subducted lithosphere may be distinctly positive, and comparable to its residual negative thermal buoyancy near the bottom of the mantle. This could mean that the lithosphere approaches neutral buoyancy, which would have implications for the dynamics of subducted lithosphere in the deepest mantle, but there are still substantial uncertainties in all quantities at present.

5.4 Three-dimensional seismic structure

Resolving the global three-dimensional structure of the mantle (that is, variations in horizontal directions as well as vertical) has been the subject of concentrated effort for over two decades, and recent progress has been gratifying. Although there had been earlier speculation that the mantle under continents ought to be different from that under oceans, clear evidence for lateral variations within the mantle were not presented until the mid-1960s [52, 53, 54]. Since then, the resolution of lateral variations in seismic mantle structure has progressed substantially, using regional and global syntheses of surface waves and body waves (e.g. [55, 56, 57]).

One major focus of this work has been to determine whether subducted lithosphere passes through the transition zone or not, since this would directly address the question of whether mantle convection occurs in two separate layers (upper mantle and lower mantle, Figure 5.2) or as a single flow through the whole mantle. However, there have been other important features revealed, especially very large-scale variations, that seem to correlate with past plate activity, and large variations in the thickness of continental lithosphere.

5.4.1 Seismic detection of subducted lithosphere

Seismologists have invested much effort in trying to establish whether or not subducted lithosphere penetrates below the transition zone. The path of subducted lithosphere is revealed in two ways, one very obvious and the other more subtle.

The occurrence of deep earthquakes was established by Wadati in 1928, and with the advent of plate tectonics they were interpreted as occurring within the anomalously cool subducted lithosphere (Chapter 3). Figure 5.12 shows a selection of cross-sections through deep seismic (Wadati–Benioff) zones. The deep earthquakes trace curving surfaces that have shallow dips near the surface (*dip* is the structural geology term for the angle a plane makes with the horizontal) and steepen with depth to dip angles between about 30° and 90°. Most of these surfaces are fairly smooth, the Tonga zone being the main exception. In several widely separated places, the Wadati–Benioff zones extend to 660–680 km, but no deeper [58].

The geometry revealed in Figure 5.12 carries important information. The profiles are asymmetric about their origins at the

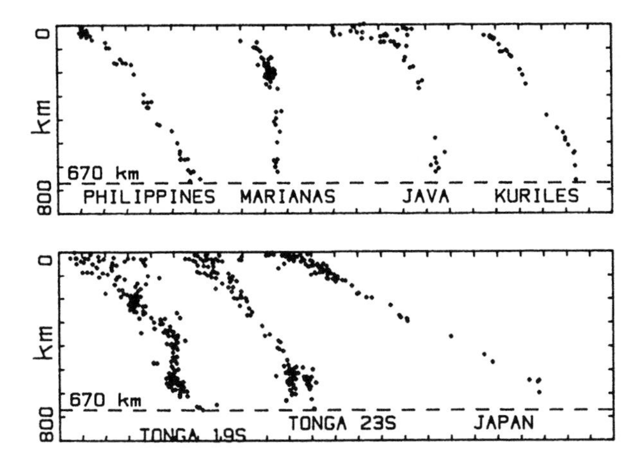

Figure 5.12. Cross-sections through several Wadati–Benioff deep seismic zones showing the locations of deep earthquakes projected onto the plane of cross section. From Davies and Richards [59]. Copyright by The University of Chicago. All rights reserved.

earth's surface (at oceanic trenches), corresponding to the fact that where two plates converge, only one descends, and the surface of contact is a reverse or thrust fault. This is behaviour expected of a brittle medium, which the lithosphere evidently is (Chapters 6, 9).

Most of the zones curve smoothly down, and those that extend deepest end close to the depth of the 660-km discontinuity (Figure 5.1), which is interpreted to be the location of major phase transformations (Figure 5.8). This conjunction has figured prominently in discussions of whether or not mantle convection can penetrate the transition zone. If the subducted lithosphere is deflected horizontally at this depth, there is little suggestion of it either in the shape of these profiles or in the density of earthquakes that might be expected during deformation of the subducted lithospheric 'slab'. The possible exceptions to this statement are the Tonga zone and one deep outlying earthquake in the Mariana zone (which does not appear in Figure 5.12).

However, earthquake locations cannot answer directly the question of slab penetration below the transition zone, because the earthquakes themselves do not extend through. This might be either because the slabs themselves do not penetrate, or because the mechanical or stress state of the slab is changed by its passage through the transition zone, so that earthquakes no longer occur within it. The plausibility of the latter interpretation is supported by current ideas that deep earthquakes are themselves triggered by phase transformations [46].

Seeking more direct evidence, seismologists have sought to map the variation of seismic velocities in the vicinity of subduction zones, in the hope of detecting differences caused by the lower temperature or different composition of subducted slabs. If slabs extend below the 660-km discontinuity, they can in principle be detected in this way. Early results were controversial [60], but more recent work supports the conclusion that many slabs do penetrate. Examples of such results are shown in Figure 5.13.

The Aegean, Tonga and Japan profiles each show a continuous zone of high seismic velocity extending from the surface through the transition zone and deep into the lower mantle. The earthquakes fall within these zones. There is also a deep high-velocity zone extending from the Central American seismic zone, but it is not continuous from the surface. This is inferred to be the signature of the former Farallon plate (Chapter 9). The apparent gap near the surface corresponds with where the recently subducted lithosphere has been young and thin, and therefore hard to detect seismically.

In the Tonga and Japan zones the earthquakes extend to near 660 km depth (compare with Figure 5.12), and the extension of the velocity anomaly below the cut-off of earthquakes is evident,

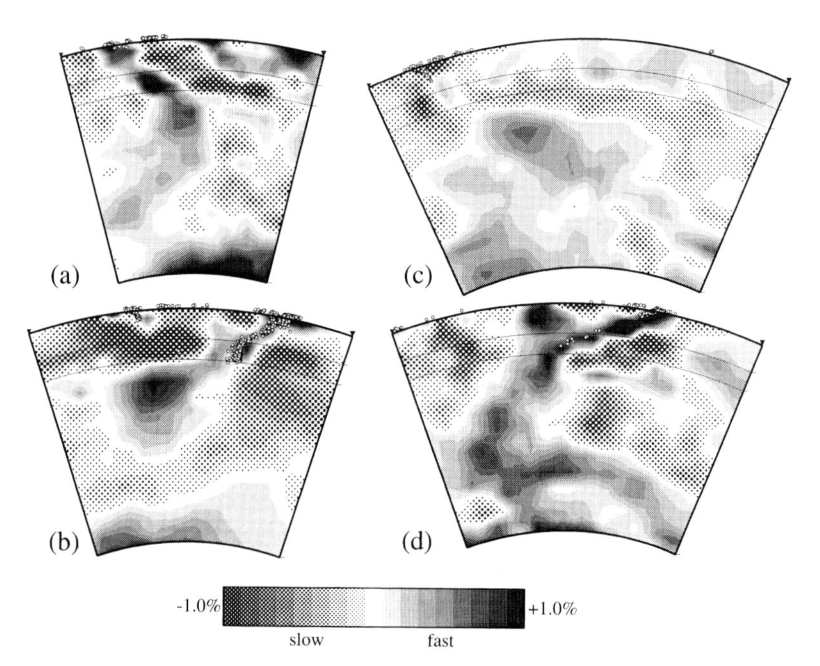

Figure 5.13. Profiles through seismic tomography models of subduction zones. (a) Aegean, (b) Tonga, (c) Central America, (d) Japan. The shading shows variations in the shear wave velocity. Small circles show earthquake locations. From the model S_SKS120 of Widiyantoro [61].

particularly under Japan. There are no deep earthquakes under the Aegean, but the velocity anomaly is clear. Neither are there deep earthquakes under Central America, and in this case there is not a clear velocity anomaly, although a weak one cannot be ruled out. Thus it is clear that the absence of earthquakes does not necessarily imply the absence of a velocity anomaly and, by implication, the absence of subducted lithosphere.

Not only is the evidence for the penetration of subducted lithosphere into the lower mantle quite strong in Figure 5.13, but there are intriguing details revealed. Thus the anomaly under Japan curves back to the east in the deep mantle, possibly corresponding to older lithosphere subducted from other locations in the region. The Tonga anomaly is offset across the transition zone, plausibly corresponding to the fact that the Tonga trench has migrated to the east in the last few million years [62, 63]. Also the Tonga anomaly terminates in a large blob that might be due to subduction at the Tonga trench only having started about 40 Ma ago, with subduction previously occurring further to the west and with the opposite polarity [62]. Such interpretations are not very firm at this stage, but the fact that they are even suggested is a measure of how much the seismological resolution has improved in recent years.

5.4.2 Global deep structure

Global maps of variations of both P and S seismic wave speeds in the depth range 1200-1400 km are shown in Figure 5.14. This depth range is well into the lower mantle. Models from substantially independent data sets have revealed quite similar structures [56, 61, 64], and this has encouraged confidence that the structures are well-resolved and not artefacts of the difficult data analysis procedures.

The main features evident in both maps of Figure 5.14 are bands of high wave speed under North and Central America and extending from the Middle East to New Guinea. These bands correspond with former or present major subduction zones [65]. The sections in Figure 5.13 are through several parts of these bands, and the bands can be interpreted as the locations of subducted lithosphere in the lower mantle. The high wave speed bands correlate well with the geoid lows of Figure 4.9.

Global models also reveal relatively slow wave speeds under Africa and the central Pacific. These are perceptible in Figure 5.14, but not strongly expressed, the best resolved probably being that under southern Africa. The prominent slow blob in the south-central Pacific is in a region not well covered by the seismic tomography, so it is not yet clear if it is real.

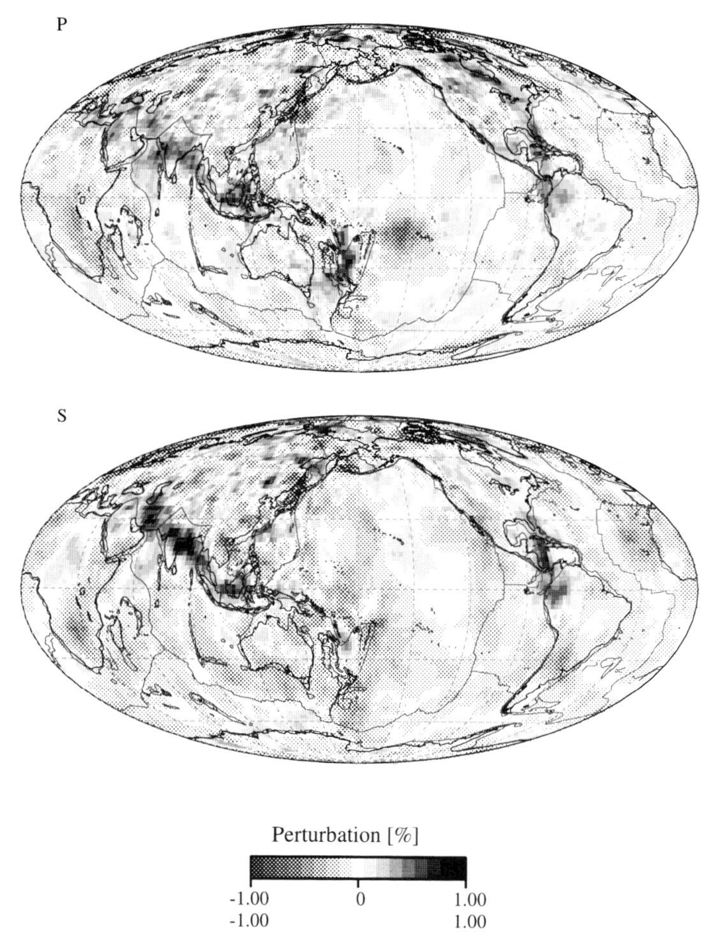

Figure 5.14. Lateral variations in seismic compressional (P) and shear (S) wave velocities averaged between 1200 and 1400 km depth. From Widiyantoro [61].

The amount of lateral heterogeneity in the mantle varies considerably with depth. The total range of variations in shear wave velocity is about $\pm 6\%$ in the uppermost mantle, decreasing to about $\pm 1\%$ in most of the interior mantle, and rising to about $\pm 2\%$ in the D$''$ layer at the bottom of the mantle. The variations in the uppermost mantle are addressed in the next section.

5.4.3 Spatial variations in the lithosphere

The radially symmetric model of the earth presented in Figure 5.1 obviously does not represent the differences between continental and oceanic crust, which have been deduced from the evidence of the gravity field for about a century (see Chapters 2 and 3). Surprisingly, it was not until the 1950s that definitive seismological evidence was adduced, from long-period surface waves and from ship-borne seismic refraction [66]. The current picture is that the oceanic crust has a remarkably uniform thickness of about 7 km, while continental crust is usually 35–40 km thick, but ranges up to 70 km thick under major mountain ranges and down to essentially zero thickness in regions that have undergone rifting.

Currently variations within continental regions are clearly evident in seismic models. An example is shown in Figure 5.15. Lateral variations are most pronounced in the upper 200–300 km of the mantle, with highest seismic velocities under continental shields, and progressively lower velocities under younger continental crust, oceans, tectonically active continental regions and midocean ridges. Low velocity anomalies have also been found under some volcanic hotspots, such as Iceland in the North Atlantic (Figure 5.15).

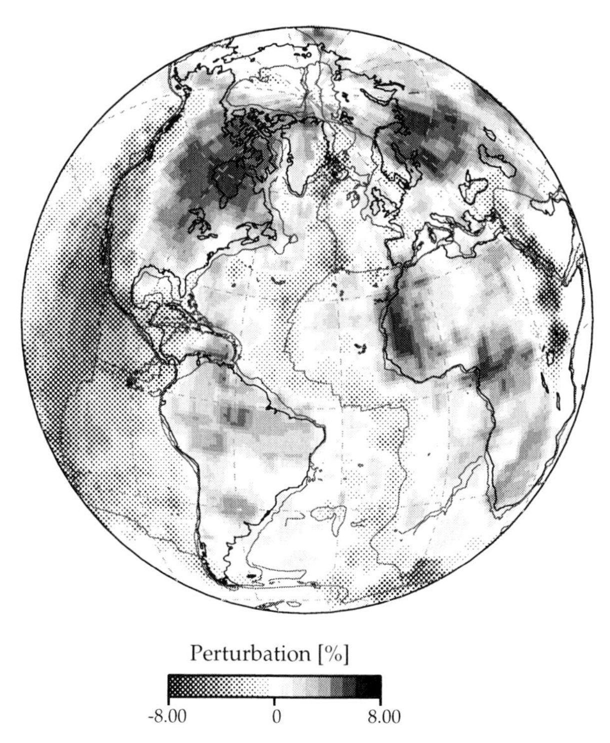

Figure 5.15. Lateral variations in seismic shear wave velocity averaged between 100 and 175 km depth. The high velocities under continental shield reveal the thick lithosphere in these regions, extending to depths of 200 km, or possibly more. From Grand [56].

5.5 References

1. C. M. R. Fowler, *The Solid Earth: An Introduction to Global Geophysics*, Cambridge University Press, Cambridge, 1990.
2. A. Mohorovičić, Das Beben vom 8.x.1909, *Jahrb. Met. Obs. Zagreb (Agram.)* **9**, 1–63, 1909.
3. R. A. Daly, *Strength and Structure of the Earth*, 434 pp., Prentice-Hall, New York, 1940.
4. K. E. Bullen, *An Introduction to the Theory of Seismology*, 381 pp., Cambridge University Press, Cambridge, 1965.
5. R. D. Oldham, Constitution of the interior of the earth as revealed by earthquakes, *Quart. J. Geol. Soc. London* **62**, 456–75, 1906.
6. B. L. N. Kennett, E. R. Engdahl and R. Buland, Constraints on seismic velocities in the earth from travel times, *Geophys. J. Int.* **122**, 108–24, 1995.
7. J. P. Montagner and B. L. N. Kennett, How to reconcile body-wave and normal-mode reference Earth models?, *Geophys. J. Int.* **125**, 229–48, 1996.
8. I. Lehmann, P′, *Bur. Centr. Seism. Internat. A* **14**, 3–31, 1936.
9. R. Jeanloz, The earth's core, *Science* **249**, 56–65, 1983.
10. P. Byerly, The Montana earthquake of June 28, 1925, *Seismol. Soc. Amer. Bull.* **16**, 209–65, 1926.
11. I. Lehmann, Transmission times for seismic waves for epicentral distances around 20°, *Geodaet. Inst. Skr.* **5**, 44, 1934.
12. H. Jeffreys and K. E. Bullen, *Seismological Tables*, British Association Seismological Committee, London, 1940.
13. M. Niazi and D. L. Anderson, Upper mantle structure from western North America from apparent velocities of P waves, *J. Geophys. Res.* **70**, 4633–40, 1965.
14. L. R. Johnson, Array measurements of P velocities in the upper mantle, *J. Geophys. Res.* **72**, 6309–25, 1967.
15. B. Gutenberg and C. F. Richter, On seismic waves, *Beitr. Geophys.* **54**, 94–136, 1939.
16. H. Jeffreys, The times of P, S, and SKS and the velocities of P and S, *Mon. Not. R. Astron. Soc.* **4**, 498–533, 1939.
17. K. E. Bullen, Compressibility-pressure hypothesis and the Earth's interior, *Mon. Not. R. Astron. Soc.* **5**, 355–68, 1949.
18. J. R. Cleary, The D″ region, *Phys. Earth Planet. Inter.* **19**, 13–27, 1974.
19. C. J. Young and T. Lay, The core-mantle boundary, *Annu. Rev. Earth Planet. Sci.* **15**, 25–46, 1987.
20. D. Loper and T. Lay, The core-mantle boundary region, *J. Geophys. Res.* **100**, 6379–420, 1995.
21. E. J. Garnero and D. V. Helmberger, A very low basal layer underlying large-scale low-velocity anomalies in the lower mantle beneath the Pacific: evidence from core phases, *Phys. Earth Planet. Inter.* **91**, 161–76, 1995.

22. E. J. Garnero and D. V. Helmberger, Seismic detection of a thin laterally varying boundary layer at the base of the mantle beneath the central Pacific, *Geophys. Res. Lett.* **23**, 977–80, 1996.
23. B. Gutenberg, On the layer of relatively low wave velocity at a depth of about 80 kilometers, *Seismol. Soc. Amer. Bull.* **38**, 121–48, 1948.
24. D. L. Anderson, Latest information from seismic observations, in: *The Earth's Mantle*, T. F. Gaskell, ed., Academic Press, New York, 1967.
25. D. L. Anderson and R. S. Hart, An earth model based on free oscillations and body waves, *J. Geophys. Res.* **81**, 1461–75, 1976.
26. A. E. Ringwood, *Composition and Petrology of the Earth's Mantle*, 618 pp., McGraw-Hill, 1975.
27. A. E. Ringwood, Phase transformations and their bearing on the constitution and dynamics of the mantle, *Geochim. Cosmochim. Acta* **55**, 2083–110, 1991.
28. I. N. S. Jackson and S. M. Rigden, Composition and temperature of the mantle: seismologial models interpreted through experimental studies of mantle minerals, in: *The Earth's Mantle: Composition, Structure and Evolution*, I. N. S. Jackson, ed., Cambridge University Press, Cambridge, 405–60, 1998.
29. F. Birch, Elasticity and constitution of the earth's interior, *J. Geophys. Res.* **57**, 227–86, 1952.
30. F. Birch, The velocity of compressional waves in rocks, Part 2, *J. Geophys. Res.* **66**, 2199–224, 1961.
31. J. D. Bass, Elasticity of minerals, glasses and melts, in: *Mineral Physics and Crystallography, A Handbook of Physical Constants*, T. J. Ahrens, ed., American Geophysical Union, Washington, D.C., 45–63, 1995.
32. A. Yeganeh-Haeri, Synthesis and re-investigation of the elastic properties of single-crystal magnesium silicate perovskite, *Phys. Earth Planet. Inter.* **87**, 111–21, 1994.
33. F. Birch, Composition of the earth's mantle, *Geophys. J. R. Astron. Soc.* **4**, 295–311, 1961.
34. F. Birch, On the possibility of large changes in the earth's volume, *Phys. Earth Planet. Inter.* **1**, 141–7, 1968.
35. D. L. Anderson and J. D. Bass, The transition region of the earth's upper mantle, *Nature* **320**, 321–8, 1986.
36. L.-G. Liu, On the 650-km seismic discontinuity, *Earth Planet. Sci. Lett.* **42**, 202–8, 1979.
37. R. Jeanloz and E. Knittle, Density and composition of the lower mantle, *Philos. Trans. R. Soc. London Ser. A* **328**, 377–89, 1989.
38. I. Jackson, Elasticity, composition and temperature of the earth's lower mantle: a reappraisal, *Geophys. J. Int.* **134**, 291–311, 1998.
39. A. M. Dziewonski and D. L. Anderson, Preliminary reference earth model, *Phys. Earth Planet. Inter.* **25**, 297–356, 1981.
40. J. D. Bernal, Discussion, *Observatory* **59**, 268, 1936.

41. A. E. Ringwood, The olivine–spinel transition in the earth's mantle, *Nature* **178**, 1303–4, 1956.
42. A. E. Ringwood and A. Major, Synthesis of Mg_2SiO_4–Fe_2SiO_4 spinel solid solutions, *Earth Planet. Sci. Lett.* **1**, 241–5, 1966.
43. L.-G. Liu, Silicate perovskite from phase transformations of pyrope garnet at high pressure and temperature, *Geophys. Res. Lett.* **1**, 277–80, 1974.
44. T. Irifune, Phase transformations in the earth's mantle and subducting slabs: Implications for their compositions, seismic velocity and density structures and dynamics, *The Island Arc* **2**, 55–71, 1993.
45. L. Liu and A. E. Ringwood, Synthesis of a perovskite-type polymorph of $CaSiO_3$, *Earth Planet. Sci. Lett.* **28**, 209–11, 1975.
46. S. H. Kirby, W. B. Durham and L. A. Stern, Mantle phase changes and deep-earthquake faulting in subducting lithosphere, *Science* **252**, 216–25, 1987.
47. C. R. Bina and G. Helffrich, Phase transition Clapeyron slopes and transition zone seismic discontinuity topography, *J. Geophys. Res.* **99**, 15 853–60, 1994.
48. H. Yusa, M. Akaogi and E. Ito, Calorimetric study of $MgSiO_3$ garnet and pyroxene: heat capacities, transition enthalpies, and equilibrium phase relations in $MgSiO_3$ at high pressures and temperatures, *J. Geophys. Res.* **98**, 6453–60, 1993.
49. S. E. Kesson, J. D. Fitz Gerald and J. M. G. Shelley, Mineral chemistry and density of subducted basaltic crust at lower mantle pressures, *Nature* **372**, 767–9, 1994.
50. S. E. Kesson, J. D. Fitz Gerald and J. M. Shelley, Mineralogy and dynamics of a pyrolite lower mantle, *Nature* **393**, 252–5, 1998.
51. A. E. Ringwood, Phase transformations and differentiation in subducted lithosphere: implications for mantle dynamics, basalt petrogenesis, and crustal evolution, *J. Geol.* **90**, 611–43, 1982.
52. M. N. Toksöz and D. L. Anderson, Phase velocities of long-period surface waves and structure of the upper mantle, *J. Geophys. Res.* **71**, 1649–58, 1966.
53. A. L. Hales and E. Herrin, Travel times of seismic waves, in: *The Nature of the Solid Earth*, E. C. Robertson, ed., McGraw-Hill, New York, 172–215, 1972.
54. R. A. Wiggins and D. V. Helmberger, Upper mantle structure under the Western United States, *J. Geophys. Res.* **78**, 1870–80, 1973.
55. W. Su, R. L. Woodward and A. M. Dziewonski, Degree 12 model of shear velocity heterogeneity in the mantle, *J. Geophys. Res.* **99**, 6945–80, 1994.
56. S. P. Grand, Mantle shear structure beneath the Americas and surrounding oceans, *J. Geophys. Res.* **99**, 11 591–621, 1994.
57. R. D. van der Hilst, B. L. N. Kennett and T. Shibutani, Upper mantle structure beneath Australia from portable array deployments, in: *Structure and Evolution of the Australian Continent*, J. Braun, J.

Dooley, B. Goleby, R. van der Hilst and C. Klootwijk, eds., American Geophysical Union, Washington, D.C., 39–57, 1998.

58. P. B. Stark and C. Frohlich, The depths of the deepest deep earthquakes, *J. Geophys. Res.* **90,** 1859–69, 1985.
59. G. F. Davies and M. A. Richards, Mantle convection, *J. Geol.* **100**, 151–206, 1992.
60. K. C. Creager and T. H. Jordan, Slab penetration into the lower mantle beneath the Mariana and other island arcs of the northwest Pacific, *J. Geophys. Res.* **91**, 3573–89, 1986.
61. S. Widiyantoro, Studies of seismic tomography on regional and global scale, Ph.D. Thesis, Australian National University, 1997.
62. R. van der Hilst, Complex morphology of subducted lithosphere in the mantle beneath the Tonga trench, *Nature* **374**, 154–7, 1995.
63. R. W. Griffiths, R. I. Hackney and R. D. van der Hilst, A laboratory investigation of effects of trench migration on the descent of subducted slabs, *Earth Planet. Sci. Lett.* **133**, 1–17, 1995.
64. S. Grand, R. D. van der Hilst and S. Widiyantoro, Global seismic tomography: a snapshot of convection in the earth., *Geol. Soc. Amer. Today* **7**, 1–7, 1997.
65. M. A. Richards and D. C. Engebretson, Large-scale mantle convection and the history of subduction, *Nature* **355**, 437–40, 1992.
66. H. W. Menard, *The Ocean of Truth*, 353 pp., Princeton University Press, Princeton, New Jersey, 1986.

CHAPTER 6

Flow

The flow of viscous fluids traditionally has not received a lot of attention in geology and geophysics curricula. The discussion of mechanics more usually focusses on elasticity and brittle fracture, with which the propagation of seismic (elastic) waves and faulting of the crust and lithosphere may be considered. The formation of folds and other kinds of distributed deformation receives some attention in structural geology, but many geologists still may not be very familiar with the mechanics of fluids. The text by Turcotte and Schubert [1] has gone a considerable way towards filling this gap, but fluid flow is so fundamental to mantle convection that it is worth developing here. By doing this I can focus the development on the particular things needed to treat mantle convection, and I can also present it at a range of mathematical levels, from the simplest possible to some more advanced aspects.

To guide readers, some of the sections are marked *Intermediate* or *Advanced*. These labels indicate the mathematical level. The essence of the chapter can be obtained just from the unlabelled sections (6.1, 6.7, 6.8.1, 6.9, 6.10). The important concepts and results are presented in those sections with minimal mathematics. The intermediate sections include mathematical formulations of stress, strain rate, viscosity and the equations governing slow flow of viscous fluids. These should not be too challenging, though some practice may be required if the notation is unfamiliar. A couple of sections summarise more advanced results that have particular relevance here, for those who may wish to see them.

It is always useful to begin with the simplest mathematical treatment that can capture a piece of physics, because then the physical concepts are the least obscured by the mathematics. This may suffice for those who want to get a clear understanding of mantle convection but who do not aspire to make any contributions to the subject themselves. For those who do aspire to go further, it is still essential to get a clear understanding of the physical concepts before proceeding to more advanced levels. Thus I begin this chapter by introducing the ideas of stress, strain, strain rate and viscosity in examples that are very simple but that permit the basic ideas and relationships to be appreciated. The basic equations of force balance and conservation of mass can also be introduced in this simple context.

These topics are then repeated, at an intermediate level, in a way that allows two- and three-dimensional problems to be treated. The equations become much messier-looking in these cases, but a concise notation retrieves a lot of the simplicity of the simple case. This 'subscript notation' may be unfamiliar to some, but the form of the equations closely parallels the simple cases, so a bit of practice with the notation is well worth the effort.

Some particular kinds of flow are then presented, the examples chosen to be relevant to mantle convection. Some of these are fairly simple, and some are more advanced. The latter are clearly marked, and those who wish may avoid them without sacrificing understanding of later chapters. It is not my intention here to present a comprehensive treatment of mantle flow, but rather to present some particularly pertinent examples, some of which are not readily accessible outside specialist fluid dynamics texts.

More detailed treatments of mantle flow often require numerical modelling. I do not present anything on numerical methods here because my focus is on developing a physical understanding in a way that is accessible to as wide an audience as possible. Analytical solutions are the most useful in this regard, because they reveal the way the fluid behaviour depends on parametres and material properties. The results of some numerical models will nevertheless be used in later chapters because known analytical solutions do not approach the realism required to demonstrate some key aspects of the behaviour of the mantle system.

This rather long chapter concludes with two sections on the mechanical properties of the mantle and crust. The first (6.9) outlines how observations of post-glacial rebound have been used to derive constraints on mantle viscosity. The second (6.10) considers the *rheology* of rocks more generally, rheology being the science of how materials respond to an applied stress. This includes brittle failure, which is characteristic of the lithosphere and central to the distinctive character of mantle convection. It also includes the dependence of viscosity on temperature and its possible dependence on stress (more correctly referred to as nonlinear rheology).

6.1 Simple viscous flow

In mechanical terms, a fluid is a material that can undergo an unlimited amount of deformation. A solid, on the other hand, may deform to a small extent, but it will break if you try to deform it too much. Another distinction is that many solids will deform only by a certain amount under the action of a particular force, and then return to their original shape if you stop applying that force. Such materials are called elastic. On the other hand, a fluid will keep deforming as long as a force is applied to it, and if the force is removed it will simply stop deforming, without returning to its original shape.

These distinctions are often very clear in our common experience, but in some circumstances they are not so clear. Thus, for example, some metals are elastic under the action of a small force, but yield and permanently deform if you apply a larger force. Malleable wire is a familiar example. A metal deforming permanently is behaving more like a fluid. The tendency to behave more like a fluid is enhanced in many materials if we heat them, and metals again provide a familiar example. Even when a material is solid for all practical purposes, it may be undergoing very slow deformation, so that we can consider it to be a fluid over hundreds or millions of years. We mentioned the example of glass in Chapter 2.

A *linear viscous fluid* is a material whose rate of deformation is proportional to the applied force. We will look here at how we can quantify that statement. I included the term 'linear' in the statement because in more general fluids the rate of deformation may be a more complicated function of the applied force. Linear viscous fluids are also known as *Newtonian* fluids. Strictly speaking, the term 'viscous' applies to materials in which the proportionality is linear, although the term is sometimes used more loosely. More general behaviour, such as that of malleable wire, is called variously *ductile*, *malleable* or *nonlinear*. Strictly speaking, ductile refers to materials with sufficient strength under tension that they can be stretched or drawn. Malleable would be a more appropriate term for many geological materials, but the term ductile is commonly used.

In order to quantify our definition of a viscous fluid, we need ways to characterise deformation and applied force. We can do this in the very simple situation depicted in Figure 6.1a. This shows a layer of fluid between two plates. It may help to think of the fluid being 'stiff', 'thick' or 'gooey' like honey or treacle (molasses). The top plate is moving to the right with velocity V, and the bottom

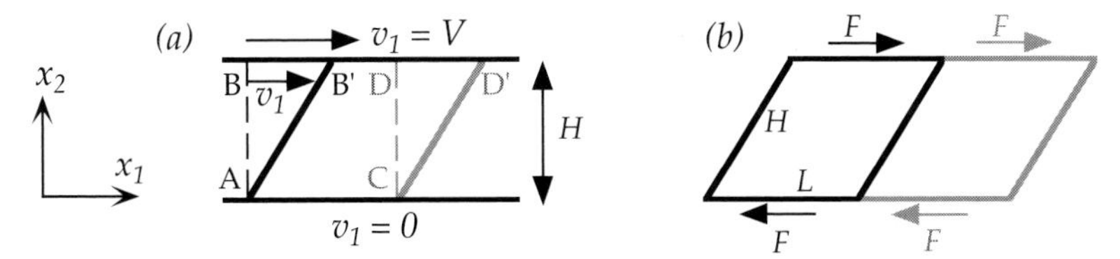

Figure 6.1. Shear flow in a layer of viscous fluid.

plate is stationary. Coodinates x_1 and x_2 are shown. If we could quickly inject a line of dye along the line AB, it would at a later time become inclined like the line AB$'$. Similarly the line CD will be carried into line CD$'$.

The box defined by ABDC becomes *deformed* into the parallelogram AB$'$D$'$C. This change in shape of the box is a measure of the deformation of the fluid. One way to measure the deformation of the box is with the ratio of lengths BB$'$/AB. If the time interval that has elapsed beween when the dye is at AB and when it is at AB$'$ is Δt, then we might write

$$\text{shape change} = V\Delta t/H$$

where H is the layer thickness. The *rate* of change of the shape or deformation is then measured by

$$\text{rate of deformation} = V/H$$

You can see that this quantity is a spatial gradient of velocity. In order to connect with the formal treatment in following sections, we will use the technical terminology. A quantity that measures deformation is called strain. Thus a quantity that measures rate of deformation is called a *strain rate*. Here I use the symbol s for strain rate. Also for consistency with later sections, I include a factor of one half in the definition of strain rate for Figure 6.1a:

$$s = \frac{V}{2H} \tag{6.1.1}$$

This quantity can serve as our measure of rate of deformation.

Now let us turn to the force causing the deformation. A force must be applied to the top plate in order to keep it moving. The moving plate then imparts a force into the adjacent fluid. The force F imparted into the top of the deformed box is depicted in Figure 6.1b as F. The magnitude of this force depends on the length, L, of the box. A second, adjacent box would also have a force F imparted into it, and the total force imparted into both boxes would be $2F$. However the deformation of each box is the same. Therefore what counts is the *force per unit area* that is applied to the fluid. We are familiar with pressure being a force per unit area, but here I want to acknowledge that pressure is a special case of the more general concept of *stress*, so I will use that term here. We need to note at this point that Figure 6.1 is implicitly a cross-section through a structure that extends into the third dimension (out of the page). We can make this explicit by assuming that the box has a width W in the third dimension. Then the stress, τ, imparted to the top of the fluid is

$$\text{stress} = \frac{\text{force}}{\text{area}} = \tau = \frac{F}{LW} \tag{6.1.2}$$

This quantity will serve as our measure of the applied force causing deformation.

We can now define a viscous fluid as one in which strain rate is proportional to stress. To be consistent with the formal development to follow, I will again include a factor of two in the definition:

$$\tau = 2\mu s \tag{6.1.3}$$

The constant of proportionality, μ, is called the *viscosity*. Since strain rate has a dimension of 1/time and stress has dimensions of force/area, or pressure, the units of viscosity are pascal seconds or Pa s. (1 pascal = 1 newton/m^2) A fluid with a high viscosity requires a greater stress to produce a given rate of deformation. Honey at room temperature has a viscosity in the range 10–100 Pa s. Water has a viscosity of about 0.001 Pa s. As we will see later, the mantle has a viscosity of the order of 10^{21} Pa s.

Equation (6.1.3) is a constitutive equation that describes the mechanical properties of a material. In order to use this in a study of convection, we need to draw upon some other basic principles: Newton's laws of motion, conservation of mass and conservation of energy. The latter will arise in Chapter 7. Here I will note how Newton's laws of motion and conservation of mass can be invoked for the situation in Figure 6.1.

The force F imparted by the top plate into the fluid induces a reaction of the fluid on the plate (Newton's first law, of action and reaction). The force will also be transmitted through the fluid to the bottom, where it will impart a force on the bottom plate, which in turn will induce an opposing reaction on the bottom of the fluid, shown as the lower force F. Newton's second law says that the acceleration of the fluid is proportional to the net force acting on it. Without saying so explicitly, I have been assuming so far that the fluid is *not* accelerating, but is flowing with constant velocity. This requires that *the net force on the fluid is zero*. The forces acting on the fluid in the box in Figure 6.1b are the force F imparted from the top plate and the opposing reaction of the bottom plate to the motion of the fluid. I have shown these as having equal magnitude in anticipation of the requirement that they must sum to zero. Writing this out,

$$\text{Net force} = F + (-F) = 0 = \text{mass} \times \text{acceleration} = \text{mass} \times 0$$

This point deserves to be emphasised. I will state it a little more generally than I have illustrated so far:

In steady, slow viscous flow, all forces sum to zero everywhere in the fluid.

We are so used to thinking of forces producing accelerations that it is easy to overlook the implication of Newton's law in this context. In mantle convection, velocities are so small that accelerations are utterly negligible. In the slow viscous flow of the mantle, applied force is balanced by viscous resistance. Another way to say this is that *momentum is completely negligible in the mantle*. For example, the uplift produced by a plume rising through the mantle (Chapter 11) is caused not by the upward momentum of the plume material but by the buoyancy of the plume material. Sometimes the expression of Newton's second law is called the momentum equation, but here I will call it the force balance equation.

I will mention conservation of mass only briefly here. Two other unstated assumptions about Figure 6.1 are that the fluid velocity is independent of horizontal position, x_1, and that the fluid is incompressible. It is then fairly obvious that the rate at which fluid flows into the box from the left is equal to the rate at which it flows out to the right. There is then no net accumulation of material and mass is conserved. If the fluid were compressible, then any imbalance of the flows into and out of the box would have to be balanced by a change in the density of the fluid in the box. For most purposes in this book we can treat the mantle as an incompressible fluid. The main context in which its compressibility is evident is in the increase of density with depth due to the great pressures in the interior (Chapter 5). However, the effect of this can be subtracted out of the equations of fluid motion to a good approximation. The equations for compressible fluids will be noted in passing in following sections.

To summarise this section, the mathematical description of flow in the mantle is done in terms of the concept of *strain rate*. The flow is driven by buoyancies, whose effect is represented as *stresses*, and these stresses cause strains to change with time. The proportionality between stress and strain rate (for materials in which simple proportionality applies) is expressed as the *viscosity* of the fluid. For a viscous fluid undergoing very slow flow, accelerations are negligible, and driving forces are everwhere in balance with viscous resisting forces. For most purposes in this book, the mantle can be approximated as an incompressible fluid. The following four sections develop each of these aspects more generally.

6.2 Stress [*Intermediate*]

When forces act on the surface of a body, their effects are transmitted through the body. This means that if you picture an imaginary surface inside the body, the material on one side of the surface will exert a force on the material on the other side. The magnitude and direction of this force may depend on the orientation of the surface. For example, in Figure 6.2a, you can readily appreciate that there will be a normal force across the surface (i), but not across the surface (ii). Also the force may have any orientation relative to the surface across which it acts, that is it need not be normal to the surface: in a solid or a viscous fluid a tangential or shearing force component may also act. A *stress* is a force component *per unit area* acting across an arbitrarily oriented surface such as (iii) in Figure 6.2a. Stress thus has the same dimensions as pressure. Following engineering usage, I will denote stress as $\boldsymbol{T}$ (for tension).

The full specification of a state of stress may require several stress components to be specified. For example, in Figure 6.2a we expect that there will be a normal stress across the surface (i) due to

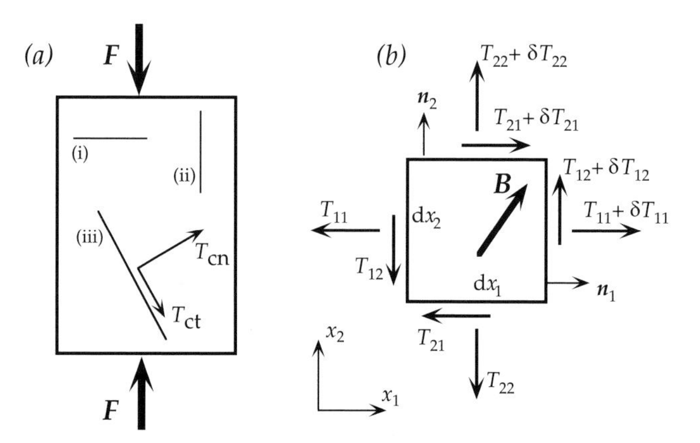

Figure 6.2. (a) Transmission of forces through a material. (b) Definitions of stress components.

the forces $\boldsymbol{F}$ shown, and we should also specify that the normal stress is zero across the surface (ii), because no horizontal forces are applied. For the record, stress is a second-order tensor, but we need not worry unduly about what a tensor is: the relevant properties will become apparent in due course.

Stresses acting across the surface (iii) in Figure 6.2a are depicted as T_{cn} and T_{ct}. These two components are sufficient to specify any possible force acting across (iii), in two dimensions. The naming convention here is that each stress component is labelled with two subscripts, the first denoting the surface across which it acts, and the second denoting the direction of the stress component itself. Thus T_{cn} is a stress component acting across surface 'c' in the normal (n) direction, while T_{ct} acts across the same surface in the tangential direction. We will now use this in a more formal way.

A systematic way of specifying stresses is to refer their components to a coordinate system. This is done in Figure 6.2b, which has coordinates x_1 and x_2. I will give the development, here and subsequently, in two dimensions where it is sufficient to demonstrate the concepts, since it is less messy than in three dimensions. The following can be generalised readily to three dimensions. Figure 6.2b depicts a small imaginary box inside a material, with faces oriented normal to the coordinate directions. Each face can be identified by its outward normal ($\boldsymbol{n}_1$ and $\boldsymbol{n}_2$). On each face stress components act, due to forces on the box exerted by surrounding material. For example, following the naming convention explained above, the component $(T_{11} + \delta T_{11})$ acts across the face whose normal n_1 is in the positive x_1 direction, and this component is also in the positive x_1 direction (I will explain the presence of the δT_{11} shortly).

The sign convention used here is that tensions are taken to be positive and compressions are taken to be negative. Further, stress components are positive when both they and the normal to the surface across which they act are in the positive coordinate direction. (Sometimes, particularly in the context of the earth's interior, pressure is taken to be positive, but this leads to confusion when shear stresses need to be considered, as we are about to do, so we will avoid this convention here.) If either the stress or the normal is in the negative coordinate direction, then the component is negative. If both are in the negative coordinate direction, then the component is positive.

The other component acting across the positive n_1 face is $(T_{12} + \delta T_{12})$ (in the direction of x_2). This is a positive shear stress component. On the left face, the component T_{11} is positive as shown, since it is in the negative x_1 direction across a face whose normal is also in the negative x_1 direction. Similarly T_{12} on the left face is a positive component.

Another purpose of Figure 6.2b is to consider the force balance on the box in the situation where the stresses vary with position. Thus the normal stress on the right face, $(T_{11} + \delta T_{11})$, is different from the normal stress on the left face, T_{11}. If the box is not accelerating, then all the forces must balance. In two dimensions, there is the possibility of rotation, and so we must also consider torques or moments: these must also balance. Consider first the balance of torques about the centre of the box. First, the force exerted by the stress T_{12} is $(T_{12} \cdot \mathrm{d}x_2)$, since stress is force per unit area, and the area over which T_{12} acts is $\mathrm{d}x_2$, assuming the box has unit length in the third dimension. Then the torque exerted about the centre is $(T_{12} \cdot \mathrm{d}x_2)(\mathrm{d}x_1/2)$. Considering each face in turn, the total torque in the clockwise direction is thus

$$\begin{aligned}(T_{21} + \delta T_{21}) \cdot \mathrm{d}x_1 \cdot \mathrm{d}x_2/2 - (T_{12} + \delta T_{12}) \cdot \mathrm{d}x_2 \cdot \mathrm{d}x_1/2 \\ + T_{21} \cdot \mathrm{d}x_1 \cdot \mathrm{d}x_2/2 - T_{12} \cdot \mathrm{d}x_2 \cdot \mathrm{d}x_1/2 = 0\end{aligned} \tag{6.2.1}$$

Dividing by $\mathrm{d}x_1 \cdot \mathrm{d}x_2$ and taking the limit as the box size approaches zero, this yields

$$T_{12} = T_{21} \tag{6.2.2}$$

This is a fundamental property of the stress tensor: it is symmetric with respect to changes in the order of the indices.

Now consider the force balance in the x_1 direction. It will be useful as we do this to include a body force, depicted as $\boldsymbol{B}$ in Figure 6.2b, which is a force per unit volume. $\boldsymbol{B}$ will be a vector with components B_1 and B_2. Then again considering each face in turn, and remembering that the tangential stresses on the top and bottom faces exert forces in the x_1 direction, the force balance condition is

$$\begin{aligned}&(T_{11} + \delta T_{11}) \cdot \mathrm{d}x_2 - T_{11} \cdot \mathrm{d}x_2 \\ &+ (T_{21} + \delta T_{21}) \cdot \mathrm{d}x_1 - T_{21} \cdot \mathrm{d}x_1 + B_1 \cdot \mathrm{d}x_1 \cdot \mathrm{d}x_2 = 0\end{aligned} \tag{6.2.3}$$

Dividing by $\mathrm{d}x_1 \cdot \mathrm{d}x_2$ and taking the limit, this yields

$$\frac{\partial T_{11}}{\partial x_1} + \frac{\partial T_{21}}{\partial x_2} + B_1 = 0 \tag{6.2.4a}$$

Similarly the force balance in the x_2 direction yields

$$\frac{\partial T_{12}}{\partial x_1} + \frac{\partial T_{22}}{\partial x_2} + B_2 = 0 \qquad (6.2.4b)$$

In Box 6.B1 I introduce a notation that takes advantage of the repetitive forms of Equations (6.2.4) in order to reduce them to a more compact form. This is called the subscript notation with summation convention. Taking note of the symmetry of the stress tensor (Equation (6.2.2)), Equations (6.2.4a) and (6.2.4b) can be written concisely as

$$\frac{\partial T_{ij}}{\partial x_j} + B_i = 0 \qquad (6.2.5)$$

This equation expresses the conservation of momentum, which in this context of no acceleration is equivalent to the equations of mechanical equilibrium or force balance. These equations show that the *gradients* of the stresses must obey these relationships if the material is to be in mechanical equilibrium. The presence of a body force modifies these relationships as shown. If the forms of these equations are unfamiliar, remember that they are simply the expression of the force balance in each coordinate direction.

Box 6.B1 Subscript notation and summation convention

The subscript notation permits concise expressions that would otherwise become large and clumsy, but it requires some familiarisation. I will briefly introduce it here, and provide some exercises at the end of the chapter.

You are probably familiar with subscripts being used to denote components of vectors and matrices. Thus a three-component vector can be written variously as

$$\boldsymbol{a} = \underline{a} = (a_1, a_2, a_3) = \{a_i\} \rightarrow a_i \qquad (6.B1.1)$$

The form $\{a_i\}$ stands for the set of a_i, for all values of i. The last form is not strictly equivalent, since it stands for a_i, for *any* value of i. Thus a general component of $\boldsymbol{a}$ stands for any component. This is the form we will use here.

The *summation convention* is that if a subscript is repeated in a term or product, it is implied that there is a summation over all values of that subscript. Thus the scalar product of two vectors can be written

$$\boldsymbol{a} \cdot \boldsymbol{b} = a_1 b_1 + a_2 b_2 + a_3 b_3 = \sum_{i=1}^{3} a_i b_i = a_i b_i \qquad (6.B1.2)$$

The last form employs the summation convention, since the subscript is repeated within the product. In effect the summation sign can be dropped because you know (usually) from the context which values the subscript can take. Occasionally there are situations where this is not true, and the explicit summation must be shown.

Summations are implicit in the following examples.

$$a_{ii} = a_{11} + a_{22} + a_{33} \tag{6.B1.3}$$

$$a_{ij}b_j = a_{i1}b_1 + a_{i2}b_2 + a_{i3}b_3 \tag{6.B1.4}$$

$$\frac{\partial a_i}{\partial x_i} = \frac{\partial a_1}{\partial x_1} + \frac{\partial a_2}{\partial x_2} + \frac{\partial a_3}{\partial x_3} = \nabla \cdot \boldsymbol{a} \tag{6.B1.5}$$

However there is no implied summation in

$$a_i + b_i \tag{6.B1.6}$$

which stands simply for the sum of any corresponding pair of components of $\boldsymbol{a}$ and $\boldsymbol{b}$, such as $a_2 + b_2$. This is because the index is not repeated *within* a term or product. Sometimes you need to turn the summation convention off. Thus if you want to refer to *any* diagonal component of a_{ij}, you must say explicitly 'a_{ii} (no summation)'.

A repeated index is, in effect, an internal dummy index that does not appear in the total expression. Thus, in Equation (6.B1.4), the end result is a vector component with index i, the j having been summed out. This means also that the name of the summed index is internal. Thus it is quite valid to write

$$a_{ii} = a_{kk} \tag{6.B1.7}$$

Correspondingly, a summation reduces the *order* of the term, that is the number of unsummed subscripts. Thus, in Equation (6.B1.3), $\boldsymbol{a}$ is a second-order tensor, but a_{ii} is a scalar (a zero-order tensor).

Just as a scalar cannot be added to a vector, all terms in an expression must be of the same order. Thus

$$a_i b_i + c_i$$

is not valid, but

$$a_i b_i + d$$

is valid.

The role of the Kronecker delta is worth spelling out. It is defined as

$$\delta_{ij} = 1 \text{ if } i = j$$

$$\delta_{ij} = 0 \text{ if } i \neq j \tag{6.B1.8}$$

and is analogous to a unit matrix $\boldsymbol{I} = [\delta_{ij}]$. When it occurs in a sum, its effect is to select out one term from the sum. Thus

$$a_i\delta_{i2} = a_1\delta_{12} + a_2\delta_{22} + a_3\delta_{32} = 0 + a_2 + 0 = a_2 \qquad (6.B1.9)$$

and

$$b_{ijk}\delta_{km} = b_{ijm} \qquad (6.B1.10)$$

6.2.1 Hydrostatic pressure and deviatoric stress

In the special case where the state of stress is a hydrostatic pressure, the normal components, like T_{11}, are all equal and the tangential components are zero. Thus, in three dimensions,

$$\begin{aligned} T_{11} &= T_{22} = T_{33} = -P \\ T_{12} &= T_{13} = T_{23} = 0 \end{aligned} \qquad (6.2.6a)$$

where I have taken pressure to be positive in compression, whereas T is positive in tension. Another way to write this, using the subscript notation and the Kronecker delta (Box 6.B1), is

$$T_{ij} = -P\delta_{ij} \qquad (6.2.6b)$$

The use of this sign convention for T may seem inappropriate for the earth's interior, where the state of stress is one of minor deviations from overwhelming pressure, but the equations are simpler with this convention. As well, for most of our purposes here, the large hydrostatic pressure can be subtracted out. This is because flow is not driven by hydrostatic pressure, but depends on deviations from hydrostatic pressure. This motivates the idea of *deviatoric stress*, below.

First, we can generalise the idea of pressure by defining P in a general state of stress (that is, other than that defined in Equations (6.2.6)) as

$$P = -(T_{11} + T_{22} + T_{33})/3 = -T_{ii}/3 \qquad (6.2.7)$$

In other words, pressure is defined as the negative of the average of the normal stress components.

We can now define a *deviatoric stress*, τ_{ij}, as the total stress minus the average of the normal stress components, so that

$$\tau_{ij} = T_{ij} - T_{kk}\delta_{ij}/3 = T_{ij} + P\delta_{ij} \qquad (6.2.8)$$

The pressure term in the last form is positive because of the different sign conventions of pressure and stress. A different subscript, k, is used in the summation in the first form so there is no confusion with the subscripts i and j, which can take arbitrary values in this equation. The effect of the Kronecker delta is that only the diagonal components of the stress are modified. In explicit matrix form, τ_{ij} is equivalent to

$$\tau_{ij} \Rightarrow \begin{bmatrix} T_{11}+P & T_{12} & T_{13} \\ T_{21} & T_{22}+P & T_{23} \\ T_{31} & T_{32} & T_{33}+P \end{bmatrix}$$

The deviatoric stress is that part of a general state of stress that differs from hydrostatic pressure or isotropic stress, and it is the part that can generate fluid flow.

6.3 Strain [*Intermediate*]

Strain is a measure of deformation. There are in fact many different measures that might be used to characterise deformation, and it is a matter of convenience which one is chosen. We will make choices here that are convenient for the present purpose. When deformation occurs, different parts of a body are displaced by different amounts. In other words there are spatial gradients of displacement. Displacement relates two different positions of a body. For example, Figures 6.3a and 6.3b depict a body in different positions at different times. Suppose the initial position of a point in the body is x_i^0 and the final position is x_i. Then the displacement of the point is defined as

$$u_i = x_i - x_i^0 \tag{6.3.1}$$

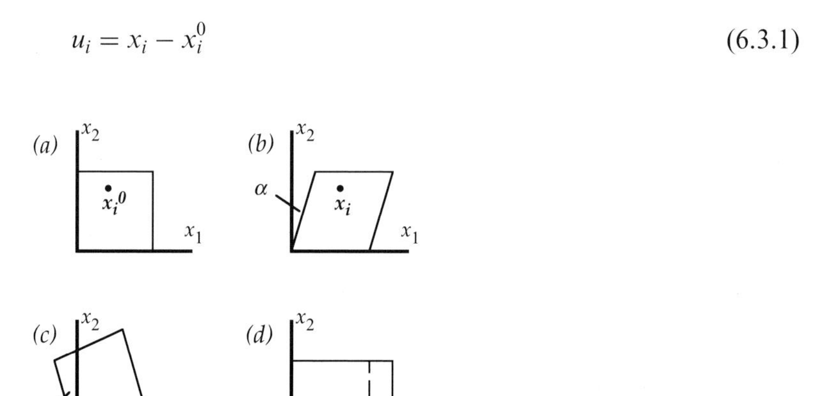

Figure 6.3. Definitions of displacement and strain.

Different parts of the body in Figure 6.3b are displaced by different amounts relative to their initial position. The variation of the two components of displacement with position is described by

$$u_1 = bx_2, \qquad u_2 = 0$$

where b is a constant. In other words the displacement in the x_1 direction increases with increasing x_2, and there are no displacements in the x_2 direction. Thus there is a gradient of displacement u_1:

$$\frac{\partial u_1}{\partial x_2} = b = -\tan\alpha$$

where α is the (counter-clockwise) angle through which the body has been sheared in this deformation. In Section 6.1 we used a displacement gradient like this to characterise the deformation of the fluid layer, and we can use the same idea here, as we will see.

If there is a shearing in the other orientation as well, parallel to x_2 as illustrated in Figure 6.3c, then

$$\tan\alpha_1 = \frac{\partial u_2}{\partial x_1}$$
$$\tan\alpha_2 = -\frac{\partial u_1}{\partial x_2}$$

taking counter-clockwise rotations to be positive.

Now if $\alpha_1 = \alpha_2$, then the body has simply rotated with no deformation. Since rotation does not involve any internal deformation of the body, we need a way to distinguish deformation from solid-body rotation. For example, $(\alpha_1 - \alpha_2)$ is zero if there is only rotation and no deformation ($\alpha_1 = \alpha_2$), and non-zero if there is deformation, so it can serve as a measure of deformation. On the other hand, if $\alpha_2 = -\alpha_1$ then there is no net rotation, in which case $(\alpha_1 + \alpha_2)$ is zero and the body undergoes pure shear. If $(\alpha_1 + \alpha_2)$ is non-zero there is rotation, so $(\alpha_1 + \alpha_2)$ can serve as a measure of rotation. These ideas are used to define a strain tensor and a rotation tensor. However, instead of the angles of rotation, we will use their tangents, which are the displacement gradients noted above. In the example of Figure 6.3c, a measure of rotation is

$$\omega_{12} = \tfrac{1}{2}(\tan\alpha_1 + \tan\alpha_2) = \tfrac{1}{2}\left(\frac{\partial u_2}{\partial x_1} - \frac{\partial u_1}{\partial x_2}\right) \tag{6.3.2}$$

and a measure of deformation is

$$e_{12} = \frac{1}{2}(\tan\alpha_1 - \tan\alpha_2) = \frac{1}{2}\left(\frac{\partial u_1}{\partial x_2} + \frac{\partial u_2}{\partial x_1}\right) \tag{6.3.3}$$

Now let us look at a different kind of deformation. Figure 6.3d depicts a stretching deformation (relative to Figure 6.3a). It is described by

$$u_1 = cx_1, \qquad u_2 = 0$$

where c is a constant, and the associated displacement gradient is $\partial u_1/\partial x_1 = c$. In this case there is no rotation to worry about, so the displacement gradient will serve as it stands as a measure of this deformation:

$$e_{11} = \frac{\partial u_1}{\partial x_1} \tag{6.3.4}$$

Using the gradient of the displacement in this case distinguishes deformation from simple solid-body translation: in a simple translation to the right, u_1 is constant, so $c = 0$ and $e_{11} = 0$.

We can now collect these ideas together concisely by defining a *strain tensor*

$$e_{ij} = \frac{1}{2}\left(\frac{\partial u_i}{\partial x_j} + \frac{\partial u_j}{\partial x_i}\right) \tag{6.3.5}$$

and an *infinitesimal rotation tensor*

$$\omega_{ij} = \frac{1}{2}\left(\frac{\partial u_j}{\partial x_i} - \frac{\partial u_i}{\partial x_j}\right) \tag{6.3.6}$$

The latter is called infinitesimal because strictly it measures angles of rotation only for small angles.

It is obvious from the definition that e_{ij} is symmetric. You can easily see that the definition of e_{ij} includes the case of Equation (6.3.3). In the case of the stretching deformation of Figure 6.3d, it yields

$$e_{11} = \frac{1}{2}\left(\frac{\partial u_1}{\partial x_1} + \frac{\partial u_1}{\partial x_1}\right) = \frac{\partial u_1}{\partial x_1}$$

so it serves for this case too. Examples of other kinds of deformation and the strains that measure them are given as exercises at the end of the chapter, and I recommend that you work through these to gain some familiarity with how this strain tensor works. We will not be concerned much further with rotation.

There is one special case of deformation worth spelling out, namely a change of volume. If there is stetching in each of three dimensions, then a small cube would expand, and its new volume would be

$$V = V_0(1 + e_{11})(1 + e_{22})(1 + e_{33}) \approx V_0(1 + e_{11} + e_{22} + e_{33})$$

The relative change in volume is then

$$\begin{aligned}(V - V_0)/V_0 &= e_{11} + e_{22} + e_{33} \\ &= e_{ii} = \partial u_i/\partial x_i = \nabla \cdot \boldsymbol{u} \equiv \Theta\end{aligned} \tag{6.3.7}$$

This quantity Θ is called the *dilatation*, and it is just the divergence of $\boldsymbol{u}$.

By analogy with the definition of deviatoric stress, we can define a *deviatoric strain*. Instead of subtracting out an average isotropic stress (that is, a pressure), we subtract out an average isotropic strain, that is, a dilatation. In this case our sign convention for dilatation is the same as for general strains, so we get

$$\xi_{ij} = e_{ij} - \Theta\delta_{ij}/3 = e_{ij} - e_{kk}\delta_{ij}/3 \tag{6.3.8}$$

You can see the analogy with Equation (6.2.8). This has the property that its diagonal terms sum to zero:

$$\xi_{ii} = e_{ii} - e_{kk} = 0$$

Thus ξ_{ii} does not register a change in volume, only a change in shape. It will be useful in discussing the viscosity of fluids in Section 6.5.

6.4 Strain rate [*Intermediate*]

It is easy to extend the definition of strain to its rate of change with time. In this case, the rate of displacement of a point in a body is just its velocity, $\boldsymbol{v}$, so differentiation of Equation (6.3.5) with respect to time yields

$$s_{ij} \equiv \frac{\partial e_{ij}}{\partial t} = \frac{1}{2}\left(\frac{\partial v_i}{\partial x_j} + \frac{\partial v_j}{\partial x_i}\right) \tag{6.4.1}$$

and s_{ij} is a *strain rate* tensor. This is analogous to Equation (6.1.1), in which we used a velocity gradient to measure the rate of shear of the fluid layer in Section 6.1. The *rate of dilatation* is the divergence of the velocity:

$$\frac{\partial \Theta}{\partial t} = s_{kk} = \frac{\partial v_k}{\partial x_k} = \nabla \cdot \boldsymbol{v}$$

and a *deviatoric strain rate* tensor is

$$\zeta_{ij} = s_{ij} - s_{kk}\delta_{ij}/3 \tag{6.4.2}$$

A good way to think of this quantity is that, with volume changes removed, it measures the rates of shearing deformations, or rates of changes of shape at constant volume.

6.5 Viscosity [*Intermediate*]

A viscous fluid is one that resists shearing deformations. Strictly speaking, it is one for which there is a linear relationship between strain rate and stress. Such fluids are sometimes called Newtonian or linear viscous fluids. You will see in Section 6.10 that more general relationships occur. The fluids of common experience are viscous, though for air and water the viscosity is quite low. Honey and treacle (molasses) are more viscous, especially when cold.

The simplest explication of viscosity is in a situation where the fluid is undergoing simple shear, as was depicted in Figure 6.1. The top plate is moving to the right, the bottom plate is stationary, and the line AB is displaced into the line AB'. The only non-zero velocity gradient is $\partial v_1/\partial x_2$, and the non-zero strain rate components are, using the definition Equation (6.4.1)

$$s_{12} = s_{21} = \frac{1}{2}\frac{\partial v_1}{\partial x_2}$$

In a linear viscous fluid, the non-zero deviatoric stress components would then be

$$\tau_{12} = \tau_{21} = \mu\frac{\partial v_1}{\partial x_2} = 2\mu s_{12} \tag{6.5.1}$$

where the constant of proportionality is μ, the *viscosity*. This is equivalent to Equation (6.1.3) derived earlier.

The viscosity μ is defined here following the convention used by Batchelor [2] in which it is the ratio of the stress component to the velocity gradient, which leaves a factor of 2 in the ratio of stress to strain rate. Sometimes a viscosity is defined by the ratio of stress to strain rate; for this I will use the symbol η. It differs from μ by a factor of 2:

$$\tau_{12} = \eta s_{12}, \qquad \eta = 2\mu \tag{6.5.2}$$

The definition of viscosity in cases with more general stresses and strains than the simple shearing depicted in Figure 6.1 requires some care at this point. It is usually assumed that fluids exhibit viscous behaviour only with respect to shearing deformations. Shearing deformations are measured by the deviatoric strain rate defined by Equation (6.4.2). It is conceivable that a fluid might also exhibit a viscous resistance to volume changes (in addition to its elastic resistance to compression). That is to say, the resistance to compression might depend on the *rate* of compression (viscous resistance), as well as on the degree of compression (elastic resistance). We could then define a *bulk viscosity*, by analogy with the bulk modulus of elasticity. However, I follow the usual practice of assuming that the bulk viscosity is negligible. The purpose of this digression has been to motivate the particular general form of the relationship between stress and strain rate that I am about to present.

If we simply generalise Equation (6.5.1) to all components, $\tau_{ij} = 2\mu s_{ij}$, there are two potential problems. First, it would imply that a bulk viscosity exists. Second, it would imply that the bulk viscosity is the same as the shear viscosity, and this is not necessarily so (the molecular mechanisms resisting deformation, if any, might well be different in compression from those that operate in shear). To avoid these problems, we can define viscous behaviour to apply only between *deviatoric stress* and *deviatoric strain rate*. Then neither pressure nor volume changes appear in the relationship. Thus

$$\begin{aligned} \tau_{ij} &= 2\mu\zeta_{ij} \\ &= 2\mu(s_{ij} - s_{kk}\delta_{ij}/3) \end{aligned} \tag{6.5.3}$$

This is a general constitutive relationship for a *compressible linear viscous fluid.*

Sometimes the compressibility of a fluid is negligible, and it can be treated as incompressible. In this case $\partial\Theta/\partial t = s_{kk} = 0$, and Equation (6.5.3) simplifies to

$$\tau_{ij} = 2\mu s_{ij} \quad \text{[incompressible]} \tag{6.5.4}$$

Although the earth's mantle material is compressed about 30% by volume near its base, the effect of compression can be subtracted out, to a sufficient approximation for many purposes (Chapter 7). The mantle can then be treated as incompressible, and Equation (6.5.4) can be used.

6.6 Equations governing viscous fluid flow [*Intermediate*]

In order to quantify the dynamics of viscous fluid flow, we must combine the constitutive relation of the fluid with equations expressing conservation of mass, momentum and energy. As we discussed in Section 6.1, acceleration and inertia are negligible in the mantle, so conservation of momentum reduces to a force balance. In the context of mantle convection, conservation of energy involves heat, which will be considered in Chapter 7.

6.6.1 Conservation of mass

For most purposes in this book, we can assume that the mantle is an incompressible fluid. For this case, conservation of mass becomes conservation of fluid volume. Then the rate at which fluid flows into a small volume like that depicted in Figure 6.4 must equal the rate at which it flows out. The volume of fluid that flows through the left side of the box in a time interval $\mathrm{d}t$ is equal to $v_1 \cdot \mathrm{d}t \cdot \mathrm{d}x_2$. The contributions through all four sides should sum to zero:

$$[v_1 \mathrm{d}x_2 + v_2 \mathrm{d}x_1 - (v_1 + \mathrm{d}v_1)\mathrm{d}x_2 - (v_2 + \mathrm{d}v_2)\mathrm{d}x_1]\mathrm{d}t = 0$$

Dividing this by $\mathrm{d}t \cdot \mathrm{d}x_1 \cdot \mathrm{d}x_2$ yields

$$\frac{\partial v_1}{\partial x_1} + \frac{\partial v_2}{\partial x_2} = \frac{\partial v_i}{\partial x_i} = \nabla \cdot \boldsymbol{v} = 0 \quad \text{[incompressible]} \tag{6.6.1}$$

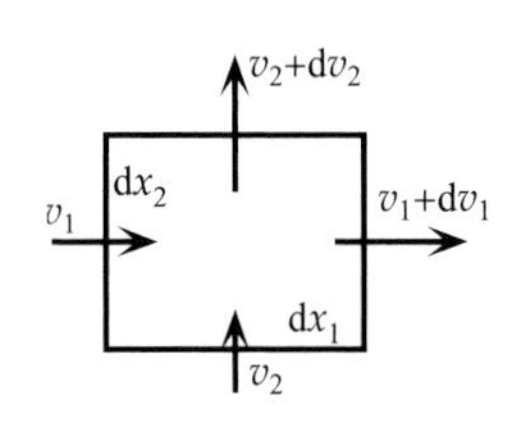

Figure 6.4. Flows into and out of a small region, used to derive the equation for conservation of mass.

In other words the divergence of the velocity is zero for an incompressible fluid. This is often called the *continuity equation*.

If the fluid is compressible, we must allow for the density, ρ, to vary with time and with position. Since we are not considering compressible fluids much here, I simply quote the result, which can be derived using the same approach [2] :

$$\frac{\partial \rho}{\partial t} + v_i \frac{\partial \rho}{\partial x_i} + \rho \frac{\partial v_i}{\partial x_i} = 0 \tag{6.6.2}$$

In vector notation, the middle term is $\boldsymbol{v} \cdot \nabla\rho$.

6.6.2 Force balance

We have seen already in Section 6.2 that the general condition for stress components and body forces to be balanced, so that fluid elements do not undergo acceleration, is expressed by Equation (6.2.5). The deviatoric stresses arising from the flow of a viscous fluid are expressed by Equation (6.5.3) or (6.5.4), and the deviatoric stresses are related to total stress through Equation (6.2.8). We can now combine these into a more specific equation. Here I just follow the incompressible case. The total stress is (Equations (6.2.8), (6.5.4))

$$\begin{aligned} T_{ij} &= \tau_{ij} - P\delta_{ij} \\ &= 2\mu s_{ij} - P\delta_{ij} \end{aligned}$$

and the general force balance for an *incompressible* viscous fluid is then (Equation (6.2.5))

$$2\frac{\partial(\mu s_{ij})}{\partial x_j} - \frac{\partial P}{\partial x_i} + B_i = 0 \tag{6.6.3}$$

This equation simplifies if the viscosity is independent of position. The assumption of constant viscosity is common outside of the mantle flow context, and it is useful for some purposes in this book, so I note some special forms of the equations for this case as we go along. If μ is independent of position, then Equation (6.6.3) becomes

$$2\mu\frac{\partial s_{ij}}{\partial x_j} - \frac{\partial P}{\partial x_i} + B_i = 0$$

This can be put in terms of velocity gradients using the definition (6.4.1) of s_{ij}. First, you can see that

$$\begin{aligned} 2\frac{\partial s_{ij}}{\partial x_j} &= \frac{\partial}{\partial x_j}\left(\frac{\partial v_i}{\partial x_j} + \frac{\partial v_j}{\partial x_i}\right) = \frac{\partial^2 v_i}{\partial x_j \partial x_j} + \frac{\partial}{\partial x_i}\left(\frac{\partial v_j}{\partial x_j}\right) \\ &= \frac{\partial^2 v_i}{\partial x_j \partial x_j} \\ &= \nabla^2 v_i \end{aligned}$$

where the second line follows from the continuity equation (6.6.1), which says that $(\partial v_j / \partial x_j) = 0$ for an incompressible fluid. The third line defines what is called the *Laplacian operator:*

$$\begin{aligned} \nabla^2 = \nabla \cdot \nabla &= \frac{\partial}{\partial x_k} \frac{\partial}{\partial x_k} \\ &= \frac{\partial^2}{\partial x_1^2} + \frac{\partial^2}{\partial x_2^2} + \frac{\partial^2}{\partial x_3^2} \end{aligned} \tag{6.6.4}$$

The denominator term $(\partial x_k \partial x_k)$ is written in this repeating form so that the summation convention is seen explicitly to apply. The term ∂x_k^2 would be ambiguous in this respect.

Now, finally, the force balance equation for an *incompressible, constant-viscosity* viscous fluid becomes

$$\mu \nabla^2 v_i - \frac{\partial P}{\partial x_i} + B_i = 0 \tag{6.6.5a}$$

or

$$\mu \frac{\partial^2 v_i}{\partial x_j \partial x_j} - \frac{\partial P}{\partial x_i} + B_i = 0 \tag{6.6.5b}$$

6.6.3 Stream function (incompressible, two-dimensional flow)

A further simplification of the equations is possible when the fluid is incompressible and the flow is two-dimensional, that is to say when one velocity vector component is zero. It is then possible to define a function that allows the continuity and force balance equations to be put into other mathematically useful forms. In this case, the continuity equation (6.6.1) is

$$\frac{\partial v_1}{\partial x_1} + \frac{\partial v_2}{\partial x_2} = 0 \tag{6.6.6}$$

If we define a *stream function* ψ such that

$$v_1 = \frac{\partial \psi}{\partial x_2}, \qquad v_2 = -\frac{\partial \psi}{\partial x_1} \tag{6.6.7}$$

then you can see by substitution that the continuity equation is satisfied identically.

In two dimensions, the horizontal and vertical force balance equations for an incompressible, constant viscosity fluid are (Equation (6.6.5a))

$$\mu\nabla^2 v_1 - \frac{\partial P}{\partial x_1} + B_1 = 0$$
$$\mu\nabla^2 v_2 - \frac{\partial P}{\partial x_2} + B_2 = 0$$

If the horizontal equation is differentiated with respect to x_2, the vertical equation differentiated with respect to x_1, and the second subtracted from the first, the result is

$$\mu\nabla^2\left(\frac{\partial v_1}{\partial x_2} - \frac{\partial v_2}{\partial x_1}\right) + \left(\frac{\partial B_1}{\partial x_2} - \frac{\partial B_2}{\partial x_1}\right) = 0$$

and the pressure terms have cancelled out. Substitution from the definition (6.6.7) of ψ then yields

$$\mu\nabla^2(\nabla^2\psi) + \left(\frac{\partial B_1}{\partial x_2} - \frac{\partial B_2}{\partial x_1}\right) = 0 \tag{6.6.8}$$

If there are no body forces

$$\nabla^4\psi = 0 \tag{6.6.9}$$

where $\nabla^4 = \nabla^2\nabla^2$ is called the *biharmonic operator*, and Equation (6.6.9) is called the *biharmonic equation*.

Equations (6.6.8) and (6.6.9) ensure that both the continuity equation and the force balance equations are satisfied. Thus the stream function allows the flow equations to be expressed in a very compact form. You will see below that it also leads to some useful analytic solutions to the flow equations.

The usefulness of the stream function does not stop there. Its name derives from the fact that lines of constant ψ are lines along which fluid flows. To see this, consider the difference, $d\psi$, between two close points, P and Q, depicted in Figure 6.5a:

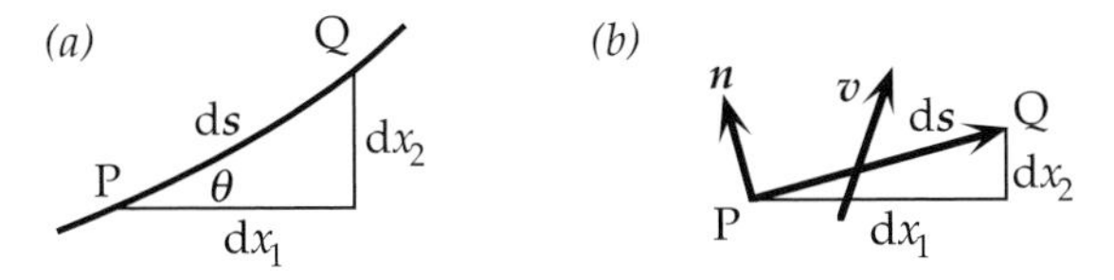

Figure 6.5. Geometric relationships to elucidate stream functions.

$$\begin{aligned} d\psi &= \frac{\partial\psi}{\partial x_1}dx_1 + \frac{\partial\psi}{\partial x_2}dx_2 \\ &= -v_2 dx_1 + v_1 dx_2 \\ &= -v_2 ds \cos\theta + v_1 ds \sin\theta \end{aligned}$$

Now if the line element **ds** is chosen to be parallel to the velocity $\boldsymbol{v}$, then

$$v_1 = v\cos\theta, \qquad v_2 = v\sin\theta$$

which implies, upon substitution, that $d\psi = 0$. In this case, ψ would have the same value at P and Q. It follows that if **ds** is part of a curve that is parallel to the local velocity along its length, then ψ is a constant along this curve.

Another property of the stream function is that the velocity is proportional to the local gradient of the stream function. This means that if streamlines are defined at equal intervals of ψ, like topographic contours, the velocity is inversely proportional to their spacing. This property can be shown using Figure 6.5b. The volumetric rate of flow dV through the surface defined by the line **ds** joining P and Q and extending a unit distance in the third dimension (out of the page) is $\boldsymbol{v}\cdot\boldsymbol{n}ds$ where $\boldsymbol{n}$ is the unit normal to the surface. The vector $\boldsymbol{n}$ has components

$$\boldsymbol{n} = (-dx_2, dx_1)/ds$$

Thus the flow rate is

$$dV = -v_1 dx_2 + v_2 dx_1 = -d\psi$$

the latter step being from the definition (6.6.7) of ψ. The volume flux ϕ is the volume flow rate per unit area:

$$\phi = dV/ds = -d\psi/ds$$

and if **ds** is chosen to be oriented normal to the local velocity, this is just the vector gradient of ψ.

6.6.4 Stream function and force balance in cylindrical coordinates [*Advanced*]

It will be useful for considering mantle plumes later to have the flow equations in a form convenient for solving problems with axial symmetry. Since my focus here is on presenting the central physical arguments in the most direct possible way, rather than on mathematical elaborations, I give only an abbreviated development here, fuller treatments being available elsewhere [2, 3].

The stream function defined by Equations (6.6.7) can be viewed as one component of a vector potential $(0, 0, \psi)$. The Cartesian velocities are then given by $\boldsymbol{v} = \nabla \times (0, 0, \psi)$. An analogous form can be used when there is axial symmetry. However with axial symmetry there are two possibilities. The first is to carry the so-called *Lagrangian* stream function ψ directly over. This preserves the relationship between velocity and derivative of the stream function. The second is to include a factor of $1/r$ to preserve the relationship between the stream function and the volume flux, which is proportional to (rv). The latter approach yields the *Stokes* stream function, Ψ, defined such that

$$\boldsymbol{v} = -\nabla \times \left(\frac{\Psi}{r} \boldsymbol{i}_\varphi \right) \tag{6.6.10}$$

where $\boldsymbol{i}_\varphi$ is a unit vector in the cylindrical coordinate system (r, φ, z) depicted in Figure 6.6.

The velocity components are

$$v_r = \frac{1}{r}\frac{\partial \Psi}{\partial z}, \qquad v_z = -\frac{1}{r}\frac{\partial \Psi}{\partial r} \tag{6.6.11}$$

To express the force balance equation in cylindrical coordinates, it is useful to define a *vorticity*

$$\boldsymbol{\Omega} = \nabla \times \boldsymbol{v} \tag{6.6.12}$$

With axial symmetry there is only one non-zero component: $\boldsymbol{\Omega} = (0, \Omega, 0)$. Ω is twice the rate of change of the rotation tensor ω_{12} defined by Equation (6.3.2); the factor of 2 is for convenience here and is often omitted from the definition. Substitution of the

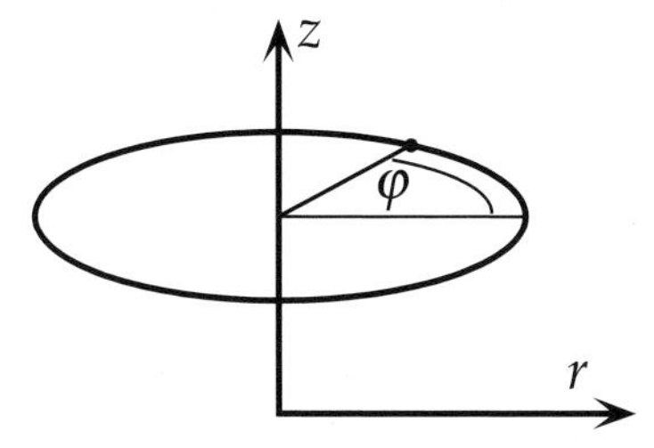

Figure 6.6. Cylindrical coordinates for axially symmetric problems.

velocity components of Equation (6.6.11) into this definition yields, after some manipulation

$$\Omega = \frac{1}{r} \mathrm{E}^2 \Psi \tag{6.6.13}$$

where E^2 is a differential operator related to the Laplacian operator ∇^2 of Equation (6.6.4):

$$\begin{aligned} \mathrm{E}^2 &= r \frac{\partial}{\partial r}\left(\frac{1}{r}\right)\frac{\partial}{\partial r} + \frac{\partial^2}{\partial z^2} \\ &= \nabla^2 - \frac{2}{r}\frac{\partial}{\partial r} \end{aligned} \tag{6.6.14}$$

For the incompressible fluid being considered here, vector identities yield

$$\nabla \times \boldsymbol{\Omega} = -\nabla^2 \boldsymbol{v}$$

Then the force balance equation (6.6.5a), with no body forces, can be written

$$\mu \nabla \times \boldsymbol{\Omega} = -\nabla P$$

and taking the curl yields

$$\nabla^2 \boldsymbol{\Omega} = \boldsymbol{i}_\varphi \left(\nabla^2 \Omega - \frac{\Omega}{r^2} \right) = 0 \tag{6.6.15}$$

and

$$\mathrm{E}^2 \Omega + \frac{2}{r}\frac{\partial \Omega}{\partial r} - \frac{\Omega}{r^2} = 0$$

Finally this can be manipulated into the form

$$\mathrm{E}^4 \Psi = \mathrm{E}^2 \mathrm{E}^2 \Psi = 0 \tag{6.6.16}$$

The analogy with Equation (6.6.9) for the Cartesian case is evident. Again we have the continuity and force balance equations put into a compact form that will be used in Section 6.8.

6.7 Some simple viscous flow solutions

Some flow solutions in relatively simple situations will help you to gain more physical insight into how viscous flow works. Additional exercises are provided at the end of the chapter.

6.7.1 Flow between plates

In the situation depicted earlier in Figure 6.1, flow is driven by the top moving plate. There are no body forces and there is no pressure gradient. In this situation the force balance Equation (6.6.5b) becomes

$$\frac{\partial^2 v_1}{\partial x_2^2} = 0$$

With the boundary conditions depicted, the solution to this is $v_1 = Vx_2/H$. This solution actually justifies the assumption implicit in Figure 6.1 that the velocity variation across the layer is linear.

Suppose now that both plates are stationary but there is a horizontal pressure gradient specified by

$$P = P_0 - P'x_1$$

as depicted in Figure 6.7. Then Equation (6.6.5b) becomes

$$\frac{\partial^2 v_1}{\partial x_2^2} = -\frac{P'}{\mu} \tag{6.7.1}$$

with the solution

$$v_1 = \frac{P'}{2\mu}(H - x_2)x_2 \tag{6.7.2}$$

Thus the velocity profile is parabolic, with a maximum at the centre of the layer. It will be useful for later to calculate the volumetric flow rate, Q, through this layer:

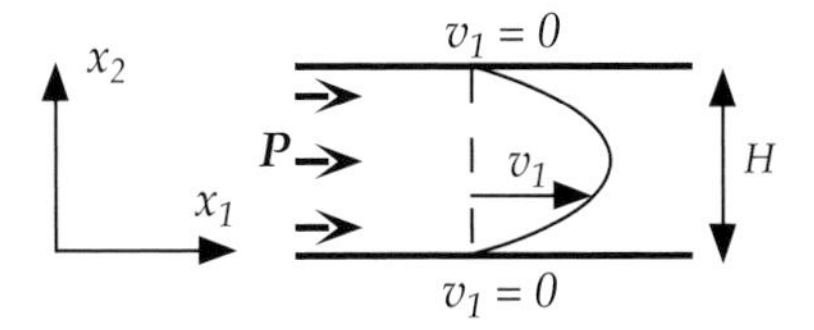

Figure 6.7. Flow between plates driven by a pressure gradient.

$$Q = \int_0^H v_1 \mathrm{d}x_2 = \frac{P' H^3}{12\mu} \tag{6.7.3}$$

Thus Q is proportional to the cube of the layer thickness.

This solution illustrates the fundamental point made earlier about slow viscous flow, that the flow is determined by a local balance between a driving pressure gradient and viscous resistance.

6.7.2 Flow down a pipe

It will be useful for later to derive the analogous flow through a pipe. I will present this problem from first principles rather than starting from the rather mathematical approach of the cylindrical stream function equation of Section 6.6.4. This will reveal even more directly the local balance between the driving force and the viscous resistance.

Here I assume that the pipe is vertical and the flow is driven by the weight of the fluid, rather than by a pressure gradient. This situation is directly analogous to convection, in which there is a balance between buoyancy forces and viscous resistance. It will have particular application in the theory of mantle plumes of Chapter 11.

Figure 6.8 depicts a fluid of density ρ flowing down a pipe (radius a) under the action of its own weight. A fluid element of length $\mathrm{d}z$ and radius r, like that shown, has weight

$$W(r) = \pi r^2 \cdot \mathrm{d}z \cdot \rho g$$

This is balanced by viscous resistance R acting on the sides of the element. If the flow is steady, there will be no net force on the top and bottom of the element. The viscous stress will be proportional to the local radial gradient of the vertical velocity: $\mu \cdot \partial v / \partial r$. The total resisting force is this stress times the surface area, $2\pi r \cdot \mathrm{d}z$, over which it acts. Thus

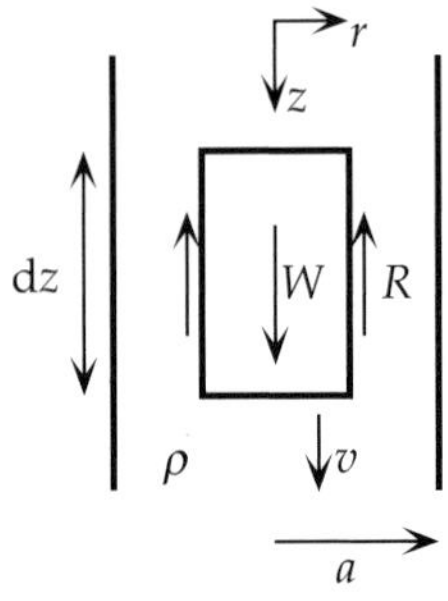

Figure 6.8. Force balance and viscous flow down a pipe.

$$R(r) = 2\pi r \cdot \mathrm{d}z \cdot \mu \cdot \partial v / \partial r$$

A balance of forces requires $R + W = 0$, which yields

$$\frac{\partial v}{\partial r} = -\frac{\rho g}{2\mu} r$$

Comparison with Equation (6.7.1) shows that the weight of the fluid here is playing the same role, through the term (ρg), as the pressure gradient in the plate problem of Figure 6.7.

The solution for this problem in which the fluid velocity is zero at the walls of the pipe is

$$v = \frac{\rho g}{4\mu}\left(a^2 - r^2\right) \tag{6.7.4}$$

and the volumetric flow rate is

$$Q = \frac{\pi \rho g}{8\mu} a^4 \tag{6.7.5}$$

The velocity profile is parabolic, as in the plate problem, but Q depends on a higher power of the size of the conduit than in the planar case, since the fluid is resisted all around in the pipe, but only from the top and bottom between the plates.

6.8 Rise of a buoyant sphere

A blob of buoyant fluid rising slowly through a viscous fluid, with negligible momentum, adopts the shape of a sphere. Drops and bubbles are commonly approximately spherical in shape, but in common situations the reason is mainly because of surface tension. The effect of momentum is also involved with water drops and bubbles of air in water, which usually causes distortions. One can observe some cases where drops and bubbles are more nearly spherical, such as air bubbles in honey, or the buoyant blobs in a 'lava lamp'. The mathematical analysis by Batchelor [2] shows more rigorously that the preferred shape is spherical.

The rise of a buoyant sphere is relevant to the mantle because there is good reason to believe that a new plume begins as a large spherical 'head', as we will see in Chapter 11. It is instructive to consider this case because it is relatively simple in concept, and because again it illustrates the balance between buoyancy and viscous resistance. It also is an appropriate example to demonstrate the usefulness of rough estimates. Not only can these give reasonable numerical estimates, but they reveal the scaling properties in the problem, by which I mean the way the behaviour would change if parametres or material properties were different.

6.8.1 Simple dimensional estimate

Let me begin by posing the question of how long it would take a plume head to rise through the mantle. In the absence of any prior indication, we might not know whether it would take ten thousand years or a billion years. Almost any kind of rough estimate would improve on this level of our ignorance. To obtain an initial estimate, consider the sphere sketched in Figure 6.9. *Buoyancy*, technically, is the *total force* arising from the action of gravity on the density difference between the sphere and its surroundings. Thus the buoyancy of the sphere is

$$B = -4\pi r^3 g \Delta\rho/3 \tag{6.8.1}$$

This force will cause the sphere to rise if the density of the sphere is less than that of its surroundings, so that $\Delta\rho$ is negative. The velocity, v, at which the sphere rises will be such that the viscous resistance from the surrounding material balances this buoyancy force. This velocity is measured relative to fluid at a large distance.

We can estimate the viscous resistance as follows. Viscous stress is proportional to strain rate, as described in Sections 6.1 and 6.5. Strain rate is proportional to velocity gradients. If we assume that the upward flow velocity in the fluid is about v near the sphere and decreases to a fraction of v over a distance of one sphere radius, then the velocity gradient will be of the order of v/r. More importantly, if v or r is changed, the velocity gradients will change in proportion. Thus, even without knowing the details of the flow and of the velocity gradients, by taking the strain rate to be of the order of v/r we can incorporate the idea that it will be ten times larger if v is ten times larger or if r is ten times *less*.

Now viscous stress, τ, is viscosity times strain rate, so

$$\tau = c\mu v/r$$

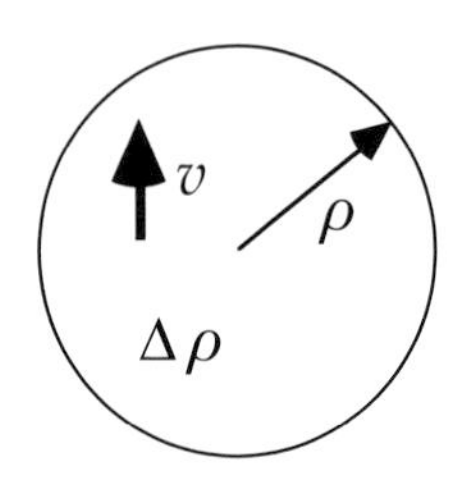

Figure 6.9. A buoyant sphere of density contrast $\Delta\rho$ rising with velocity v.

where μ is viscosity and c is a constant whose value will be of the order of 1 if the above logic is appropriate. Stress is force per unit area, so the total resisting force acting on the sphere is approximately stress times the surface area of the sphere:

$$R = -4\pi r^2 \cdot c\mu v/r = -4\pi c r \mu v \tag{6.8.2}$$

where the minus sign comes from taking upwards to be positive. The forces on the sphere will be balanced, and hence its velocity will be constant, if $B + R = 0$. So, using Equations (6.8.1) and (6.8.2),

$$4\pi c r \mu v + 4\pi r^3 g \Delta\rho/3 = 0$$

The value of the steady velocity will thus be

$$v = -g\Delta\rho r^2/3c\mu \tag{6.8.3}$$

If v is less than this value, the resistance will be less than the buoyancy, and the sphere will accelerate. If v is greater, the sphere will decelerate. Thus this value of v is a stable equilibrium value to which v will tend after any perturbation of the sphere's motion.

A more rigorous theory for a solid sphere is presented below. A solution for a fluid sphere with a viscosity, μ_s, different from the surrounding fluid can be obtained by a similar approach. These theories confirm the form of Equation (6.8.3), and yield

$$c = \frac{\mu + 1.5\mu_s}{\mu + \mu_s} \tag{6.8.4}$$

The value of c ranges between 1 and 1.5, thus justifying our hope that it would be of the order of 1. The limit of 1.5 is obtained when μ_s is infinite, and this corresponds to a solid sphere. The limit of 1 is obtained when $\mu_s = 0$, that is the fluid sphere is inviscid.

From Equation (6.8.3) you can see that the rise velocity of the sphere is proportional to its density deficit and inversely proportional to the viscosity of the surrounding material, and neither of these dependences is surprising. From Equation (6.8.4), you can see that the viscosity inside the sphere is not very important: an inviscid sphere rises only 50% faster than a solid sphere. This implies that the main resistance to the sphere's rise comes from the surrounding viscous fluid that it has to push through in order to rise.

With the other factors held constant, Equation (6.8.3) also says that a larger sphere rises faster, in proportion to r^2. This results from competing effects. On the one hand, the buoyancy is proportional to r^3 if $\Delta\rho$ is held constant. Against this, the resistance is proportional to the surface area, which varies as r^2. But the resistance is also proportional to the strain rate, which is proportional to v/r, as noted above. Thus a larger sphere generates smaller strain rates at a given velocity, and thus smaller viscous stresses. The net dependence of the resistance is thus on r (Equation (6.8.2)), and the net dependence of the velocity is on r^2.

Let us now apply Equation (6.8.3) to a mantle plume head with a radius of 500 km (Chapter 11) and a density deficit of 30 kg/m^3 (corresponding to a temperature excess of about 300 °C: Chapter 7). Assume a viscosity of the surrounding mantle of 10^{22} Pa s, typical of mid-mantle depths. Then Equation (6.8.3) gives $v = 2.5 \times 10^{-9}$ m/s = 80 mm/a = 80 km/Ma. At this rate the plume head would rise through 2000 km of mantle in 25 Ma. Thus we can get a useful idea of how long it might take a new plume head to reach the surface from deep in the mantle. Just as importantly, we know also how this estimate depends on the assumptions we have made, such as that the deep mantle viscosity is 10^{22} Pa s. If this viscosity is uncertain by, say, a factor of 3, then our estimate of the rise time is also uncertain by a factor of 3: it might be anything between about 8 Ma and 80 Ma.

6.8.2 Flow solution [*Advanced*]

I will present here the rigorous solution for a solid sphere rising though a very viscous fluid. This was first developed by Stokes [4]. Versions of it are presented by Happel and Brenner [3] (p. 119) and Batchelor [2] (p. 230). Their versions are developed in more general contexts for mathematicians and fluid dynamicists. Here I outline an approach that is more direct in the present context.

The situation is sketched in Figure 6.10, which depicts a buoyant solid sphere of radius a rising slowly, with velocity U, through a viscous fluid of viscosity μ. The problem is symmetric about the vertical axis, and it is convenient to use spherical coordinates (r, θ, ϕ), where ϕ is the azimuthal angle about the axis. Sometimes it is also useful to use the cylindrical coordinates (ϖ, ϕ, z). I explained in Section 6.6.4 that with axial symmetry it is possible to define Stokes' stream function, Ψ, and that the force balance equations reduce to

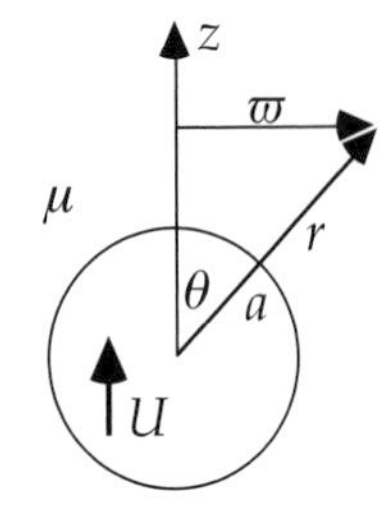

Figure 6.10. Rising buoyant sphere with spherical coordinates (r, θ, ϕ) and cylindrical coordinates (ϖ, ϕ, z).

$$\mathrm{E}^4 \Psi = 0 \tag{6.8.5}$$

where E is a differential operator given by Equation (6.6.14). In spherical coordinates, E has the form

$$\mathrm{E}^2 = \frac{\partial^2}{\partial r^2} + \frac{\sin\theta}{r^2}\frac{\partial}{\partial\theta}\left(\frac{1}{\sin\theta}\frac{\partial}{\partial\theta}\right)$$

and $\mathrm{E}^4 = \mathrm{E}^2(\mathrm{E}^2)$.

The boundary conditions are that the fluid velocity $\boldsymbol{u}$ equals $\boldsymbol{U}$ on the surface of the sphere and $\boldsymbol{u}$ approaches zero at infinity. These can be expressed as follows.

At $r = a$: $\quad u_r = U\cos\theta$

so

$$\Psi = -0.5Ua^2\sin^2\theta \tag{6.8.6a}$$

and

$$u_\theta = -U\sin\theta$$

so

$$\frac{\partial\Psi}{\partial r} = -Ua\sin^2\theta \tag{6.8.6b}$$

At r $= \infty$: $\quad \dfrac{\Psi}{r^2} \to 0 \qquad (6.8.6c)$

A common method for solving equations such as (6.8.5) is separation of variables, which can often be used if the boundary conditions are compatible with the solution being a product of separate functions of each independent variable. The spherical geometry suggests using functions of r and θ, and the form of the boundary conditions suggests trying

$$\Psi = \sin^2\theta F(r) \tag{6.8.7}$$

where F is an unknown function. Substitution into the definition of E^2 above yields

$$E^2\Psi = \sin^2\theta\left(F'' - \frac{2F}{r^2}\right) \equiv f(r)\sin^2\theta \tag{6.8.8}$$

where a prime denotes differentiation and $f(r)$ is another unknown function. Another application of E^2 yields

$$E^4\Psi = \sin^2\theta\left(f'' - \frac{2f}{r^2}\right)$$

so from Equation (6.8.5)

$$f'' - \frac{2f}{r^2} = 0 \tag{6.8.9}$$

This equation has a solution of the form

$$f = Ar^2 + \frac{B}{r}$$

so from Equation (6.8.8)

$$F'' - \frac{2F}{r^2} = Ar^2 + \frac{B}{r}$$

This has a particular solution of the form $Ar^4/10 - Br/2$, to which a homogeneous solution of the same form as that for f should be added:

$$F = \frac{Ar^4}{10} - \frac{Br}{2} + Cr^2 + \frac{D}{r} \tag{6.8.10}$$

The boundary condition (6.8.6c) requires $A = C = 0$, while (6.8.6a,b) require $B = 3Ua/2$ and $D = Ua^3/4$. Substitution into Equation (6.8.7) yields finally

$$\Psi = \frac{1}{4} Ua^2 \left(\frac{a}{r} - 3\frac{r}{a} \right) \sin^2\theta \tag{6.8.11}$$

From this stream function we can deduce the fluid velocities and other quantities. In particular we want an expression for the viscous resistance to the sphere, and for this it is convenient to have expressions for the pressure and vorticity. The velocities can be found directly from the definition of Ψ in Section 6.6.4:

$$u_r = -\frac{U}{2}\left[\left(\frac{a}{r}\right)^3 - 3\left(\frac{a}{r}\right)\right]\cos\theta$$

$$u_\theta = -\frac{Ua}{4r}\left[\left(\frac{a}{r}\right)^2 + 3\right]\sin\theta$$

and I also showed there that the ϕ-component of the vorticity is

$$\zeta = \frac{1}{\varpi} \mathrm{E}^2 \Psi = \frac{3Ua}{2r^2} \sin\theta \tag{6.8.12}$$

The pressure can be found most easily by putting the force balance equation in the form

$$\nabla p = \mu \nabla^2 \boldsymbol{u} = -\mu \nabla \times \zeta$$

Substitution from Equation (6.8.12) and integration with respect to r and θ yields

$$p = p_\infty + \frac{3\mu U a}{2r^2} \cos\theta \qquad (6.8.13)$$

where p_∞ is the pressure at infinity.

To get the force on the sphere with minimal manipulation, we need a general result for stresses on a no-slip surface. This is derived in Box 6.B2, where it is shown that the normal and tangential stress components can be written in terms of the pressure and vorticity on the boundary, as given by Equations (6.B2.1) and (6.B2.2). Since these are scalar quantities, the result is independent of the coordinate system, and can be transferred to the surface of the sphere, a portion of which is sketched in Figure 6.11. We want the net force on the sphere in the positive z direction, which we get by adding the z-components of the surface stresses and integrating them over the surface of the sphere. The net z-component is

$$T_z = T_{rr} \cos\theta - T_{r\theta} \sin\theta$$

From the result in Box 6.B2, we get $T_{rr} = -p$ and $T_{r\theta} = \mu\zeta$, noting that the sign of ζ in the x–z coodinates of Figure 6.B2 is opposite to its sign in the r–θ coordinates of Figure 6.11, assuming that the coordinate systems are right-handed. Substituting for p and ζ from Equations (6.8.12) and (6.8.13), we get the simple expression

$$T_z = -p_\infty \cos\theta - 3\mu U a/2r^2$$

The net force on the sphere is obtained by integrating over strips of the sphere between θ and $\theta + \mathrm{d}\theta$, so that

$$F_z = \int_0^\pi T_z \cdot 2\pi a \sin\theta \cdot a \cdot \mathrm{d}\theta$$

with the final result

$$F_z = -6\pi\mu U a \qquad (6.8.14)$$

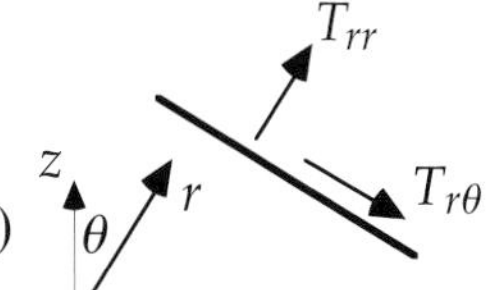

Figure 6.11. Stress components on the surface of a portion of a sphere.

The contribution from the p_∞ term is zero, as is expected for the net force from a uniform pressure.

This result has the same form as the dimensional estimate, Equation (6.8.2), and they are the same if $c = 3/2$ there, which is the value obtained from Equation (6.8.4) for a solid sphere. The formula for the velocity of the sphere follows directly as before (Equation (6.8.3)).

The analysis for a fluid sphere proceeds in the same way, except that now a solution for the flow inside the sphere must be matched to a solution for the flow outside the sphere. Thus the boundary conditions are different. The interior and exterior solutions both have the general form given by Equations (6.8.7) and (6.8.10). The calculation of the net force does not simplify in the same way, since the result from Box 6.B2 does not apply in this case. A derivation is given by Batchelor [2] (p. 235).

Box 6.B2 Stresses on a no-slip boundary

The result we need is most easily obtained in Cartesian coordinates, as sketched in Figure 6.B2. From the boundary condition, you can see that the strain component $s_{xx} = \partial u_x/\partial x = 0$. From the conservation of mass for an incompressible fluid, this implies also that $\partial u_y/\partial y = s_{yy} = 0$ on the boundary. Then from the constitutive relation for a viscous fluid,

$$T_{yy} = -p + 2\mu s_{yy} = -p \tag{6.B2.1}$$

From the boundary condition, we also have that $\partial u_y/\partial x = 0$, so that $s_{yx} = 0.5\partial u_x/\partial y$. But also, the vorticity is

$$\zeta = \left(\frac{\partial u_y}{\partial x} - \frac{\partial u_x}{\partial y}\right) = -\frac{\partial u_x}{\partial y}$$

Then the shear stress component becomes

$$T_{yx} = 2\mu s_{yx} = -\mu\zeta \tag{6.B2.2}$$

Thus on the no-slip boundary, the normal and tangential stress components take the simple forms (6.B2.1) and (6.B2.2).

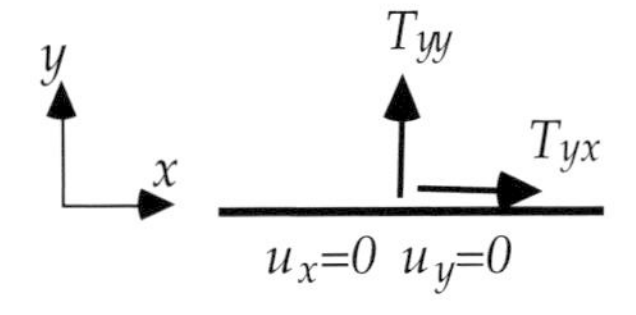

Figure 6.B2. Stress components on a no-slip boundary.

6.9 Viscosity of the mantle

There are a number of observations that indicate that on geological time scales the mantle deforms like a fluid, and these can be used also to deduce something about the relevant rheological properties of the mantle. Usually it is assumed that the mantle is a linear viscous fluid, and the material is characterised in terms of a viscosity. In Chapter 3 I discussed the origins of the idea that the mantle is deformable, which came particularly from evidence from the gravity field that the earth's crust is close to a hydrostatic (or isostatic) balance, on large horizontal scales, as would be expected if the interior is fluid. I briefly mentioned there that by the 1930s observations of the upward 'rebound' of the earth's surface after the melting of ice-age glaciers had been used to estimate the viscosity of the mantle. This approach, and results from recent versions of it, will now be presented. I will also discuss constraints from the gravity field over subduction zones and from small variations in the earth's rotation. The former provides some additional constraints on the variation of viscosity with depth.

6.9.1 Simple rebound estimates

The land surfaces of Canada and of Scandinavia and Finland (Fennoscandia) have been observed to be rising at rates of millimetres per year relative to sea level. The main observation on which this inference is based is a series of former wave-cut beach levels raised above present sea level. These have been dated in a number of places to provide a record which is usually presented as relative sea level versus time, an example of which is shown in Figure 6.12a.

The inferred sequence of events is sketched in Figure 6.12b. An initial reference surface (6.12b(i)) is depressed a distance u by the weight of glacial ice during the ice age (6.12b(ii)). (The ice load peaked about 18 ka and ended about 10 ka before present.) After melting removed the ice load, the reference surface rose back towards its isostatically balanced level (6.12b(iii)). That rising continues at present with velocity v.

A very simple analysis will illustrate the approach to deducing a mantle viscosity and give a rough estimate of the result. The removal of the ice load generates a stress in the underlying mantle which we can think of for the moment simply as a pressure deficit due to the remaining depression in the earth's surface, which is filled by air or water. This stress is, approximately

$$\tau_p \approx \Delta\rho g u$$

where $\Delta\rho$ is the density contrast between the mantle and the air or water. This stress is resisted by viscous stresses in the mantle.

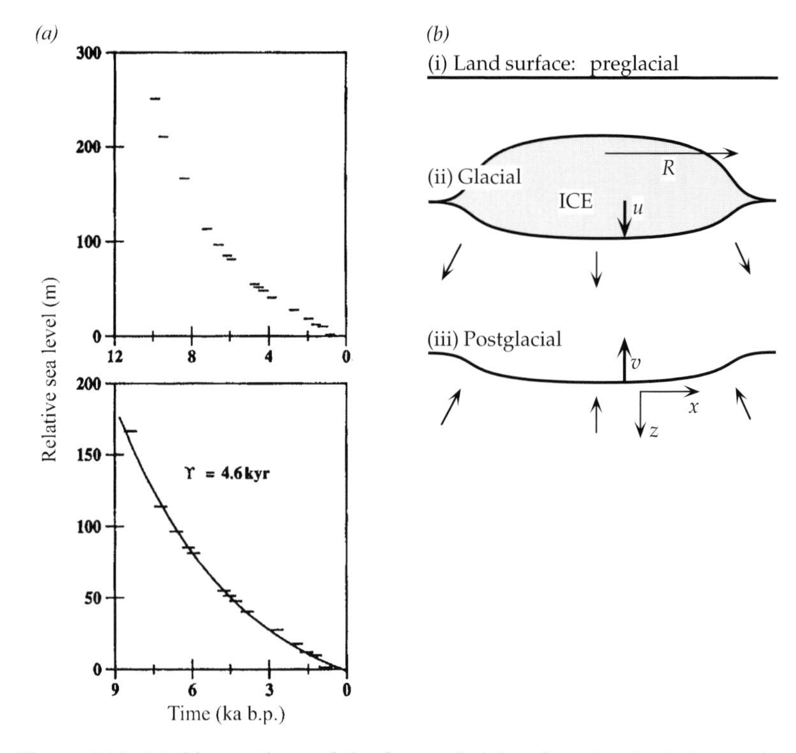

Figure 6.12. (a) Observations of the former height of sea level relative to the land surface at the Angerman River, Sweden. From Mitrovica [5].
(b) Sketch of the sequence of deformations of the land surface (i) before, (ii) during, and (iii) after glaciation.

Viscous stress is proportional to strain rate, which is proportional to velocity gradient. A representative velocity gradient is v/R, where v is the rate of uplift of the surface and R is the radius of the depression. Thus (compare with Equation (6.5.1)) the viscous stress τ_r will be approximately

$$\tau_r \approx \mu \frac{v}{R} = -\frac{\mu}{\mathrm{R}} \frac{\partial u}{\partial t}$$

where the last step follows because v is minus the rate of change of u.

Equating τ_p and τ_r and rearranging gives the differential equation

$$\frac{1}{u} \frac{\partial u}{\partial t} = -\frac{\Delta \rho g R}{\mu} \equiv -\frac{1}{\tau} \tag{6.9.1}$$

where the last identity defines a time scale τ. This has the simple solution

$$u = u_0 e^{-t/\tau}$$

In other words, the depth of the depression decays exponentially with time. The observations of relative sea level fit such an exponential variation well. In Figure 6.12a the decay has a time scale of 4.6 ka.

This observed time scale can be used in Equation (6.9.1) to estimate a viscosity of the mantle: $\mu = \Delta\rho g R\tau$. Taking $\Delta\rho$ to be 2300 Mg/m^3, the difference between the densities of water and the mantle, R to be 1000 km and g to be about 10 m/s^2 yields $\mu = 3 \times 10^{21}$ Pa s. A more rigorous analysis by Haskell [6] in 1937 yielded 10^{21} Pa s. Our estimate here is very rough, but clearly it gives the right order of magnitude, and makes the physics clear.

A more rigorous, though still simplified, analysis can be done by considering a sinusoidal perturbation of the earth's surface. You can think of this as the longest-wavelength Fourier component of the depression in Figure 6.12b, with wavelength $\lambda = 4R$ and wavenumber $k = 2\pi/\lambda = \pi/2R$. Thus suppose that after the ice has melted there is a component of the perturbation to the surface topography of the form

$$u(x, 0) = U \cos kx$$

where the coordinates here will be denoted (x, z), with x horizontal and z vertically downward. The rate of change of this displacement, $v = -\partial u/\partial t$, can be matched by a stream function of the form

$$\psi(x, z) = VZ(z) \sin kx$$

where $V = \partial U/\partial t$ and Z is an unknown function of z. Substitution of this into the biharmonic Equation (6.6.9) then yields

$$\left(\frac{d^2}{dz^2} - k^2\right)^2 Z = 0$$

which has a general solution of the form

$$Z = a_1 e^{kz} + a_2 e^{-kz} + a_3 z e^{kz} + a_4 z e^{-kz}$$

Requiring the solution to decrease at great depth implies $a_1 = a_3 = 0$. Two other boundary conditions are that the surface vertical velocity amplitude be V and the horizontal velocity be zero. Using the definition of the stream function (Equation 6.6.7) and application of these conditions then yields

$$\psi = -\frac{V}{k}(1 + kz)e^{-kz} \sin kx \tag{6.9.2}$$

This prescribes everything about the solution to this problem, but relating it to the rebound problem still requires the vertical stress, T_{zz}, at the surface. This stress can be related to the amplitude of the surface displacement, and hence to the restoring stress at the surface, because the high parts exert an excess downward normal stress due to the extra weight of the topography. The low parts exert a (notional) upward normal stress. It is the differences between the weight of the topography in different places that drive the rebound, and these should also match T_{zz}. Thus the stress at $z = 0$ exerted by the topography is

$$W = \Delta\rho g u = \Delta\rho g U \cos kx$$

This must be balanced by the viscous stress, T_{zz}, calculated from Equation (6.9.2). From Equations (6.2.8) and (6.5.4), $T_{zz} = 2\mu s_{zz} - P$. The pressure P can be obtained from the force balance equations, (6.6.5). The calculations are somewhat tedious. It turns out that $s_{zz} = 0$ and

$$T_{zz} = -2V\mu k \cos kx$$

Equating T_{zz} and W,

$$\frac{\partial U}{\partial t} = -\frac{\Delta\rho g}{2\mu k} U$$

which has the solution

$$U = U_0 e^{-t/\tau}$$

where U_0 is the initial value of U and

$$\tau = 2\mu k/\Delta\rho g = \mu\pi/\Delta\rho g R \tag{6.9.3}$$

This result differs from Equation (6.9.1) by the factor π, and so it will yield a viscosity of $\mu = 10^{21}$ Pa s using the same numbers as used above. This is the same result as obtained by Haskell, even though the problem has been rather idealised.

6.9.2 Recent rebound estimates

A full analysis of postglacial rebound could involve the time and space history of the ice load, the changes in the volume of the oceans as ice accumulates on the continents, the resulting changed magnitude of the ocean load and changed distribution of the ocean load near coastlines, the self-gravitation of the changing mass distributions at large scales, the elasticity of the lithosphere, lateral variations in lithosphere thickness, especially at continental margins, and possible lateral variations in mantle viscosity (e.g. [7]). The full problem is thus very complicated, and has absorbed a great deal of effort.

It turns out that there are certain observations that probe the mantle viscosity more directly, without being greatly affected by the complications introduced by the other factors. One of these is the time scale of rebound at the centre of a former ice sheet after it has all melted. The case in Figure 6.12a is an example of this. Mitrovica [5] has analysed the sensitivity of the inferred viscosity to the ice load history and the assumed thickness of the lithosphere and shown it to be small. He has also analysed the depth-resolution of the observation, that is the sensitivity of the observed rebound time scale to differences in viscosity structure at various depths. This showed that the time scale depends mainly on the average viscosity of about the upper 1400 km of the mantle, a result that is consistent with the intuitive expectation that the deformation due to the ice load will penetrate to a depth comparable to its radius.

The viscosity of 10^{21} Pa s inferred by Haskell from similar data thus represents an average viscosity to a depth of about 1400 km. Mitrovica showed that it is possible for the upper mantle viscosity to be less than the average and the viscosity of the upper part of the lower mantle to be more than the average, with a contrast of an order of magnitude or more, so long as the average value is preserved. Observations from North America support this inference. The North American ice sheet was larger than in Fennoscandia, and hence its rebound is sensitive to greater depth, about 2000 km. A similar analysis of observations from the southern part of Hudson Bay, near its centre, showed that the top of the lower mantle has a higher viscosity. Combining these two analyses suggests a lower mantle viscosity of about 6×10^{21} Pa s and a corresponding upper mantle viscosity of about 3×10^{20} Pa s. Neither of these observations constrains the viscosity in the lower third of the mantle, 2000–3000 km depth, which may be higher still.

These results are consistent with two other types of study, one of geoid anomalies over subduction zones, discussed in the next section, and the other of postglacial relative sea level changes far from ice sheets. The latter are a second special case that seem to be less sensitive to complicating factors. The idea here is that far from ice sheets relative sea level is controlled not by the ice load, whose effects are negligible, but by changes in the volume of ocean water as ice accumulates on distant continents and then melts again. This causes relative sea level away from the ice loads to be low during glaciations, and to rise as the ice melts, the reverse of the sequence within glaciated areas that is depicted in Figure 6.12a.

It is observed, however, that far from ice sheets the relative sea level rise has not been monotonic, but has overshot by some metres before dropping to present levels. An example is shown in Figure 6.13. The reason for the overshoot is that the ocean basins are not static during the process, because the change in ocean volume changes the load on the sea floor. Consequently, as water is withdrawn the sea floor rises slightly, and as the water is returned it is depressed again. This process happens with a time lag because the mantle is viscous, which means that immediately after all the water has been put back, the sea floor has not completely subsided to its isostatic level, and the water floods slightly onto the continents. Subsequently, as the sea floor completes its delayed subsidence, the water retreats from the continents by a few metres. Thus these so-called Holocene highstands are a measure of the delayed response of the seafloor level to the changing ocean load, and hence

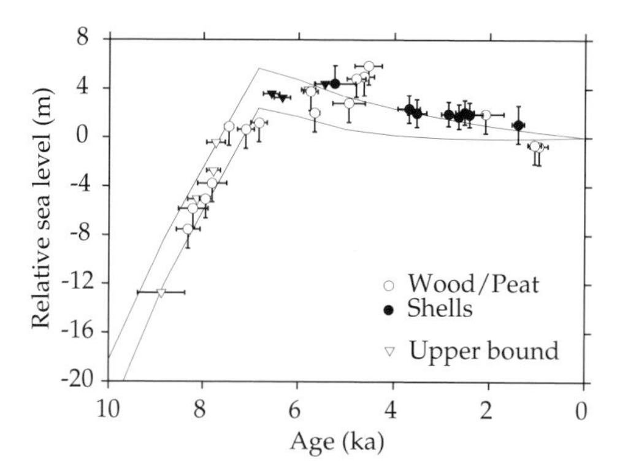

Figure 6.13. Variation of relative sea level with time before present at west Malaysian Peninsula. This example shows a small overshoot of the postglacial rise in sea level, due to the delayed response of the sea floor to the increasing water load. The curves are the envelope of models that plausibly fit the observations from many sites. The steep rise is due to the addition of water until about 7 ka ago, and the subsequent slow fall is due to continued adjustment of the sea floor to the increased water load. From Fleming *et al.* [8]. Copyright by Elsevier Science. Reprinted with permission.

of mantle viscosity [8]. Because the ocean basins are large in horizontal extent, the effects penetrate to great depth in the mantle, and it is expected that the inferred viscosity is an average essentially of the whole depth of the mantle.

Analyses of these observations by Lambeck and coworkers [7, 9, 10] have led to conclusions very similar to those quoted above, a representative result being an upper mantle viscosity of 3×10^{20} Pa s and a lower mantle viscosity of 7×10^{21} Pa s.

So far these approaches have not been combined into a single study of depth resolution, so it is not yet clear whether there is more information to be gained about the lowest third of the mantle. In particular it is not clear whether there is direct evidence for the possibility that the deep mantle has an even higher viscosity, as will be suggested in Section 6.10 on the basis of rock rheology.

6.9.3 Subduction zone geoids

A completely different kind of observation has been used to constrain the relative variation of mantle viscosity with depth, though it does not constrain the absolute values of the viscosity. We saw in Chapter 4 that there are positive gravity and geoid anomalies over subduction zones (Figure 4.9). It was also noted there that the geoid is sensitive to density variations to greater depths than is gravity, and so it is the more useful for probing the mantle. The idea is that these geoid anomalies reflect the presence of higher-density subducted lithosphere under subduction zones. However, the net effect on the gravity field is not as simple as might seem at first sight, because the density variation also causes vertical deflections of the earth's surface and of internal interfaces, which in turn contribute perturbations to the gravity field. The net perturbation to the gravity field depends on a delicate balance of these contributions, and is sensitive to the vertical variation of viscosity in the mantle.

I will explain qualitatively the principles involved in this approach, but without going into details or quantitative analysis. This is partly because the analysis is not simple, and partly because not all aspects of this problem are fully understood at the time of writing. Although the observed geoid can be matched by some models, the accompanying surface perturbations do not match observations as well.

The key physics is shown in Figure 6.14. If there is a high-density anomaly in the mantle, of total excess mass m, then its gravitational attraction will make an extra positive contribution to the geoid. If the mantle were rigid, this would be the only con-

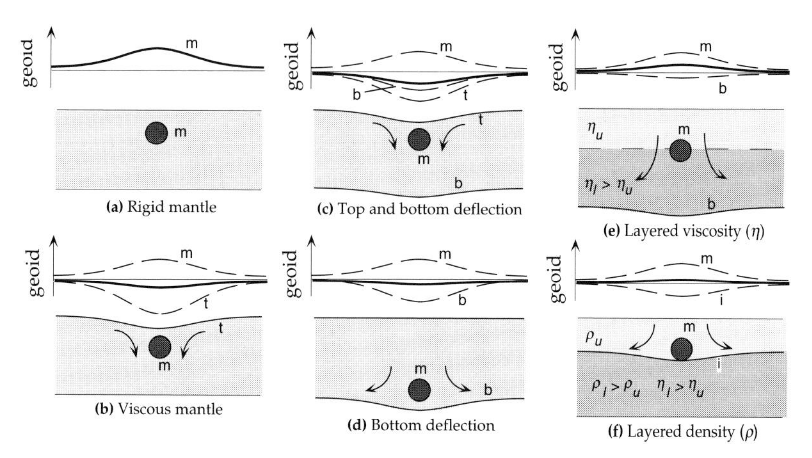

Figure 6.14. Sketches illustrating the ways an internal mass, *m*, may deflect the top and bottom surfaces or internal interfaces of the mantle, and their contributions to the geoid. Dashed curves are the geoid contributions from mass anomalies correspondingly labelled. Solid curves are total geoid perturbation.

tribution, and the result would be a positive geoid anomaly, as depicted in Figure 6.14a. If however the mantle is viscous, then the extra mass will induce a downflow, and this will deflect the top surface downwards by some small amount (Figure 6.14b). In effect, the viscosity of the mantle transmits some of the effect of the internal load to the surface via viscous stresses. The depression of the surface is a negative mass anomaly (rock is replaced by less dense air or water), and this will make a negative contribution to the geoid. The net geoid will depend on the relative magnitudes of the contributions from the internal mass and from the surface deflection. Actually, both the top and bottom surfaces of the mantle will be deflected, and each deflection will create a negative mass anomaly (Figure 6.14c,d).

The magnitude and sign of the net geoid anomaly depend on the relative magnitudes of the top and bottom deflections, and these depend on the depth of the mass anomaly and any stratification of viscosity or composition that might exist in the mantle. Two principles are at work here. One is that the mass anomalies of the surface deflections balance the internal mass anomaly: it is the same thing as an isostatic balance. The second is that the geoid contribution of a mass anomaly decreases inversely as its distance from the surface. (The geoid is related to gravitational *potential*, which falls off as 1/distance.)

Now if the internal mass anomaly is near the top, then the depression of the bottom surface and its gravity signal are negligible (Figure 6.14b). The top depression has a total mass anomaly in this case that nearly balances the internal mass, but it is closer to the surface (being *at* the surface) than the internal mass, so its geoid signal is stronger. Consequently the net geoid anomaly is small and *negative* (Figure 6.14b). This actually remains true for all depths of the internal mass (Figure 6.14c,d), so long as the mantle is uniform in properties, though this result is less easy to see without numerical calculations. It was demonstrated by Richards and Hager [11].

If the lower mantle has a higher viscosity than the upper mantle and the mass is within the lower mantle, it couples more strongly to the bottom surface. As a result, the bottom deflection is greater than for a uniform mantle, whereas the top deflection is smaller (Figure 6.14e). Because the geoid signal from the bottom depression is reduced by distance, it turns out that it is possible for the positive contribution from the internal mass to exceed the sum of the negative contributions from the deflections, and the result is a small *positive* net geoid (Figure 6.14e; [11, 12]).

Richards and Hager [11, 12] also considered the possibility that there is an increase in intrinsic density within the mantle transition zone (Figure. 6.14f). The effect of such an internal interface is to reduce the magnitude of the net geoid, because much of the compensation for the internal mass anomaly occurs through a deflection of the internal density interface. Since they are close together, their gravity signals more nearly cancel. The result is that although it is possible for the net geoid to be positive in this case, it is harder to account for the observed amplitude of the geoid anomalies over subduction zones.

A full consideration of subduction zone geoids requires using slab-shaped mass anomalies and spherical geometry, which affects the fall-off of geoid signal with the depth of the mass anomaly. Analyses by Richards, Hager and coworkers [11–14] yielded three important conclusions.

The first is that there is an increase in mantle viscosity with depth, located roughly within the transition zone, by a factor between 10 and 100, with a preferred value of about 30. This is reasonably consistent with the more recent inferences from post-glacial rebound discussed in Section 6.9.2.

The other two conclusions are of less immediate relevance to mantle viscosity structure, but are important for later discussion of possible dynamical layering of the mantle (Chapter 12). The second conclusion is that it is difficult to account for the observed magnitudes of geoid anomalies if there is an intrinsic density interface within the mantle (other than the density changes associated with phase transformations; Section 5.3). The third conclusion is that subducted lithosphere must extend to minimum depths of about 1000 km to account for the magnitude of the geoid anomalies.

As I noted at the beginning of this section, the geoid does not constrain the absolute magnitude of the viscosity, only its relative depth dependence. This is because the viscous stresses are proportional to the internal mass anomaly, not to the viscosity. A lower viscosity would be accommodated by faster flow, and the stresses would be the same. Consequently the surface deflections would be the same and the geoid analysis would be unaffected. The geoid anomaly depends on the *instantaneous* force balance, into which time does not enter explicitly, rather than on flow *rates*, whereas the glacial rebound effect involves flow rates and time explicitly in the observations and in the physics.

6.9.4 Rotation

The changing mass distribution of the earth during the process of glaciation and deglaciation changes the moments of inertia of the earth, and hence its rotation. Since the mass rearrangements that result from glacial cycles are delayed by mantle viscosity, there is in principle important information about mantle viscosity in these adjustments, and observations do show continuing changes both in the rate of rotation and in the pole of rotation of the earth. A potential advantage of these constraints is that they depend on the largest-scale components of the mass redistribution, and so are sensitive to the entire depth of the mantle.

According to Mitrovica [5], models of these processes are rather sensitive to poorly constrained details of the ice load history and to lithosphere thickness. As well, some models have taken the Haskell viscosity to represent the mean only of the upper mantle, rather than of the upper 1400 km of the mantle, and consequently they do not properly reconcile the two kinds of constraint. At present it is not clear that reliable additional information has been extracted from this approach, but work is continuing.

6.10 Rheology of rocks

Rheology is the study of the ways materials deform in response to applied stresses. Rocks exhibit a range of responses to stress. The response depends on the rock type, temperature, pressure and level of deviatoric stress. It ranges from elastic–brittle near the surface, where pressures and temperatures are low, to ductile or viscous behaviour at the high temperatures and pressures of the interior. The relationship between stress and rate of deformation may be linear or nonlinear. A linear relationship between stress and strain rate defines a Newtonian viscous fluid. Brittle failure is an example of an extremely nonlinear rheology. It is plausible, though not conclusively demonstrated, that at the low deviatoric stresses associated with mantle convection, the mantle behaves as a linear viscous fluid. However, there is evidence from laboratory experiments that at slightly higher stresses the relationship may become moderately nonlinear.

These features will be briefly summarised here. There are two principal points to be highlighted. First, a brittle–ductile transition occurs in mantle material within depths less than about 50 km. Second, in the ductile range, the viscosity (or effective viscosity in nonlinear flow) is strongly dependent on temperature, changing by up to a factor of 10 for a 100 °C change in temperature. Two other points are also quite significant. The effect of pressure may also be substantial over the depth range of the mantle, and small amounts of water may decrease viscosity by about one order of magnitude.

There remain great uncertainties about the details of mantle rheology. This is because experiments in the pertinent ranges of pressure and temperature are quite difficult, because the time scales and strain rates of the earth are orders of magnitude different from what can be attained in experiments, and because the rheology can be sensitive to the many details of rock and mineral composition and structure, especially to grain size. These uncertainties will be briefly indicated at the end of this section.

6.10.1 Brittle regime

The transition from brittle behaviour near the surface to ductile behaviour at depth in the mantle has a crucial influence on mantle convection that distinguishes it from most other convecting fluid systems, as was indicated in Chapter 3 and will be elaborated in Chapter 10.

I will use the term brittle here loosely for a regime in which deformation is concentrated along faults or narrow shear zones, and in which the behaviour is grossly like that described below using the Mohr–Coulomb theory. As you might expect, the processes controlling failure in aggregates of crustal minerals of a wide range of compositions and in a wide range of conditions are complex [15]. However, a general behaviour emerges in which faults occur and in which they have characteristic orientations relative to the stress field, and these are the essential points I want to present.

Much of the shallow crust is pervasively fractured, but some of it is not, and presumably in the deep crust fractures tend to heal.

Thus we should consider both the brittle failure of intact rock and the sliding of adjacent rock masses along pre-existing fractures. Suppose a piece of rock is subjected to a shear stress, σ_s, and to a confining normal stress, σ_n, as sketched in Figure 6.15a. If there is a pre-existing fracture, suppose that it is parallel to the direction of shearing. Whether there is a pre-existing fracture or not, there is a critical shear stress at which fracture or fault slip occurs, depending either on the strength of the intact rock or on the frictional property of the pre-existing fracture.

It turns out that for either case the shear stress necessary to cause slip is proportional to the normal stress acting across the fault surface. This is in accord with common experience in the case of frictional sliding, in which it is harder to make blocks slide past each other if they are pressed together. Thus we can write

$$\sigma_s = \mu_f \sigma_n + C_f \tag{6.10.1}$$

where μ_f is a coefficient of friction, C_f is a cohesive strength, and σ_s and σ_n are the shear and compressive normal stresses, respectively. Because the engineering convention of considering stress to be positive in tension is unfamiliar and clumsy in this context, I will use the geological convention and notation, in which $\sigma = -\boldsymbol{T}$, which is positive in compression. When applied to frictional sliding with particular values of μ_f and C_f, Equation (6.10.1) is known as Byerlee's rule. When applied to fracture, it is called the Mohr–Coulomb criterion [15]. For my present purpose, it is sufficient to consider C_f to be negligible. Typically $\mu_f \approx 0.6$–0.8 for rocks.

We can use Equation (6.10.1) to find the orientation in which a new fracture is most likely to occur, or the orientation of a pre-existing fracture which is most prone to slipping. To do this, we need to relate stress components on planes with different orientations. A property of the stress tensor is that there is always an orientation of mutually perpendicular planes on which the shear

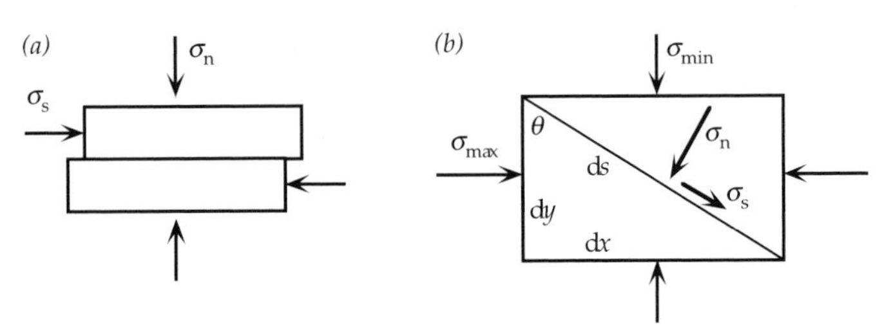

Figure 6.15. Illustration of the relationship between shear stress and normal stress in fracture or frictional sliding.

stresses vanish, leaving three non-zero normal stress components. (If coordinates are defined relative to these planes, the stress tensor contains only the diagonal components. Finding the orientation of these planes is equivalent to diagonalising the stress tensor.) The demonstration of this comes from considering the relationship between stresses on planes of oblique orientation, such as in Figure 6.15b, relative to stresses on coordinate planes. I will not give it here. It can be found in many structural geology and engineering texts, for example. The normal stresses in this orientation are called the *principal stresses*, and they can be arranged in order as the maximum, intermediate and minimum principal stresses.

Figure 6.15b portrays a particular situation that allows us to derive the relationship between the stress components on the oblique plane relative to the maximum and minimum principal stresses. First note that the areas of the orthogonal planes are $dx = ds \cdot \sin\theta$ and $dy = ds \cdot \cos\theta$. Taking the force balance first in the direction parallel to σ_n and then parallel to σ_s yields

$$\sigma_n = \sigma_{max} \cos^2\theta + \sigma_{min} \sin^2\theta$$
$$\sigma_s = (\sigma_{max} - \sigma_{min}) \sin\theta \cos\theta$$

Standard trigonometric identities then yield

$$\sigma_n = \sigma_c + \sigma_r \cos 2\theta \tag{6.10.2a}$$

$$\sigma_s = \sigma_r \sin 2\theta \tag{6.10.2b}$$

where

$$\sigma_c = (\sigma_{max} + \sigma_{min})/2$$
$$\sigma_r = (\sigma_{max} - \sigma_{min})/2$$

These relationships can be represented geometrically as in Figure 6.16. The stress components on any surface whose normal is oriented at angle θ to the direction of maximum principal stress fall on a circle in this plot, with its centre at the average stress, σ_c, and with radius equal to half the differential stress, σ_r. This circle is known as Mohr's circle.

The Mohr–Coulomb criterion for fracture, Equation (6.10.1), can also be represented on this plot as the sloping line making an angle $\phi = \tan^{-1}(\sigma_s/\sigma_n)$ with the σ_n axis. If the differential stress σ_r is large enough that the Mohr circle is tangent to this line, then the shear stress on a plane with the corresponding orientation is suffi-

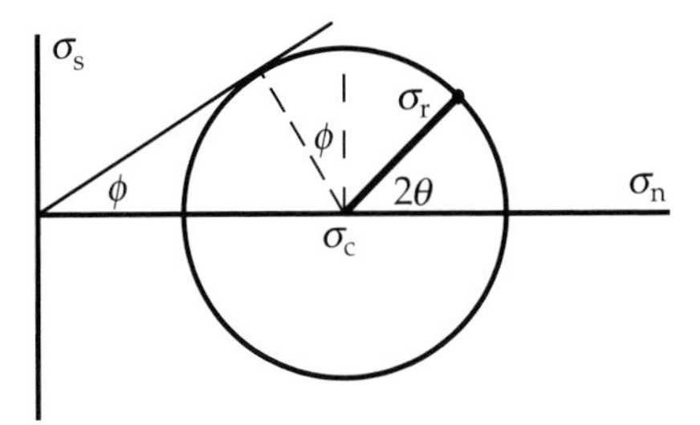

Figure 6.16. Mohr's circle: the geometric representation of stress components on planes with different orientations. The Mohr–Coulomb criterion (Equation (6.10.1)) can also be represented on this diagram by the sloping line (assuming $C_f = 0$).

cient to cause fracture. This tells us that the most likely orientation of a fracture is one whose normal makes an angle such that $2\theta = \phi + \pi/2$ ($\theta = \phi/2 + \pi/4$) with the maximum principal stress direction. It might have been thought that fracture was most likely on a surface with $\theta = \pi/4$, where the shear stress is maximal, but according to this theory the influence of the normal stress component means that a slightly different orientation is preferred, on which σ_n is less. Typically $\phi \approx 30$–$40°$, so $\theta \approx 60$–$65°$.

This simple theory of fracturing gives a reasonable first-order account both of fracturing observed in the laboratory and of faults observed in the earth's crust. It is found, for example, that normal faults are generally steeper than 45° and reverse faults less steep than 45°, as is expected from this theory. This is explained by Figure 6.17, which shows the expected relationships between maximum or minimum principal stress and the standard fault types of structural geology.

This theory also seems to work for the deeper crust and the mantle part of the lithosphere, even though the rheology there is expected to be more ductile. Evidently deformation is still sufficiently concentrated into narrow shear zones that this theory has some relevance. It is found, for example, that some reverse faults cut completely through the continental crust and into the mantle. It is found also that the major plate boundaries tend to correspond quite well with the standard fault types of Figure 6.17, as discussed in Chapters 3 and 9.

6.10.2 Ductile or plastic rheology

A fairly general relationship for rocks between strain rate, s, and stress, σ, temperature, T, grain size, d, and pressure, P, is [16]

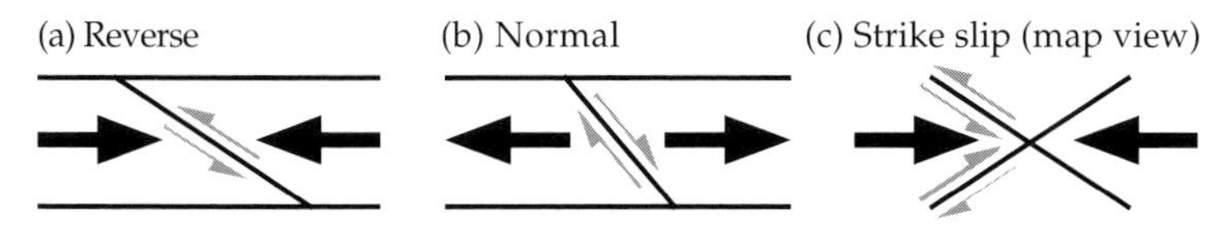

Figure 6.17. Relationship between deviatoric stress in the crust and the principal fault types of structural geology.

$$s = A\left(\frac{\sigma}{G}\right)^n \left(\frac{b}{d}\right)^m \exp\left[-\frac{E^* + PV^*}{RT}\right] \qquad (6.10.3)$$

In this equation, A is a constant, G is the elastic shear modulus, b is the length of the Burgers vector of the crystal structure (about 0.5 nm), E^* is an activation energy, V^* is an activation volume and R is the gas constant.

In mantle minerals there are likely to be two main deformation mechanisms. In *diffusion creep*, the deformation is limited by diffusion of atoms or vacancies through grains, and the stress dependence is linear ($n = 1$). There is a strong grain size dependence, with $m = 2$–3. In *dislocation creep*, the deformation is limited by the motion of dislocations through the grains, the stress dependence is nonlinear ($n = 3$–5) and there is no grain size dependence ($m = 0$). In each regime there is a strong temperature dependence, but it tends to be stronger in dislocation creep ($E^* = 400$–550 kJ/mol) than in diffusion creep ($E^* = 250$–300 kJ/mol).

Karato and Wu [16] have estimated that in the upper mantle the contributions from the two mechanisms may be of similar order, with one or the other dominating in different circumstances. The lower mantle may be mainly in the linear regime of diffusion creep. Since garnet phases tend to have lower plasticity, it has been suggested that viscosities might be higher within the transition zone. These possibilities must be balanced against the viscosities inferred from observational constraints that were discussed earlier.

There are considerable uncertainties in the absolute magnitudes of the strain rates or apparent viscosities predicted from laboratory data. Nevertheless, an important value of the laboratory work is in establishing the general form of the dependence of the strain rate on state variables and material characteristics. For example, the effect of increasing the temperature from 1600 K to 1700 K is, taking $E^* = 250$ kJ/mol and $R = 8.3$ kJ/mol K, to increase the strain rate by a factor of 3. If $E^* = 500$ kJ/mol, more appropriate for dislocation creep, the strain rate increases by a factor of 9.

Thus strain rate, and effective viscosity, is strongly dependent on temperature.

For linear rheologies, the viscosity is simply $\mu = \sigma/2s$. From Equation (6.10.3), it is then possible to write the dependence of viscosity on temperature in the form

$$\mu = \mu_r \exp\left[\frac{(E^* + PV^*)}{R}\left(\frac{1}{T} - \frac{1}{T_r}\right)\right] \tag{6.10.4}$$

where μ_r is the viscosity at the reference temperature T_r. Figure 6.18 shows the variation of viscosity with temperature for activation energies of 400 kJ/mol and 200 kJ/mol, assuming the same viscosity of 10^{21} Pa s at a reference temperature of 1300 °C.

The effect of pressure on strain rate is not well understood, because it is hard to reconcile the laboratory and observational constraints. Inferences from postglacial rebound suggest that the deep mantle viscosity is at least one order of magnitude higher than in the upper mantle, but probably not more than three orders of magnitude higher, although this has not been directly tested against the observational constraints. Laboratory estimates are that V^* is about 15–20 cm^3/mol for dislocation creep in olivine and about 5–6 cm^3/mol for diffusion creep. Even with the latter value, using the pressure of about 130 GPa at the base of the mantle, the predicted viscosity increase is about 12 orders of magnitude over the depth of the mantle. To accord with the observational indications, V^* should be no larger than about 2.5 cm^3/mol.

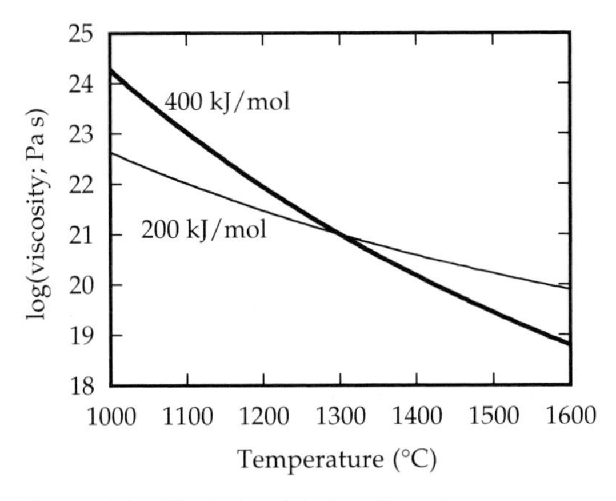

Figure 6.18. Variation of viscosity with temperature for two different activation energies. The viscosity is calculated from Equation (6.10.4), assuming a viscosity of 10^{21} Pa s at a reference temperature of 1300 °C.

Experiments on olivine show that the strain rate increases by more than one order of magnitude if the olivine is water saturated, but the effects of temperature and pressure on this behaviour are not well constrained [16]. Since grain size can be affected by deformation, it is possible that there is a feedback in diffusion creep in which higher strain rates cause smaller grain sizes which in turn cause higher creep rates. Together with the possibility that the nonlinear dislocation creep regime is sometimes entered, these possibilities account for a great deal of the uncertainty about absolute strain rates and apparent viscosities in the mantle.

It has been most commonly assumed that the rheology of the mantle is linear. To some extent this is because it is mathematically easier to analyse linear rheology. However, the observational constraints give some support to this approach, though not a compelling argument. If, for example, the rheology were nonlinear during postglacial rebound, then the mantle flow would tend to be more concentrated towards the surface load, and there would tend to be a peripheral bulge developed as mantle was squeezed more to the side than to great depth [17]. This does not appear to have happened, but conclusions from this kind of argument are sensitive to the ice load history and other complications of postglacial rebound.

Whether a linear or nonlinear rheology is assumed, a useful approach is to assume a form like Equation (6.10.3) and combine it with constraints from observations to determine some of the constants, such as A and V^*. This is the approach implicit in Equation (6.10.4). You will see in later chapters that there is a broad consistency between inferences from observations, the general linear form of Equation (6.10.3), and the basic features of mantle convection. In this book only the most basic points are being addressed, and this simple approach is therefore taken. However, mantle rheology must be recognised as one of the main uncertainties in considering mantle convection.

6.10.3 Brittle–ductile transition

The transition between brittle behaviour and ductile or malleable behaviour will occur when ductile deformation can occur fast enough to prevent the stress from becoming large enough to cause brittle failure. Since ductile deformation rates in particular are so dependent on conditions, there is no unique stress, temperature or pressure at which this will occur. Nevertheless it is useful to show some representative examples, with the understanding that other conditions would give significantly different results.

There is a problem in comparing two such different rheological responses. One approach is to plot the differential stress ($\sigma_{max} - \sigma_{min}$) that the material can sustain under particular conditions. Then the behaviour that can sustain the least stress is the one that will prevail. Figure 6.19 shows the 'strength envelopes', that is maximum stress versus depth, for representative conditions of oceanic and continental lithosphere [15].

It is necessary to assume a geotherm (temperature versus depth) for each case, and for the ductile deformation a strain rate of 10^{-15}/s is assumed. This is representative of mantle convection strain rates and some lithospheric deformation rates. For the oceanic mantle, a lithospheric age of 60 Ma and a dry olivine ductile rheology are assumed, while for the continental mantle, a wet olivine rheology is assumed. The dashed segment in each is an intermediate 'semi-brittle' regime in which deformation is by microscopic fracture pervasively through the material (that is, not concentrated along a fault).

A distinctive feature of the continental envelope is that it is bimodal. This is because the deformation rate of crustal minerals is much greater than that of mantle minerals, so the lower crust deforms much more rapidly and prevents brittle failure. However, the actual deformation rate of the lower crust is quite uncertain, and the limit of brittle behaviour might be between 300 and 600 MPa [15]. These curves indicate that the continental lithosphere

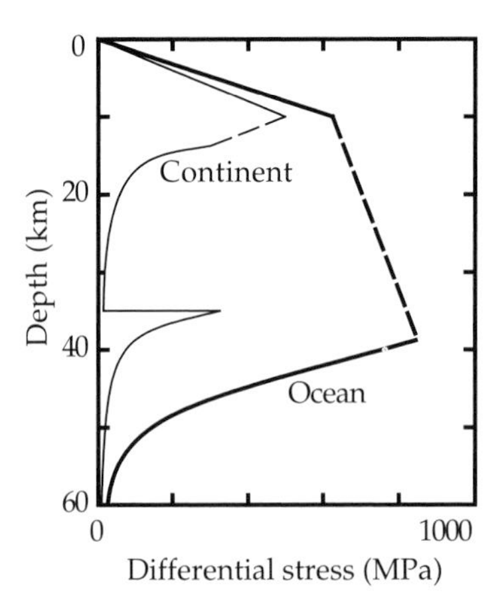

Figure 6.19. Strength envelopes estimated for representative oceanic and continental geotherms. Such estimates depend greatly on details assumed (see text). Each case comprises three regimes: brittle (straight lines), semi-brittle (dashed lines) and ductile (curves). In the continental case, the crustal ductile response changes to the mantle ductile response at 35 km depth. After Kohlstedt *et al.* [15]. Copyright by the American Geophysical Union.

is considerably weaker than the oceanic lithosphere, a fact that seems to be borne out by the observed tendencies of plate boundaries to be diffuse deformation zones within continents, but sharp boundaries within oceans (Chapter 4).

The general relevance of these estimates to mantle convection is that the mantle can be expected to behave as a ductile or viscous material deeper than a few tens of kilometres, but as a brittle material at shallower depths.

6.11 References

1. D. L. Turcotte and G. Schubert, *Geodynamics: Applications of Continuum Physics to Geological Problems*, 450 pp., Wiley, New York, 1982.
2. G. K. Batchelor, *An Introduction to Fluid Dynamics*, Cambridge University Press, Cambridge, 1967.
3. J. Happel and H. Brenner, *Low Reynolds Number Hydrodynamics*, 553 pp., Prentice-Hall, Englewood Cliffs, NJ, 1965.
4. G. G. Stokes, *Trans. Camb. Philos. Soc.* **9**, 8, 1851.
5. J. X. Mitrovica, Haskell [1935] revisited, *J. Geophys. Res.* **101**, 555–69, 1996.
6. N. A. Haskell, The viscosity of the asthenosphere, *Am. J. Sci.*, ser. 5 **33**, 22–8, 1937.
7. K. Lambeck and P. Johnston, The viscosity of the mantle: evidence from analyses of glacial rebound phenomena, in: *The Earth's Mantle: Composition, Structure and Evolution*, I. N. S. Jackson, ed., Cambridge University Press, Cambridge, 461–502, 1998.
8. K. Fleming, P. Johnston, D. Zwartz, Y. Yokoyama and J. Chappell, Refining the eustatic sea-level curve since the Last Glacial Maximum using far- and intermediate-field sites., *Earth Planet. Sci. Lett.* **163**, 327–42, 1998.
9. K. Lambeck and M. Nakada, Late Pleistocene and Holocene sea-level change along the Australian coast, *Palaeogeogr., Palaeoclimatol., Palaeoecol.* **89**, 143–76, 1990.
10. K. Lambeck, P. Johnston and M. Nakada, Holocene glacial rebound and sea-level change in northwestern Europe, *Geophys. J. Int.* **103**, 451–68, 1990.
11. M. A. Richards and B. H. Hager, Geoid anomalies in a dynamic earth, *J. Geophys. Res.* **89**, 5487–6002, 1984.
12. B. H. Hager, Subducted slabs and the geoid: constraints on mantle rheology and flow, *J. Geophys. Res.* **89**, 6003–15, 1984.
13. B. H. Hager, R. W. Clayton, M. A. Richards, R. P. Comer and A. M. Dziewonski, Lower mantle heterogeneity, dynamic topography and the geoid, *Nature* **313**, 541–5, 1985.
14. B. H. Hager and R. W. Clayton, Constraints on the structure of mantle convection using seismic observations, flow models and the

geoid, in: *Mantle Convection*, W. R. Peltier, ed., Gordon and Breach, New York, 657–763, 1989.

15. D. L. Kohlstedt, B. Evans and S. J. Mackwell, Strength of the lithosphere: constraints imposed by laboratory experiments, *J. Geophys. Res.* **100**, 17 587–602, 1995.
16. S. Karato and P. Wu, Rheology of the upper mantle: a synthesis, *Science* **260**, 771–8, 1993.
17. L. M. I. Cathles, *The Viscosity of the Earth's Mantle*, 390 pp., Princeton University Press, Princeton, 1975.

6.12 Exercises

1. Subscript notation and summation convention: note which of the following expressions are valid, and expand any summations into explicit form. Assume the two-dimensional case (that is, indices running from 1 to 2).
   (a) $a_i b_j$. (b) $a_{ij} + b_j$. (c) $a_{ij} b_j$. (d) $a + b_i c_i$. (e) $a + b_j c_j$.
   (f) $a_{ijk} b_k$. (g) $\partial a_i / \partial x_i$. (h) $\partial a_{ij} / \partial x_j$. (i) $\partial a_{ij} / \partial x_k$.

2. Sketch the deformation described by the following displacements and give the values of each component of the two-dimensional strain tensor (Equation (6.3.5)) and rotation tensor (Equation (6.3.6)).
   (a) $u_x = ay$, $u_y = 0$. (b) $u_x = 0$, $u_y = ay$. (c) $u_x = ay$, $u_y = ax$.
   (d) $u_x = ay$, $u_y = -ax$. (e) $u_x = ax$, $u_y = ay$.

3. Referring to Figure 6.7, suppose that, instead of the top surface of the fluid layer being a zero-velocity surface, it is a free-slip surface, that is the shear stress is zero on the top surface. Derive the velocity profile in this case, and a formula for the volumetric flow rate.

4. If a mantle plume has a volumetric flow rate of 400 m^3/s, a radius of 50 km, and a density deficit of 20 kg/m^3, estimate the viscosity of the material in the plume. Assume that the plume is a vertical cylinder with stationary sides.

5. Calculate the rising or sinking velocity of the following objects.
   (a) A plume head of temperature excess 300 °C and radius 500 km in a mantle of viscosity 10^{22} Pa s.
   (b) A 'drop' of iron of radius 50 km and average density excess 5 Mg/m^3 in a hot mantle with an average viscosity of 10^{20} Pa s. This example gives some quantitative feel for the idea that during the formation of the earth liquid iron would gather into large pools and sink to the core.

6. [*Advanced*]. Solve for the flow around and within a fluid sphere of viscosity μ_s rising through a fluid of viscosity μ. From this calculate the viscous resistance to the sphere, and its rise velocity. The solutions have the same forms as Equations (6.8.7) and (6.8.10), but the boundary conditions are different. The result from Box 6.B2 does not apply. You can get there with a lot of algebra. Some shortcuts are outlined by Batchelor [2] (p. 235).

7. (a) Complete the steps in the derivation of the stream function solution (6.9.2), that is, show that the general forms assumed are solutions of the relevant equations and that the solution satisfies the boundary conditions. (b) Derive the expression for the normal stress at the surface, T_{zz}, leading to Equation (6.9.3).

8. (a) Calculate a representative strain rate for the mantle assuming the horizontal velocity at the top is 100 mm/a and that it is zero at a depth of 1000 km. (b) From Equation (6.10.3) for the strain rate of a rock, and using the material constants given below, calculate the value of the constant A that would yield the strain rate calculated above. Assume the mantle temperature is 1400 °C, the pressure is (approximately) zero, the stress is 3 MPa and the grain size is 1 mm. What is the viscosity? (c) Now calculate the change in viscosity for (i) $T = 1500$ °C, (ii) grain size = 10 mm, (iii) pressure = 30 GPa (equivalent to a depth of about 1000 km).

   Material constants: $n = 1$, $\mu_e = 80\,\text{GPa}, m = 2.5$, $b = 0.5\,\text{nm}$, $E^* = 300\,\text{kJ/mol}$, $V^* = 5\,\text{cm}^3\text{/mol}$, $R = 8.3\,\text{kJ/mol K}$.

CHAPTER 7

Heat

Heat transport is an integral part of convection. Heat is transported in two principal ways in the mantle: by conduction and by advection. Advection means that heat is carried along with mass motion. Heat is also *generated* internally by radioactivity. Here we consider these processes in turn. A key feature of heat conduction is that there is a fundamental relationship between the time scale of conductive cooling (or heating) and the length scale over which the process is occurring. This is demonstrated in several ways and at different mathematical levels.

A key application of heat conduction theory is to the cooling oceanic lithosphere, and a key consequence is the subsidence of the sea floor with age. The lithosphere is a special case of a conductive thermal boundary layer, which is the source of convective motion (Chapter 8). The oceanic lithosphere and its subsidence play a central role in the discussions of Chapters 10 and 12 of what can be inferred about the form of mantle convection from observations. The role of the continents in the earth's thermal regime is considered separately, since continental lithosphere does not partake in subduction like oceanic lithosphere.

The advection of heat is a phenomenon that can be understood in quite simple terms. It is presented first in a simple way, and the idea is then used to derive a general equation that describes heat generation and transport, including both conduction and advection. Finally, thermal properties of materials are briefly considered, including their likely variations with pressure. This leads into the concept of adiabatic gradients of temperature and density.

7.1 Heat conduction and thermal diffusion

Let us start from Fourier's 'law' of heat conduction, that the rate of flow of heat is proportional to the temperature gradient:

$$q = -K\partial T/\partial x \tag{7.1.1}$$

where T is temperature, q is the rate of flow of heat per unit area (that is, the heat flux) in the positive x direction, and K is the conductivity of the material. The negative sign ensures that heat flows from hotter to cooler regions.

We need to be able to consider situations in which the temperature varies with time and in which heat is generated by radioactivity within the rocks. To do this, let us consider the thermal energy budget of the small block of material sketched in Figure 7.1. Suppose that the temperature T depends only on time and the x-coordinate. The change in heat content of the block during a time interval will be equal to the heat conducted in minus the heat conducted out plus the heat generated internally. Suppose the temperature changes by an amount $\mathrm{d}T$ within a time interval $\mathrm{d}t$. Then the change in heat content H is

$$\mathrm{d}H = \rho S\mathrm{d}x \cdot C_P \cdot \mathrm{d}T$$

where ρ is the density, S is the area of the end surfaces of the block (so that $\rho S\,\mathrm{d}x$ is the mass of the block), and C_P is the specific heat at constant pressure of the material. The specific heat measures the capacity of a material to hold heat, and for mantle minerals it has a value of the order of 1000 J/kg °C. (The subscript P is used because this is the specific heat at constant pressure. In other words it is the change in heat content, per unit mass per degree, with the pressure held constant so that thermal expansion is allowed to happen. It is possible to define the specific heat at constant volume, C_V, but we will not have any use for this here. Since the two quantities have significantly different values, it is usual to distinguish them.)

Again taking positive heat flow to be in the positive x direction, the heat added by conduction through the left side of the box during the interval $\mathrm{d}t$ is $qS\,\mathrm{d}t$, and the heat lost by conduction through the right side is $(q + \mathrm{d}q)S\,\mathrm{d}t$. If A is the rate of radioactive heat generation per unit volume, the heat generated during $\mathrm{d}t$ is $A \cdot S\mathrm{d}x \cdot \mathrm{d}t$. The total heat budget for the time interval $\mathrm{d}t$ is then

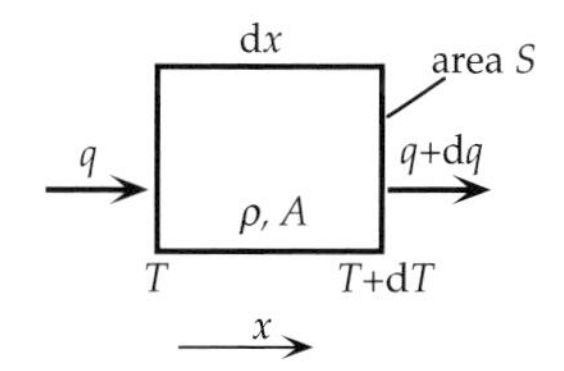

Figure 7.1. Heat budget of a small block of material with heat transported by thermal conduction and with internal heat generation (A).

$$\rho S\mathrm{d}x \cdot C_P \cdot \mathrm{d}T = qS\,\mathrm{d}t - (q + \mathrm{d}q)S\,\mathrm{d}t + A \cdot S\mathrm{d}x \cdot \mathrm{d}t$$

which yields, upon dividing by $S\,\mathrm{d}x\,\mathrm{d}t$ and taking limits,

$$\rho C_P \frac{\partial T}{\partial t} = -\frac{\partial q}{\partial x} + A \tag{7.1.2}$$

If K is uniform (that is, independent of x), this can be written, using Equation (7.1.1),

$$\frac{\partial T}{\partial t} = \kappa \frac{\partial^2 T}{\partial x^2} + a \tag{7.1.3}$$

where

$$\kappa = K/\rho C_P$$
$$a = A/\rho C_P$$

κ is called the *thermal diffusivity* and a is the rate of increase of temperature due to the radioactive heating. Notice that the dimensions of κ are length^2/time. With $K = 3\,\mathrm{W/m\,°C}$, $\rho = 3000\mathrm{kg/m^3}$ and $C_P = 1000\,\mathrm{J/kg\,°C}$, a typical value of κ for rocks is $10^{-6}\ \mathrm{m^2/s}$.

Equation (7.1.3) governs conductive heat flow in situations where the material is not moving, so that conduction is the only method of heat transport. We will use it below to consider the thermal structure of the lithosphere and of thermal boundary layers more generally. It is an example of a diffusion equation. A similar equation governs the diffusion of chemical species through solids, for example, though without the generation term a. This is why κ is called a diffusivity. As we will see, heat conduction causes temperature differences to spread out and become more uniform, in other words to diffuse. The term *thermal diffusion* is often used interchangeably with heat conduction: they mean the same thing, but in cases where the temperature is changing with time it is useful to emphasise the diffusive nature of the process by using the term thermal diffusion.

7.2 Thermal diffusion time scales

It is a very general feature of thermal diffusion (and other forms of diffusion) that the time it takes for a body to heat up or cool down is related to its size in a particular way. This general property of thermal diffusion is a key to building a simple understanding of thermal convection. It also governs a key part of mantle convection in a particularly simple way. In this section we look at this aspect of thermal diffusion in different ways, in order to provide a clear understanding of it.

7.2.1 Crude estimate of cooling time

Suppose a layer of magma with some high temperature T intrudes between sedimentary rock layers (forming a sill), as illustrated in Figure 7.2. Can we estimate how long it will take to cool? Would it, for example, take hours, or weeks, or centuries? Often it is possible to get some idea of an answer by making very crude approximations. (We have already seen examples of useful rough approximations in Sections 6.8.1 and 6.9.1.)

Suppose the sill thickness is D. At first ($t = 0$) there will be a very steep temperature gradient at the top and bottom of the sill (Figure 7.2), but after some time, t, you might expect the temperature profile to have smoothed out, as sketched. This will be justified more rigorously below. At this stage, a typical temperature difference is T, and a typical length scale over which the temperature varies by this much might be about half the thickness of the sill, $D/2$. Let us try approximating the differentials in Equation (7.1.3) with these large differences (assuming there is no heat generation, so that $a = 0$):

$$\frac{T}{t} = \kappa \frac{T}{(D/2)^2}$$

which yields $t = D^2/4\kappa$. Notice that this is independent of T.

What does this time t mean? According to Equation (7.1.3), T/t is a rough measure of the rate of change of T, so t should be a rough measure of the time it takes for the temperature to change by a significant fraction of T. Suppose D is 10 m and κ is 10^{-6} m^2/s. Then $t \approx 2.5 \times 10^7$ s, which is about 9 months. If this seems to be a surprisingly long time, it illustrates that rocks are not very good conductors of heat. Of course this may only be an approximate result, but it suggests that the cooling time for a 10 m sill is months rather than hours or centuries.

Notice now that the cooling time depends on the square of the thickness D. Thus if D is only 1 m, then $t \approx 3$ days, and if D is 10 cm then $t \approx 40$ minutes. This behaviour is quite characteristic

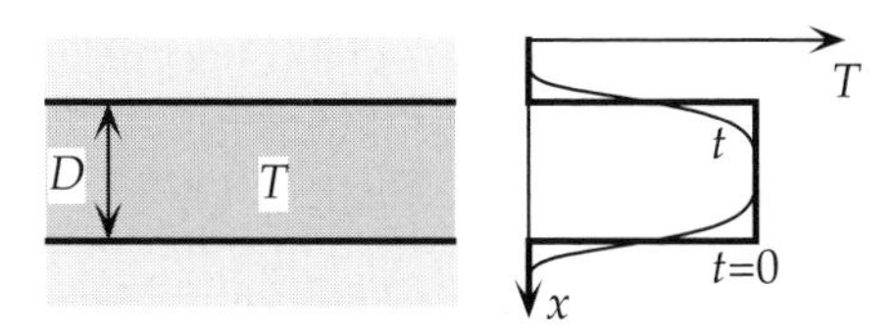

Figure 7.2. Cooling magma layer (or sill).

of diffusion processes: the time scale depends on the square of the length scale, with the diffusivity being the constant of proportionality. In fact this is built into the dimensions of the thermal diffusivity:

$$\kappa \approx \frac{D^2}{t} \tag{7.2.1}$$

This is the fundamental point to understand about time-dependent heat conduction processes. With this simple formula, we can make rough estimates of such things as the thickness of the oceanic lithosphere and the rate at which mantle convection should go. The latter will be done in Chapter 8. You will see in following sections that Equation (7.2.1) always emerges from a more rigorous analysis, with a proportionality constant of the order of 1.

7.2.2 Spatially periodic temperature [*Intermediate*]

In order to keep the mathematics from being unnecessarily complicated, let us approximate the initial temperature variation with depth (x) in Figure 7.2 as

$$T(x, 0) = T_0 \cos px \tag{7.2.2}$$

where $p = 2\pi/(2D)$ is a wavenumber, corresponding to a wavelength of $2D$. You can, if you want, regard this as the first Fourier component of the initial square temperature variation of Figure 7.2. Although this is still a crude approximation to the actual initial temperature distribution, it allows a rigorous solution of Equation (7.1.3) to be derived.

The evolution of the temperature is governed by Equation (7.1.3) with $a = 0$. If the geometry and initial conditions are appropriate, a solution to a partial differential equation such as this can often be found by assuming the solution to be a product of a function of depth, $\chi(x)$, and a function of time, $\Theta(t)$:

$$T(x, t) = \chi(x)\Theta(t) \tag{7.2.3}$$

(This method is called 'separation of variables'.) Substitution of this into Equation (7.1.3) and rearrangement leads to

$$\frac{1}{\Theta}\frac{\mathrm{d}\Theta}{\mathrm{d}t} = \kappa \frac{\mathrm{d}^2\chi}{\mathrm{d}x^2} \equiv -\frac{1}{\tau} \tag{7.2.4}$$

where τ is a constant with dimension time. The first and second parts of this equation must each be equal to a constant because t and x can be varied independently, so the only way the two expressions can remain equal is if they are each equal to the same constant, which I have written with malice aforethought as $-1/\tau$. This equation is now in the form of two ordinary differential equations, each of which can be readily solved. Thus the first and third terms of Equation (7.2.4) can be equated and integrated to yield

$$\Theta = \Theta_0 e^{-t/\tau} \tag{7.2.5}$$

while the second and third terms can be rearranged as

$$\frac{d^2\chi}{dx^2} + \frac{\chi}{\kappa\tau} = 0$$

which has a general solution of the form

$$\chi(x) = a \cos\frac{x}{\sqrt{\kappa\tau}} + b \sin\frac{x}{\sqrt{\kappa\tau}} \tag{7.2.6}$$

We want Equations (7.2.5) and (7.2.6) to combine in Equation (7.2.3) with the constants evaluated so that the solution matches the initial condition, (7.2.2). This requires $b = 0, a\Theta_0 = T_0$ and $p = 1/\sqrt{\kappa\tau}$. The solution is then

$$T(x, t) = T_0 \exp(-t/\tau) \cos px \tag{7.2.7}$$

with

$$\tau = \frac{1}{p^2\kappa} = \frac{D^2}{\pi^2\kappa} \tag{7.2.8}$$

Compare this with the crude estimate in the last section, which yielded a time scale of $D^2/4\kappa$. They differ only by a factor of about 2.5. You can see again the dependence of time scale on the square of the length scale embodied in the dimensionality of κ (Equation (7.2.1)). With this formula, the cooling time of a 10 m sill can be estimated as 4 months.

7.2.3 Why is cooling time proportional to the square of the length scale?

A simple illustration can clarify why there is this general relationship between time scale and length scale in thermal diffusion processes. Compare the two sinusoidal temperature distributions in Figure 7.3, with similar amplitudes, and wavelengths of λ and 2λ. Heat will flow from the hotter part to the colder part. The heat flux at the point where $T = 0$ is only half as great in case (*b*) as in case (*a*), because the temperature gradient is less: the same temperature difference is spread over twice the distance. Thus the rate at which the hot part loses heat to the cool part is only half as great in case (*b*): it is proportional to $1/2\lambda$.

There is another factor to be considered. So far we have accounted for time scale being proportional only to the first power of λ. We must also take account of the fact that in case (*b*) there is twice as much heat to be moved as in case (*a*), because the volume of the hot region is twice as great. Twice as much heat flowing at half the rate will take four times as long. Thus we can conclude that the time scale for a significant reduction in the temperature differences is proportional to λ^2.

7.3 Heat loss through the sea floor

At a midocean rise crest, or spreading centre, two tectonic plates pull apart. It is observed that the zone of rifting is quite narrow, only a few tens of kilometres in width. Beyond the rift zone, each plate is a rigid unit moving away from the spreading centre. If the plates are separating, then of course there must be a replenishing flow of material ascending from below. I will argue in Chapter 12 that at normal midocean rises the upwelling is passive, being simply the flow of mantle material drawn in to replace the material moving away with the plates. In cross-section then, the situation must be like that sketched in Figure 7.4a.

Hot material, at temperature T_m, rises close to the surface at the spreading centre. Some of it melts, and the magma rises to the top to form the oceanic crust, but this can be ignored for the

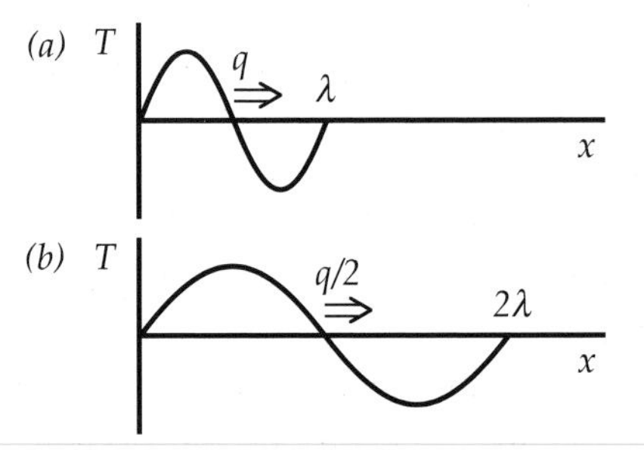

Figure 7.3. Effect of length scale on cooling time.

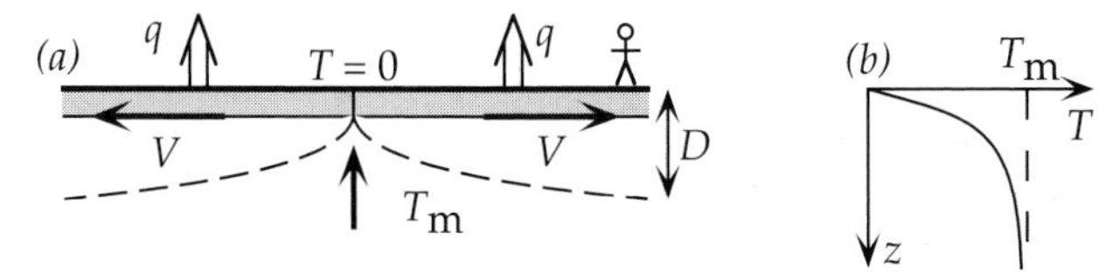

Figure 7.4. Cooling oceanic lithosphere.

moment. As the rising hot material approaches the cool surface, it will begin to cool by conduction, because the surface is maintained at a temperature close to 0 °C. If you imagine standing on one side of the spreading centre (stick figure, Figure 7.4a), you will be carried away with the plate at velocity V. The material under you will continue simply to cool by conduction. Once you are some distance from the spreading centre, its presence and your motion relative to it might be ignored, in which case the only process happening would be the conduction of heat *vertically* to the surface. We will justify this assumption in retrospect. After a time, the temperature profile with depth might look like that sketched in Figure 7.4b. As discussed in Chapter 6, the cold material will behave like an elastic/brittle material, and in this context it will not flow like the deeper (solid) material. In other words, the cooled upper part will behave like a rigid plate, consistent with what is observed.

7.3.1 Rough estimate of heat flux

We can use the results of Section 7.2 to get some idea of the thickness, D, of the cooled zone. From Equation (7.2.1), it should be approximately

$$D = \sqrt{\kappa t}$$

where t is the time for which the cooling has been proceeding. The older parts of the sea floor are over 100 Ma old. Setting $t = 100$ Ma (1 year $\approx 3.16 \times 10^7$ s) and using $\kappa = 10^{-6}$ m^2/s gives $D = 56$ km. This says that D should be roughly a few tens to 100 km.

The larger plates are typically 5000–10 000 km across, which is consistent with velocities of 50–100 mm/a (50–100 km/Ma) sustained for about 100 Ma. Thus plate widths are much greater than plate thicknesses. This justifies our assumption that heat conduction is mainly vertical, except within about 100 km of the spreading centre. Putting it another way, the vertical temperature gradients will be much greater than horizontal gradients, except near spreading centres. The plate material will be close to the spreading centre for only about 1 Ma.

We can estimate the order of magnitude of the heat flux through the sea floor to be expected from this conductive cooling process. From Equation (7.1.1) it is $q \sim K\Delta T/D \sim 70\,\text{mW/m}^2$ using $K = 3\,\text{W/m}\,^\circ\text{C}$, $\Delta T = 1300\,^\circ\text{C}$ and $D = 56\,\text{km}$. This compares with the average heat flux through the sea floor of about 100 mW/m^2, decreasing to about 50 mW/m^2 through the oldest sea floor.

7.3.2 The cooling halfspace model [*Intermediate*]

The assumptions used above are that the temperature profile at a given location moving with a plate is determined only by heat conduction in the vertical direction and that the conductivity is spatially uniform. This amounts to assuming that the mantle is a uniform infinite halfspace (that half of an infinite space below $z = 0$). With the initial condition that $T(z, 0) = T_\text{m}$, the boundary condition that $T(0, t) = 0$ and the assumption that radioactive heating can be neglected (so $a = 0$) Equation (7.1.3) has the solution

$$T(z, t) = T_\text{m}\,\text{erf}\left(\frac{z}{2\sqrt{\kappa t}}\right) \tag{7.3.1}$$

where erf stands for the *error function*:

$$\text{erf}(x) = \frac{2}{\sqrt{\pi}}\int_0^x \text{e}^{-\beta^2}\,\text{d}\beta \tag{7.3.2}$$

The derivation of this result is outlined in the next section. The error function looks like the temperature profile sketched in Figure 7.4b. It has the value 0.843 at $x = 1$.

The temperature in this solution depends on depth and time only through the combination $[z/2\surd(\kappa t)]$. Thus, for example, the temperature reaches 84% of its maximum value when $z/2\surd(\kappa t) = 1$. In other words, the depth, D, to the isotherm $T = 0.84T_\text{m}$ is

$$D = 2\sqrt{\kappa t} \tag{7.3.3}$$

This is just twice the value resulting from the rough estimate of the last section. Using $\kappa = 10^{-6}\,\text{m}^2/\text{s}$, the depth to this isotherm is thus 112 km at 100 Ma. More generally, you can see that the proportionality between length^2 and time has emerged again in this solution. It implies here that the thickness of the lithosphere should be proportional to the square root of its age. Thus D should be 56 km at 25 Ma and 11.2 km at 1 Ma. An impression of this thickening with age is included in Figure 7.4a as the dashed curves.

We can calculate the heat flux through the sea floor from this solution. For this, we need the result

$$\frac{\mathrm{d}\,\mathrm{erf}(x)}{\mathrm{d}x} = \mathrm{erf}'(x) = \frac{2}{\sqrt{\pi}}\mathrm{e}^{-x^2}$$

This follows from the fact that erf(x) depends on x only through the upper limit of the integral in Equation (7.3.2), and can be derived using basic calculus methods for differentiating integrals with variable limits. If we identify x with $z/2\sqrt{(\kappa t)}$, we can use the chain rule of differentiation:

$$\frac{\partial T}{\partial z} = T_\mathrm{m}\frac{\mathrm{d}\,\mathrm{erf}(x)}{\mathrm{d}x}\frac{\partial x}{\partial z} = T_\mathrm{m}\frac{2}{\sqrt{\pi}}\mathrm{e}^{-x^2}\frac{1}{2\sqrt{\kappa t}}$$

Then the heat flux at $z = 0$ is

$$q_0 = -K\frac{\partial T}{\partial z}\bigg|_{z=0} = -\frac{KT_\mathrm{m}}{\sqrt{\pi\kappa t}} \tag{7.3.4}$$

Thus the heat flux declines with time in proportion to $t^{-1/2}$. The minus in Equation (7.3.4) is because the heat flux is upwards, which is the negative z direction.

We saw in Chapter 4 (Figure 4.7B) that the observations of heat flow through the sea floor follow this behaviour to within the errors of measurement. The values used above yield a heat flux of 39 mW/m^2 for 100 Ma-old sea floor, compared with observed values of 40–50 mW/m^2. This very simple model, which approximates the earth below the sea floor as a uniform halfspace, thus gives a remarkably good description of the observed heat flux through the sea floor.

The physics we have considered here is the same as was considered last century by Lord Kelvin in making his estimate of the age of the earth (Chapter 2). His assumptions were that the earth had started hot and that it had been cooling by conduction to the surface ever since. He asked how long it would take for the near-surface temperature gradient (or the surface heat flux) to fall to the presently observed values. This is explored further in Exercise 4.

7.3.3 The error function solution [*Advanced*]

Since it is a central result in our understanding of oceanic lithosphere, and through that of mantle convection, I will outline the derivation of the error function solution. Another account is given, for example, by Officer [1]. A general form of solution to Equation (7.1.3) (with $a = 0$) is

$$T(z, t; \gamma) = \exp(-\kappa\gamma^2 t)[B(\gamma)\cos\gamma z + C(\gamma)\sin\gamma z] \qquad (7.3.5)$$

where the notation shows explicitly the dependence of T on the wavenumber γ. This form is just a more general version of Equation (7.2.7), and it can be derived in the same way. The coefficients B and C are also assumed to depend on γ because we can use this form to Fourier synthesise the total solution. This can be done by first Fourier analysing the initial condition $T(z > 0, 0) = T_m$, $T(0, 0) = 0$; then the time dependence of each Fourier component will have the above form. Thus the forms of B and C can be derived from the Fourier integrals of the initial condition:

$$B(\gamma) \equiv \frac{1}{\pi}\int_{-\infty}^{\infty} T(\zeta, 0)\cos\gamma\zeta\, d\zeta$$
$$C(\gamma) \equiv \frac{1}{\pi}\int_{-\infty}^{\infty} T(\zeta, 0)\sin\gamma\zeta\, d\zeta$$

The top boundary condition can be matched by assuming that the solution is antisymmetric about $z = 0 : T(z < 0, 0) = -T(z > 0, 0) = -T_m$. Substitution into these integrals then yields

$$B(\gamma) = 0$$
$$C(\gamma) = \frac{2}{\pi}T_m\int_0^{\infty}\sin\gamma\zeta\, d\zeta$$

The Fourier synthesised solution is then of the form

$$\begin{aligned} T(z, t) &= \int_0^{\infty} T(z, t; \gamma)d\gamma \\ &= \int_0^{\infty}\exp(-\kappa\gamma^2 t)C(\gamma)\sin\gamma z\, d\gamma \\ &= \frac{2T_m}{\pi}\int_0^{\infty}\int_0^{\infty}\exp(-\kappa\gamma^2 t)\sin\gamma\zeta\sin\gamma z\, d\gamma\, d\zeta \end{aligned}$$

where the expression for C has been substituted and the order of integration reversed in the third line.

The following two results allow this to be rewritten:

$$\sin A \sin B = [\cos(A - B) - \cos(A + B)]/2$$

$$\int_0^\infty \exp(-\kappa\gamma^2 t)\cos\gamma(\zeta - z)\mathrm{d}\gamma = \frac{1}{2}\sqrt{\frac{\pi}{\kappa t}}\exp[-(\zeta - z)^2/4\kappa t]$$

Then

$$T(z, t) = \frac{T_\mathrm{m}}{2\pi}\sqrt{\frac{\pi}{\kappa t}}\int_0^\infty \left[\mathrm{e}^{-(\zeta-z)^2/4\kappa t} - \mathrm{e}^{-(\zeta+z)^2/4\kappa t}\right]\mathrm{d}\zeta$$

Transform the internal variables in each of these integrals to the following:

$$\beta = \eta(\zeta - z), \qquad \beta' = \eta(\zeta + z)$$

where $\eta = 1/2\surd(\kappa t)$. Then the integrals have the same form, but over different ranges of β and β', so they can be combined to yield

$$\begin{aligned} T(z, t) &= \frac{T_\mathrm{m}}{\surd\pi}\int_{-\eta z}^{\eta z} \mathrm{e}^{-\beta^2}\mathrm{d}\beta \\ &= \frac{2T_\mathrm{m}}{\surd\pi}\int_0^{\eta z} \mathrm{e}^{-\beta^2}\mathrm{d}\beta \\ &= T_\mathrm{m}\mathrm{erf}(\eta z) \end{aligned}$$

which is the solution defined by Equations (7.3.1) and (7.3.2).

7.4 Seafloor subsidence and midocean rises

If the lithosphere cools, it will undergo thermal contraction. As a result, the surface (the sea floor) will subside, and the amount of subsidence can be roughly estimated as follows. If the temperature rises from 0 °C to about 1400 °C through the thickness of the lithosphere, then the average temperature of the lithosphere is about 700 °C. This means that the average temperature *deficit* of the lithosphere relative to the underlying mantle is ΔT = 1400 °C − 700 °C = 700 °C. This will cause the density to increase by the fraction $\Delta\rho/\rho = \alpha\Delta T$, where α is the coefficient of thermal expansion. If the lithosphere thickness is D, then a vertical column of rock of height D through the lithosphere will shorten by this fraction. In other words, the shortening, h, of the top of the column will be given by

$$h/D = \alpha \Delta T$$

This shortening h is not the actual amount by which the surface subsides, since the rock that has subsided away is replaced by water. We have to consider the isostatic balance of the column relative to a similar column at the midocean rise crest. These are illustrated in Figure 7.5. The mass per unit area in the column at the rise crest is $(d + D - h)\rho_m$, while the mass per unit area in the other column is $[d\rho_w + (D - h)\rho_l]$. Equating these, and neglecting second-order terms yields

$$d = h\rho_m/(\rho_m - \rho_w) \tag{7.4.1}$$

Old lithosphere (say 100 Ma) is about 100 km thick and $\alpha \approx 3 \times 10^{-5}/°C$. Then $h \approx 2.1$ km. Using $\rho_m = 3.3$ Mg/m^3 and $\rho_w = 1.0$ Mg/m^3, this implies $d \approx 3.0$ km. Old sea floor is indeed observed to be about 3 km deeper than midocean rise crests (Chapter 4, Figures 4.5, 4.6). This result suggests that the greater depth of the old sea floor relative to midocean rise crests may be explained simply by the thermal contraction of the lithosphere. In other words, the existence of the midocean rise topography may be explained by this cooling process.

We can test this idea more rigorously by using the solution to the thermal halfspace model obtained above. Each layer of thickness dz at depth z will have a temperature deficit of $\Delta T(z, t) = T_m - T(z, t)$. Then the total thermal contraction will be

$$h(t) = \int_0^\infty \alpha \Delta T(z, t) \mathrm{d}z$$

Using the result that

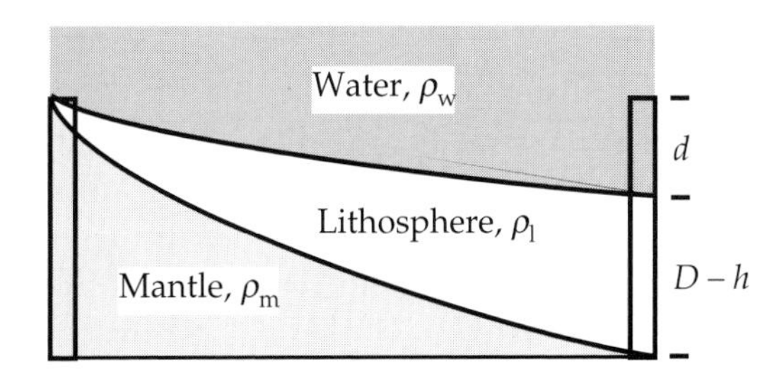

Figure 7.5. Seafloor subsidence by thermal contraction, with isostasy.

$$\int_0^\infty [1 - \mathrm{erf}(x)]\,\mathrm{d}x = \frac{1}{\sqrt{\pi}}$$

and making the appropriate variable transformation in the integral expression for h, this combines with Equation (7.4.1) to yield

$$d = \frac{\rho_\mathrm{m}\alpha T_\mathrm{m}}{(\rho_\mathrm{m} - \rho_\mathrm{w})} 2\sqrt{\frac{\kappa t}{\pi}} \tag{7.4.2}$$

Using the same values as previously, this gives $d = 3.8$ km at $t = 100$ Ma.

Equation (7.4.2) predicts that the seafloor depth should increase in proportion to the square root of its age. We saw in Chapter 4 (Figure 4.6) that the sea floor does follow this behaviour to first order, particularly for ages less than about 50 Ma. At greater ages, many parts of the sea floor are shallower than this by up to about 1 km. The possible reasons for such deviations will be discussed in Chapter 12.

Here I want to emphasise the further success of the cooling halfspace model in accounting for most of the variations of both the heat flux through the sea floor and the depth of the sea floor with age. It is a remarkably simple and powerful result, which suggests that major features of the earth's surface can be explained by a simple process of near-surface thermal conduction. It also suggests that we can usefully think of the midocean rises as standing high by default, because the surrounding sea floor has subsided, rather than the rises having been actively uplifted.

The success of the cooling halfspace model in accounting for seafloor subsidence and heat flux suggests some very important things about the earth and about mantle convection. The midocean rise system is the second-largest topographic feature of the earth after the continents, and we have seen here that it can be accounted for by a simple near-surface process: conductive heat loss to the earth's surface. *Its explanation does not require any process operating deeper than about 100 km*, and in particular it does not require a buoyant convective upwelling under midocean rises, as has often been supposed (see Chapter 3). If the midocean rise topography is not an expression of deep convection, as I am suggesting here, this leads to the question of why there is not some more obvious expression in the earth's topography of deep convection. These questions will be taken up in later Chapters 8, 10, 11 and 12, where they will lead to some important conclusions about the form of mantle convection.

Table 7.1. *Heat-producing isotopes [2].*

Element	Isotope	Half life (Ga)	Power (μW/kg of element)
Uranium	^{238}U	4.468	94.35
	^{235}U	0.7038	4.05
Thorium	^{232}Th	14.01	26.6
Potassium	^{40}K	1.250	0.0035

7.5 Radioactive heating

Radioactivity generates heat, and radioactive heat generated in the earth sustains the earth's thermal regime, as we will see in Chapter 14. There are two aspects that I want to cover here: its effect in modifying continental geotherms and its contribution to the heat budget of the mantle.

The isotopes that make the main heat contributions are ^{40}K, ^{238}U, ^{235}U and ^{232}Th. Each of these has a half life of the order of 1–10 Ga. (If they had shorter half lives, they would not still be present in significant quantities.) Their half lives and current rates of heat production are given in Table 7.1.

Geochemists find that these elements occur in similar proportions relative to each other in the crust and mantle, although their absolute concentrations differ greatly. Thus the mass ratio Th/U is usually 3.5–4 and the ratio K/U is usually $1–2 \times 10^4$. It is sufficient for our purposes to assume the particular values Th/U = 4 g/g and K/U = 10^4 g/g. (The unit g/g may seem to be redundant, but it serves to specify that this is a ratio by weight, rather than by mole or by volume, for example.) With these ratios, the total power production due to all of these isotopes, expressed per kg of uranium in the rock, is 190 μW/(kg of U). Then representative values of the concentration of uranium in different rocks allow us to estimate the total rate of heat production. Such estimates are given in Table 7.2.

These are only representative values, and there is considerable variation, especially in the continental crust. These values probably tend to be on the high side of the distribution. For the upper mantle, Jochum and others [3] have estimated on the basis of measurements of representative rocks that the likely value of the heat production rate is 0.6 pW/kg, with the value unlikely to be as great as 1.5 pW/kg.

You will see in the next section that heat production in the upper continental crust is sufficient to account for about half of typical continental heat fluxes. For example, a heat production rate

Table 7.2. *Radiogenic heat production rates, assuming Th/U = 4 kg/kg, K/U = 10^4 kg/kg. U concentrations from [4, 5].*

Region	Concentration of U	Power (pW/kg)	Density (Mg/m^3)	Power (nW/m^3)
Upper continental crust	5 μg/g	1000	2.6	2600
Oceanic crust	50 ng/g	10	2.9	30
Upper mantle	5 ng/g	1	3.3	3
Chondritic meterorites[a]	20 ng/g	4–6	3.3	12–18

[a]With K/U = 2–6 × 10^4 kg/kg.

of 2.5 $\mu W/m^3$ through a depth of 10 km will produce a surface heat flux of 25 mW/m^2, compared with typical continental values of 60–100 mW/m^2. On the other hand, the oceanic crust produces very little heat (30 nW/m^3 through 7 km gives 0.2 mW/m^2).

There is a puzzle about the amount of heat production in the mantle. It is not clear that there is enough radioactivity to account for the heat being lost at the earth's surface. Heat production is small in the upper mantle: 3 nW/m^3 through a depth of 650 km yields 2 mW/m^2. If this heat production rate applied through the whole 3000 km depth of the mantle, the surface heat flux would still be only about 10 mW/m^2. To account for the observed average oceanic heat flux of 100 mW/m^2 requires heat production in the mantle to be closer to that of oceanic crust. Some of the deficit can be accounted for by the slow cooling of the earth's interior, as will be shown in Chapter 14, and some may be explained by a greater heat production in the deeper mantle, either because the deep mantle composition is more 'primitive' (that is, closer to that of chondritic meteorites) or because there is an accumulation of subducted oceanic crust at depth, or both (Chapters 13, 14). Another contribution comes from mantle plumes, but these seem to account for less than 10% of the total (Chapter 11). In any case, there is a significant discrepancy here that has not been entirely accounted for. It is believed that none of the principal heat-producing elements would dissolve in the core in significant quantities, which implies that the discrepancy cannot be made up there. The question is addressed again in Chapters 12 and 14.

7.6 Continents

In the theory of plate tectonics, the continents are part of the lithospheric plates, carried passively as the plates move. Although the assumption that plates are non-deforming is not as good in continental areas as in oceanic areas, it is nevertheless sufficiently true to be a useful approximation. If the continents are part of the lithosphere, then heat transport within them must be by conduction rather than by convection (assuming that heat transport by percolating liquids, such as water or magma, is not important in most places most of the time). This was assumed in Section 7.3 for the oceanic lithosphere. There is, however, an important difference between the continental lithosphere and the oceanic lithosphere, and this is that the continental lithosphere is much older, since we know that most continental crust is much older. Whereas we treated the oceanic lithosphere as a transient (time-dependent) cooling problem, the continental thermal regime is more likely to be near a steady state, as we will now see.

The more stable parts of the continents, the cratons and shields, have not had major tectonic activity for periods ranging from a few hundred million years up to a few billion years. Their heat flux tends to be lower, 40–50 mW/m^2, than younger parts of the continents (Figure 4.8) It is often assumed that they are in thermal steady state, that is the heat input (from below and from radioactivity) balances the heat loss through the surface. Is this reasonable? We saw in Section 7.2 that the time scale for cooling oceanic lithosphere to a depth of 100 km is about 100 Ma. The time scale to cool or equilibrate to a depth of 200 km would then be about 400 Ma. It would thus seem to be reasonable to assume that at least the Archean shields, and perhaps the Proterozoic cratons, had approached equilibrium. I do not want to belabour this point either way. It is instructive to assume that the older continental geotherms are roughly in steady state, but on the other hand most continental areas have had some tectonic activity within the last billion years or so, and little is known about whether the lithosphere might have had a constant thickness during such periods.

The typical heat flux out of continents of about 60 mW/m^2 is due partly to heat generated within the continental crust and partly to heat conducting from the mantle below. The relative proportions of these contributions are not known very accurately, but they seem to be roughly comparable. The heat-producing elements tend to be concentrated in the upper crust, and a common and useful assumption is that their concentration decreases exponentially with depth, with a depth scale of about $h \approx 10$ km. At the surface the heat production rate is of the order of 1 nW/kg, so that the heat production rate per unit volume is $A_0 \approx 2.5\,\mu\text{W/m}^3$. The crust is very heterogeneous, so you should understand that these are merely representative numbers. Thus we might assume that the heat production rate as a function of depth is

$$A = A_0 e^{-z/h} \tag{7.6.1}$$

Let us assume that a piece of continental crust is in thermal steady state, which implies that the lithosphere thickness, the heat production rate and the distribution of heat production with depth have all been constant for a sufficiently long time. Let us also assume that heat is transported only by conduction, which excludes transport by percolating water or magma and transport by deformation of the crust and lithosphere. By steady state I mean simply that the temperature at a given depth is not changing, so that $\partial T/\partial t = 0$. Then Equation (7.1.3), which governs the evolution of temperature by conduction, is

$$\frac{\partial^2 T}{\partial z^2} = -\frac{A}{K} \tag{7.6.2}$$

where K is the conductivity.

We know that the temperature at the earth's surface is about 0 °C and its gradient is constrained by the surface heat flux. From Equation (7.1.1), the surface gradient $T_0' = -q_0/K \approx 20\,°\mathrm{C/km}$, taking $q_0 \approx -60\,\mathrm{mW/m^2}$ and $K = 3\,\mathrm{W/m\,°C}$. Equation (7.6.2) is a differential equation and these are the boundary conditions:

$$T_0 = 0\,°\mathrm{C}, \qquad T_0' = 20\,°\mathrm{C/km} \tag{7.6.3}$$

Suppose, for the moment, there were no radioactive heat production in the crust: $A = 0$. Then the solution to Equation (7.6.2) with these boundary conditions is

$$T = T_0 + T_0' z \tag{7.6.4}$$

With the values of Equations (7.6.3), the temperature at the base of the crust, about 40 km deep, would be 800 °C, at which temperature the crust would be likely to be melting, depending on its composition. The temperature at 60 km depth would be 1200 °C, at which depth the mantle would almost certainly be melting. Since seismology tells us that the mantle is largely solid, this suggests that at least one of our assumptions becomes invalid at some depth of the order of 60 km.

Now let us return to the assumption that there is radioactive heat generation, and that its variation with depth is given by Equation (7.6.1). Then the solution to Equation (7.6.2) with the same boundary conditions is

$$
\begin{aligned}
T &= T_0 + \frac{A_0 h^2}{K}\left(1 - e^{-z/h}\right) + \left(T_0' - \frac{A_0 h}{K}\right) z \\
&= T_0 + T_h\left(1 - e^{-z/h}\right) + T_m' z
\end{aligned}
\tag{7.6.5}
$$

The solutions (7.6.4) and (7.6.5) are sketched in Figure 7.6. Equation (7.6.5) approaches an asymptote at depth, line (a), given by

$$
T = (T_0 + T_h) + T_m' z \tag{7.6.6}
$$

At 40 km depth the term $e^{-z/h}$ is already as small as 0.018, so for greater depths the asymptote is a good approximation. Using this, it is easy to calculate that the temperature at 40 km depth is 550 °C, compared with 800 °C from Equation (7.6.4) without radioactive heating. The depth at which 1200 °C is reached is 96 km, compared with 60 km without radioactive heating. The mantle below the lithosphere is believed to be at a temperature of 1300 °C to 1400 °C. Assuming the latter, the lithosphere could be no more than 113 km thick using the values assumed here. This is relatively thin, and might be appropriate for a relatively young continental province.

It might seem paradoxical that lower temperatures have been calculated when radioactive heating has been included, in Equation (7.6.5), compared with temperatures from Equation (7.6.4) with no radioactive heating. In order to clarify this, I will spend some time explaining some aspects of this solution. The reason for the lower

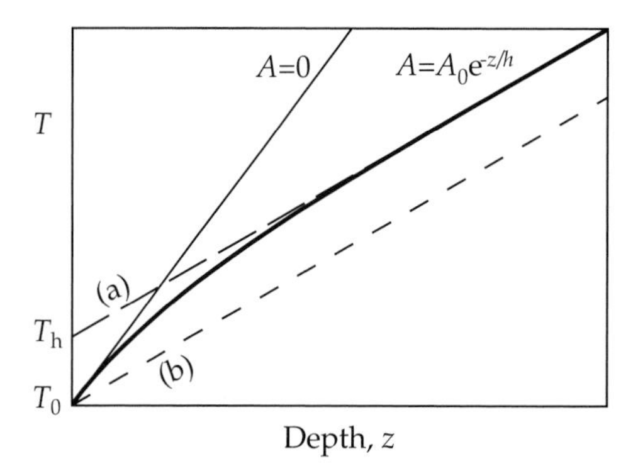

Figure 7.6. Sketch of calculated continental geotherms. Geotherm with no radioactivity ($A = 0$, light solid line), with radioactivity given by Equation (7.6.1) (heavy solid). Line (a) is the asymptote of the heavy curve. Line (b) is the geotherm, with $A = 0$, that would match the temperature at the surface and the temperature gradient below the zone of radioactive heating.

temperatures is that the temperature gradient was fixed at the surface. Physically you can think of it as follows. If the temperature gradient is fixed at the surface, then so is the surface heat flux. If there is no crustal radioactive heat generation, then all of that heat flux must be coming from below the lithosphere. This requires that the temperature gradient must equal the surface value right through the lithosphere (assuming for simplicity that the conductivity is constant). On the other hand, if some of the heat comes from radioactivity near the surface, then below the heat-generating zone the heat flux will be less, so the temperature gradient will also be smaller. This is why the geotherm in this case bends down (Figure 7.6) and reaches lower temperatures at depth than the geotherm with no heat generation (but with the same temperature and temperature gradient at the surface).

If we required instead to keep the same heat flux into the *base* of the lithosphere, then with no heat generation the result would be line (b), with a slope of $T'_{\mathrm{m}} = T'_0 - A_0 h/K$. This matches the surface temperature and has the same temperature gradient at the base of the lithosphere. You can see that including heat generation raises the temperature relative to line (b). This comparison is more in accord with simple intuition. It also illustrates what is sometimes called thermal blanketing: the geotherm with heating is hotter by the amount $T_{\mathrm{h}} = A_0 h^2/K$. With the values used above, $T_{\mathrm{h}} = 83\,^{\circ}\mathrm{C}$.

The heat flux through the base from the mantle, using the above values, is $q_{\mathrm{m}} = KT'_{\mathrm{m}} = 35\,\mathrm{mW/m^2}$. The heat flux due to radioactive heating is in this case $q_{\mathrm{h}} = A_0 h = 25\,\mathrm{mW/m^2}$. Thus you can see that in this case the total surface heat flux of $60\,\mathrm{mW/m^2}$ is the sum of $35\,\mathrm{mW/m^2}$ from the mantle and $25\,\mathrm{mW/m^2}$ from radioactive heating. If the lithosphere were thicker, the heat flux from the mantle would be less, as would the total surface heat flux.

From the point of view of considering mantle convection, an important question is how much heat escapes from the mantle through the continents, since this heat is not then available to drive mantle convection. This is the amount of heat entering the base of the continental lithosphere, and it is determined essentially by the thickness of the continental lithosphere, as modified by the thermal blanketing effect of radioactivity in the upper crust. Thus the long-term or steady-state conducted heat flux is determined by the temperature gradient $(T_{\mathrm{m}} - T_{\mathrm{h}})/D$. The mantle temperature T_{m} does not vary much in comparison with variations in the lithosphere thickness D, and you have just seen an estimate that the thermal blanketing effect of radioactivity is to raise the effective surface temperature by about $80\,^{\circ}\mathrm{C}$, which also is small in comparison with the total temperature difference. Thus if we take the effective temperature difference across continental lithosphere to be about 1200 °C and a typical range of thickness to be 100–250 km, we get a range of heat flux out of the mantle of 14–36 mW/m^2, with an average of perhaps 20–25 mW/m^2.

Continental crust covers about 40% of the earth's surface, with an area of about 2×10^{14} m^2, so the total heat loss from the mantle through continents is about 4–5 TW, about 10% of the global heat budget. Thus the mantle heat loss through the continents is not a large fraction of the total. In fact it is such a small fraction that we can regard the continents as insulating blankets. The complementary role of oceanic heat loss will be taken up in Chapter 10.

7.7 Heat transport by fluid flow (Advection)

So far we have looked at heat transported by thermal conduction, but heat can also be transported by the motion of mantle material. Suppose, for example, that mantle material with a temperature of $T_{\mathrm{h}} = 1500\,°\mathrm{C}$ replaces normal mantle at $T_{\mathrm{m}} = 1400\,°\mathrm{C}$. Then the local heat content per unit volume is increased by $\rho C_P \Delta T = \rho C_P (T_{\mathrm{h}} - T_{\mathrm{m}})$. With $\rho = 3300\,\mathrm{kg/m^3}$ and $C_P = 1000\,\mathrm{J/kg\,°C}$, the increase is $3.3 \times 10^8\,\mathrm{J/m^3}$. Now, referring to Figure 7.7, suppose the hot material is flowing up a vertical pipe of radius $R = 50$ km at a velocity $v = 1$ m/a $\approx 3 \times 10^{-8}$ m/s. Then the volume of hot material that flows past a point on the pipe within unit time is $V = \pi R^2 v \approx 250\,\mathrm{m^3/s}$. The amount of extra heat that has been carried past this point within unit time is

$$Q = V\rho C_P \Delta T = \pi R^2 v \rho C_P \Delta T \tag{7.7.1}$$

With the above values, $Q \approx 8 \times 10^{10}\,\mathrm{J/s} = 8 \times 10^{10}$ W: this is the heat flow rate. The heat flux is

$$q = Q/\pi R^2 = v\rho C_P \Delta T \tag{7.7.2}$$

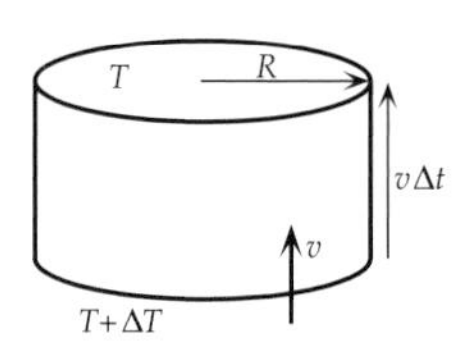

Figure 7.7. Heat carried by fluid flowing with velocity v.

Then $q \approx 10\,\mathrm{W/m^2}$. This heat flux is a much higher value than the conducted heat fluxes we discussed above. The heat flow, Q, is about 0.2% of the global heat budget, despite the small area of the pipe in comparison with the surface area of the earth. This example has been tailored to approximate a mantle plume, and these will be discussed in more detail in Chapter 11.

This process of heat transport by mass motion is called *advection*. It usually accounts for most of the heat transport in *convection*. In fact, you will see in Chapter 8 that in a sense convection only occurs when conduction is inadequate to transport heat. You can see from the above example that advection can transport much greater heat fluxes than conduction in some situations.

The distinction between advection and convection is that the term *advection* is used to refer to heat transport by mass motion regardless of the source of the mass motion, whereas in *convection* the motion is due specifically to the internal buoyancies of the material. Thus when you stir your coffee with a spoon, you force fluid motion that transports heat around in the cup by *advection*. On the other hand, if you let the cup sit, the top of the coffee will cool and sink, producing *convection* that will also advect heat.

7.8 Advection and diffusion of heat

7.8.1 General equation for advection and diffusion of heat

The approach just used in Section 7.7 can be used to derive an equation that governs the evolution of temperature in the presence of both conduction (diffusion) and advection. Advection occurs when there is fluid motion and when there are temperature differences within the fluid: if the temperature is homogeneous, then there is no net heat transport. In Section 7.1 we considered heat conduction in one dimension (the x direction). If we now suppose that in addition to the other things happening in Figure 7.1 there is a flow with velocity v in the positive x direction, then we should add two terms to the right-hand side of the heat budget for the little box:

$$Sv\,\mathrm{d}t \cdot \rho C_P T - Sv\,\mathrm{d}t \cdot \rho C_P (T + \Delta T)$$

You can recognise these as the heat advected into the box through the left-hand side, within the time interval $\mathrm{d}t$, minus the heat advected out through the right-hand side. Dividing again by $S \cdot \mathrm{d}x \cdot \mathrm{d}t$ and taking the limit yields the extra term on the right-hand side of Equation (7.1.2)

$$-\rho C_P v \frac{\partial T}{\partial x}$$

Equation (7.1.3) can then be generalised to

$$\frac{\partial T}{\partial t} + v\frac{\partial T}{\partial x} = \kappa \frac{\partial^2 T}{\partial x^2} + a \qquad (7.8.1)$$

Here the advection term is placed on the left-hand side, and you can see it depends on there being both a fluid velocity and a temperature gradient.

This equation can readily be generalised to three dimensions just by considering heat transport in the other two coordinate directions of Figure 7.1. The result is

$$\frac{\partial T}{\partial t}+v_i\frac{\partial T}{\partial x_i}=\kappa\frac{\partial^2 T}{\partial x_i \partial x_i}+a \tag{7.8.2}$$

where I have used the summation convention (Box 6.B1). This equation governs the evolution of temperature in the presence of advection, diffusion (conduction) and internal heat generation.

We now have the conceptual and mathematical tools to consider convection. We have looked at viscous fluid flow, including examples driven by buoyancy forces, and at heat transport. Convection involves the combination of these processes. Their general mathematical description is embodied in Equations (6.6.1), (6.6.3) and (7.8.2). Convection will be discussed in Part 3.

7.8.2 An advective-diffusive thermal boundary layer

Here is a relatively simple illustration of the simultaneous occurrence of advection and diffusion. We will see in Chapter 11 that mantle plumes are believed to transport material from the base of the mantle, where mantle material is heated by heat flowing out of the core, which is believed to be hotter. This heat will generate a hot thermal boundary layer at the base of the mantle. The thickness of this boundary layer will depend on the rate at which material flows through it, and also on the form of that flow. If the flow is basically horizontal, like the bottom of a large-scale convection cell, then the boundary layer thickness will depend on the time for which mantle material is adjacent to the hot core. The theory of Section 7.3 will then apply. In this case the bottom thermal boundary layer would be analogous to the top thermal boundary layer (the lithosphere), and its thickness would be proportional to the square root of the time spent at the bottom of the mantle. It is conceivable, however, that the large-scale, plate-related flow penetrates only minimally to the bottom of the mantle (Chapters 10, 12), and that the dominant flow near the bottom is a vertical downwards flow that balances the material flowing upwards in plumes. In this case the relationship between the advection and diffusion of heat would be different from that in Section 7.3. We now look at this possibility.

The situation is sketched in Figure 7.8. Material flows slowly downwards with velocity $v = -V$. Hot, low-viscosity material flows rapidly sideways within a thin layer adjacent to the core, and then into narrow plumes where it rises rapidly upwards. Away from the plumes and above the thin channel at the base, the flow can be approximated as being vertical. We now derive an expression for the temperature in this region, as a function of height, z, above the core.

If we assume that there is a steady state and no heat generation, then Equation (7.8.1) reduces to

$$v\frac{\mathrm{d}T}{\mathrm{d}z} = \kappa\frac{\mathrm{d}^2T}{\mathrm{d}z^2} \tag{7.8.3}$$

Note first of all that if we use a representative temperature scale ΔT and a representative length scale h, then rough approximations to the differentials in this equation yield

$$h = \kappa/v \tag{7.8.4}$$

Thus this ratio contains an implicit length scale, h, which we can also see from the dimensions of κ and v.

Using Equation (7.1.1) for the heat flux, q, Equation (7.8.3) can be rewritten as

$$\frac{1}{q}\frac{\mathrm{d}q}{\mathrm{d}z} = \frac{v}{\kappa} = -\frac{V}{\kappa} \equiv -\frac{1}{h}$$

which defines the length scale $h = \kappa/V$ for the particular problem in Figure 7.8. This equation can be integrated to give

$$q = q_{\mathrm{b}}\exp\left(-\frac{z}{h}\right)$$

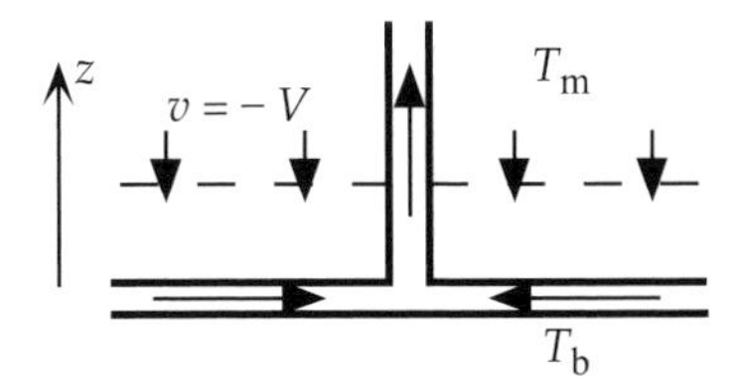

Figure 7.8. Sketch of the flow associated with a mantle plume drawing hot material from the base of the mantle. The temperature at the bottom boundary is T_{b}, and the temperature of the ambient mantle is T_{m}. Away from plumes, mantle material is assumed to flow vertically downwards with velocity $v = -V$. A thermal boundary layer (dashed line) forms above the core–mantle boundary.

where q_b is the heat flux into the base. This in turn can be integrated to give

$$T = T_m + \frac{hq_b}{K} \exp\left(-\frac{z}{h}\right) \tag{7.8.5}$$

where K is the conductivity. The temperature at the boundary, $z = 0$, is then

$$T_b = T_m + hq_b/K \tag{7.8.6}$$

In Chapter 11 we will see that the total heat flow carried by plumes is about 3.5 TW, roughly 10% of the heat flowing out of the top of the mantle. This heat is inferred to be flowing out of the core. Since the surface area of the core is only about $\frac{1}{4}$ of the surface area of the earth, the heat flux out of the core is then about 40% of the surface heat flux, or about 30 mW/m^2. In Exercise 12, later, you can deduce that the downward velocity is about $V = 1.3 \times 10^{11}$ m/s (0.4 mm/a). Then taking the density at the base of the mantle to be 5600 kg/m^3 and other quantities from Table 7.3, below, we obtain $h = 115$ km and $T_b - T_m = 385$ °C.

The physics described by this solution is that mantle material slowly flows down towards the hot interface with the core and heat conducts upwards against this flow. Thus upwards thermal diffusion is competing against downwards advection of heat. In the steady state, the temperature declines exponentially towards the ambient mantle temperature as a function of height above the interface. This thermal boundary layer has a characteristic thickness of the order of 100 km and a temperature increase across it of about 400 °C, according to the numerical values we have used.

7.9 Thermal properties of materials and adiabatic gradients

7.9.1 Thermal properties and depth dependence

We have already encountered most of the important thermal properties of materials that we will be needing in this book. It is useful to summarise them here, with some typical values. This is done in Table 7.3. However there is one aspect that we have not yet encountered, and that is the variation with depth of some of these properties, and of the temperature in the convecting mantle. Although we will not be much concerned with these depth variations, because the effects are secondary to the main points I want to demonstrate, they are nevertheless significant and worth noting.

Table 7.3. *Representative thermal properties of the mantle [6].*

Quantity	Symbol	Value $P = 0$	Value CMB	Units
Specific heat at constant pressure	C_P	900	1200	J/kg °C
Thermal conductivity	K	3	9	W/m °C
Thermal diffusivity	κ	10^{-6}	1.5×10^{-6}	m^2/s
Thermal expansion coefficient	α	3×10^{-5}	0.9×10^{-5}	/°C
Grüneisen parameter	γ	1.0–1.5	0.9	

Some estimates of values at the base of the mantle (CMB or core–mantle boundary) are included in Table 7.3. These are modified from Stacey's [6] values, mainly by using higher values of the thermal expansion coefficient.

7.9.2 Thermodynamic Grüneisen parameter

The thermodynamic relationships governing the depth dependence of the temperature can be expressed most concisely in terms of a parameter known as the thermodynamic Grüneisen parameter, γ. It is related to the thermal expansion coefficient, and this relationship is most directly evident if we define γ as [6]

$$\gamma = \frac{1}{\rho C_V}\left(\frac{\partial P}{\partial T}\right)_V \tag{7.9.1}$$

where C_V is the specific heat at constant volume. This definition shows that γ is a measure of the rate at which pressure increases as heat is input while volume is held constant. For comparison, the thermal expansion coefficient is

$$\alpha = \frac{1}{V}\left(\frac{\partial V}{\partial T}\right)_P \tag{7.9.2}$$

Thus α is a measure of the rate at which volume increases as heat is input while pressure is held constant, and α and γ are complementary measures of the effect of heating. Another way to think of γ is that it measures the pressure required to prevent thermal expansion.

Two other useful expressions for γ can be derived with the help of thermodynamic identities. The latter are complicated, and can be found in standard thermodynamics texts. A concise summary is provided by Stacey [6]. I will just quote the results here. The first form is

$$\gamma = \frac{\alpha K_S}{\rho C_P} = \frac{\alpha v_b^2}{C_P} \tag{7.9.3}$$

where $K_S = \rho(\partial P/\partial \rho)_S$ is the adiabatic bulk modulus, subscript S indicates constant entropy, and v_b is the bulk sound speed (Section 5.1.4). The second form is

$$\gamma = -\frac{V}{T}\left(\frac{\partial T}{\partial V}\right)_S = \frac{\rho}{T}\left(\frac{\partial T}{\partial \rho}\right)_S \tag{7.9.4}$$

where V is specific volume (that is, volume per unit mass: $V = 1/\rho$).

7.9.3 Adiabatic temperature gradient

As mantle material rises and sinks in the course of mantle convection, thermal diffusion is so inefficient at large scales that through most of the mantle it can be neglected. At the same time, there are large changes of pressure accompanying the vertical motion. A process of compression with no heat exchange with surroundings is called *adiabatic compression*. If it happens slowly, so that it is reversible, it is characterised by having constant entropy. A parcel of mantle that sinks slowly through the mantle experiences such adiabatic compression. During adiabatic compression, although there is no heat exchange with surroundings, the increasing pressure does work on the material as it compresses, and this increases the internal energy of the material, which is expressed as a rise in temperature. We will now estimate this adiabatic increase in temperature with depth in the mantle.

The Grüneisen parameter provides a convenient way to make this estimate. The Grüneisen parameter in the mantle can be estimated most reliably from Equation (7.9.3), since K_S, ρ and v_b are known from seismology (Section 5.1.4). C_P does not vary much with pressure. The thermal expansion coefficient is the least well constrained, and it is likely to decrease substantially under pressure [7], as indicated in Table 7.3. This is counteracted by the increase of v_b with depth (Figure 5.3). The result is that γ does not vary greatly with depth, being about 1–1.5 in the peridotite and transition zones and decreasing to slightly less than 1 at the bottom of the mantle.

If γ does not vary greatly through the mantle, then the assumption that it is constant will be a reasonable approximation. In this case, Equation (7.9.4) can be integrated to yield

$$\frac{T_l}{T_u} = \left(\frac{\rho_l}{\rho_u}\right)^{\gamma} \qquad (7.9.5)$$

where the subscripts l and u refer to lower mantle and upper mantle, respectively. With $\rho_l = 5500\,\mathrm{kg/m^3}$, $\rho_u = 3300\,\mathrm{kg/m^3}$, and $\gamma = 1.0$–1.5, this yields $T_l/T_u = 1.7$–2.15. However, about 800 kg/m^3 of the density increase through the mantle is due to phase transformations, through which Equation (7.9.4) does not apply. If we take instead $\rho_l = 4700\,\mathrm{kg/m^3}$, then $T_l/T_u = 1.4$–1.7. With $T_u = 1300\,°\mathrm{C}$, this indicates that the adiabatic increase of temperature through the mantle is about 500–900 °C, and $T_l =$ 1800–2200 °C.

A schematic temperature profile through the earth is shown in Figure 7.9. A more quantitative version is not given here, both because we are not concerned with details, and because the uncertainties are so large that greater detail is hardly justified. For example, various estimates put the temperature jump across the lower thermal boundary layer of the mantle at anything between 500 °C and 1500 °C, with some estimates even higher [6, 7], so that $T_b = 2300$–$3700\,°\mathrm{C}$. However, it is hard to reconcile these higher values with the dynamics of plumes (Chapter 11), even taking account of the likelihood of a layer of denser material at the base of the mantle (Chapter 5). Stacey [6] estimates the adiabatic temperature increase through the core to be about 1500 °C, so that the temperature at the centre of the earth might be $T_c = 3800$–$5000\,°\mathrm{C}$.

7.9.4 The super-adiabatic approximation in convection

Although the adiabatic increase of temperature through the mantle is quite large, it is not of great concern to us in this book. This is because convection will only occur if the actual temperature gradient exceeds the adiabatic gradient, as I will explain in a moment. We can therefore focus on this *super-adiabatic* gradient. An effective way to do this is to subtract the adiabatic gradient out of the mantle temperature profile for convection calculations, or in other words to neglect this effect of pressure.

To see that convection requires a super-adiabatic gradient, suppose that the interior of the mantle has an adiabatic gradient, as sketched in Figure 7.9. You might suppose at first that since the deeper mantle is hotter than the shallow mantle, it will be buoyant and therefore drive convection. However, if a small portion of this deep mantle rises vertically, it will decompress adiabatically as it rises and its temperature will follow the adiabatic profile. Thus it

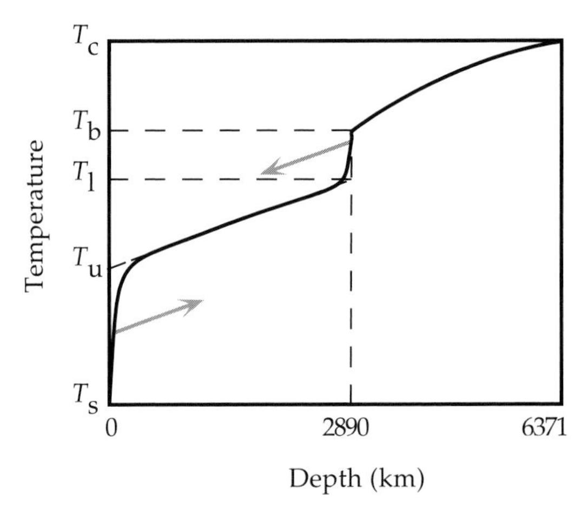

Figure 7.9. Schematic temperature profile through the earth. Thermal boundary layers are assumed at the top of the mantle (the lithosphere) and the bottom of the mantle. Numerical values of the temperatures are quite uncertain (see text). The grey arrows show adiabatic compression and decompression paths of material from the thermal boundary layers.

will remain at the same temperature as its surroundings and no thermal buoyancy will be generated.

In order to have buoyancy that will drive convection, a deep portion of the mantle must start hotter than its surroundings, as would for example material from the lower thermal boundary layer. It may then follow an adiabatic decompression path that is sub-parallel to the mantle adiabat, and consequently remain hotter and buoyant as it rises. Such a path is illustrated in Figure 7.9. An analogous path is also shown for descending, cool, negatively buoyant lithospheric material. Of course these portions of the mantle may exchange heat with the surrounding mantle by thermal diffusion, in which case their paths will tend to converge towards the mantle adiabat, but their initial buoyancy will be approximately preserved within a larger volume of material.

7.10 References

1. C. B. Officer, *Introduction to Theoretical Geophysics*, 385 pp., Springer-Verlag, New York, 1974.
2. V. M. Hamza and A. E. Beck, Terrestrial heat flow, the neutrino problem, and a possible energy source in the core, *Nature* **240**, 343, 1972.
3. K. P. Jochum, A. W. Hofmann, E. Ito, H. M. Seufert and W. M. White, K, U and Th in midocean ridge basalt glasses and heat production, *Nature* **306**, 431–6, 1986.

4. R. L. Rudnick and D. M. Fountain, Nature and composition of the continental crust: a lower crustal perspective, *Rev. Geophys.* **33**, 267–309, 1995.
5. W. F. McDonough and S.-S. Sun, The composition of the Earth, *Chem. Geol.* **120**, 223–53, 1995.
6. F. D. Stacey, *Physics of the Earth*, 513 pp., Brookfield Press, Brisbane, 1992.
7. R. Boehler, A. Chopelas and A. Zerr, Temperature and chemistry of the core–mantle boundary, *Chem. Geol.* **120**, 199–205, 1995.

7.11 Exercises

1. Use Equation (7.2.1) to estimate the time it would take for a sill of thickness 100 m to cool substantially.
2. During the ice age, glaciers kept the surface of Canada cooler than at present. The glaciers had melted by about 10 000 years ago. To about what depth in the crust would the subsequent warming of the surface have penetrated?
3. [*Intermediate*] Complete the derivation of Equation (7.2.7). Either integrate the equations or show that the forms used are solutions of the relevant equations. Apply the initial condition to evaluate the constants of integration.
4. Use Lord Kelvin's argument to estimate the age of the earth from the fact that the rate of temperature increase with depth in mines and bore holes is about 20 °C/km and assuming the upper mantle temperature to be 1400 °C. Comment on the relationship between your answer and the age of oceanic lithosphere.
5. [*Advanced*] Derive the general solution (7.3.5), using the same approach as in Exercise 3.
6. Using values in Table 7.2, calculate the thicknesses of layers composed of (i) upper continental crust, (ii) oceanic crust, and (iii) chondritic meteorites required to produce the average heat flux of 80 mW/m^2 observed at the earth's surface. What constraints does this impose on the composition of the continental crust and the mantle?
7. Calculate, by integration from the surface to great depth, the total rate of heat production per unit surface area implied by Equation (7.6.1).
8. Derive the solution (7.6.4) for temperature versus depth from Equations (7.6.2) and (7.6.3).

9. [*Intermediate*] Derive the solution (7.6.5) in the same way as for Exercise 8. Produce a quantitative version of Figure 7.6.

10. Calculate the steady-state surface heat flux that would be observed for combinations of the following values of quantities used in Section 7.6: $D = 100, 150, 250\,\text{km}$, $A_0 = 1, 3\,\mu\text{W/m}^3$, $h = 10\,\text{km}$.

11. For the Hawaiian plume, the heat flow estimated from surface topography (Chapter 11) is about 0.4 TW. Estimate the velocity of flow up the plume assuming it has a radius of 50 km and a temperature difference of (i) 100 °C, (ii) 300 °C.

12. In Chapter 11 it is estimated that plumes transport heat upwards at the rate of 3.5 TW. (a) If the temperature difference between plumes and surrounding mantle is 300 °C, what is the total volumetric flow rate of mantle material in plumes? (b) If the downwards return flow that balances the upward plume flow is uniform, what is its velocity at the bottom of the mantle. Take the surface area of the core to be $1.3 \times 10^{14}\,\text{m}^2$.

13. [*Intermediate*] Complete the derivation of Equation (7.8.5). Calculate the numerical values of the thickness of the thermal boundary layer and the temperature difference across it, using results from Exercise 12 and Table 7.3.

PART 3

ESSENCE

In Part 3 we come to the core of the subject of this book. We look at convection as a general phenomenon, at the features and properties of the mantle that make convection in the mantle distinctive, at the observations that enable us to infer the general form of mantle convection, at the two major identified modes of convection in the mantle, and at the current picture as best I can assess it.

Convection involves fluid flow and heat transport, and the way convection works in general is presented in Chapter 8. This includes a relatively simple way to estimate the rate of convection, useful ways to characterise convection quantitatively and ways to understand some basic features of a convecting system. The topography generated by convection is a key observable, and the seafloor subsidence discussed in Chapter 7 is put in a more general context.

Mantle convection is an unusual kind of fluid dynamics because the non-fluid plates seem to be an integral part of the convection process. We therefore need to look more explicitly at the way plates move and evolve, and this is done in Chapter 9. Some of the basic ideas are simple, but their consequences are not so obvious to the uninitiated, so some attention to the material and the Exercises is warranted.

Two main modes of mantle convection can be identified, one associated with the plates and one involving mantle plumes. These are discussed in turn in Chapters 10 and 11. Chapter 12 concludes Part 3 with an assessment of the main conclusions that can be drawn at present about the form of mantle convection, including discussions of the main controversies and questions and of some misconceptions.

CHAPTER 8

Convection

Convection is a kind of fluid flow driven by internal buoyancy. In general, the buoyancy that drives convection derives from horizontal density gradients. In the mantle, the main sources of density gradients are horizontal thermal boundary layers. Convection is driven when the buoyancy (positive or negative) of a thermal boundary layer causes it to become unstable, so that fluid from it leaves the boundary of the fluid and rises or falls through the interior of the fluid. This statement may seem to be labouring the obvious, but there has been a lot of confusion about the nature of mantle convection, and much of this confusion can be avoided by keeping these basic ideas clearly in mind.

In general the buoyancy driving convection may be of thermal or compositional origin. We will be concerned mainly with thermal buoyancy, but compositional buoyancy is also important in the mantle. It is best to consider first thermal convection, that is convection driven by thermal buoyancy. Some aspects of compositional buoyancy will be considered in Chapter 14.

Here I describe sources of buoyancy, give a simple example of thermal convection, and show how there is an intimate relationship between convection and the surface topography that it produces. This establishes some basic concepts that will be applied more explicitly to the mantle in subsequent chapters.

In the course of doing this, I show how convection problems scale, how the Rayleigh number encapsulates this scaling, why convection occurs only if the fluid is heated or cooled strongly enough, and how the mode of heating (from below or internally) governs the nature of the thermal boundary layers. In principle there may be two thermal boundary layers in a fluid layer, one at the top and one at the bottom, or there may be only one, depending on the way the fluid is heated and cooled.

8.1 Buoyancy

Buoyancy arises from gravity acting on density differences. Technically, buoyancy is used to describe a *force*. Thus it is not the same as a density difference. Rather, it is the product of a density difference, $\Delta\rho$, a volume, V, and the gravitational acceleration, g:

$$B = -gV\Delta\rho = -g\Delta m \tag{8.1.1}$$

where Δm is the mass anomaly due to a volume V with a density difference $\Delta\rho = \rho_V - \rho$ from its surroundings. The minus is used because, in common usage, buoyancy is positive upwards, whereas gravity and weight are positive downwards. Thus for a density excess, $\Delta\rho$ is positive and B is negative, that is downwards.

It is buoyancy rather than just density difference that is important in convection. A large density difference within a small volume may be unimportant. For example, you might expect intuitively that a steel ball-bearing, 1 cm in diameter, embedded in the mantle would not sink rapidly to the core, despite a density difference of over 100%. On the other hand, a plume head with a density contrast of only about 1% would have a significant velocity if its diameter were 1000 km, as we saw in Section 6.8.

With thermal buoyancy, density differences arise from thermal expansion. This is described by

$$\rho = \rho_0[1 - \alpha(T - T_0)] \tag{8.1.2}$$

where ρ is density, α is the volume coefficient of thermal expansion, T is temperature, and ρ_0 is the density at a reference temperature T_0. With α typically about $3 \times 10^{-5}/°\mathrm{C}$ (Table 7.3), a temperature contrast of 1000 °C gives rise to a density contrast of about 3%. In the lower mantle, where α may be only about $1 \times 10^{-5}/°\mathrm{C}$ due to the effect of pressure, the corresponding density difference would be only about 1%.

There are some density differences in the earth larger than these thermal density differences, and these are due to differences in chemical or mineralogical composition. For example the oceanic crust has a density of about 2.9 Mg/m^3, compared with an upper mantle density of about 3.3 Mg/m^3, so it has a density deficit of about 400 kg/m^3 or 12%. The total density change through the mantle transition zone is about 15%. Much or all of this is believed to be due to pressure-induced phase transformations of the mineral assemblage (Chapter 5), and so it is not necessarily a source of buoyancy. However, locally all of the density differences associated with particular transformations may be operative because the depth of the transformation is changed by temperature, as was discussed in Chapter 5. Apart from this, if the density increase through the transition zone is not all due to phase transformations, the maximum that could be attributed to a difference between the composition of the upper mantle and the lower mantle is a small percentage, according to the seismological and material property constraints discussed in Chapter 5.

It is useful to have some idea of the magnitudes of buoyancies of various objects. For example, a ball bearing would exert a buoyancy force of about −0.02 N (taking buoyancy to be positive upwards), while a plume head 1000 km in diameter with a temperature difference of 300 °C would have a buoyancy of about 2×10^{20} N. Subducted lithosphere extending to a depth of 600 km exerts a buoyancy of about −40 TN per metre of oceanic trench, that is per metre horizontally in the direction of strike of the subducted slab.

If the subducted lithosphere extended to the bottom of the mantle, about 3000 km in depth, its buoyancy would be about −200 TN/m. Comparing this with a plume head, it takes a piece of subducted lithosphere about 1000 km wide and 3000 km deep to equal in magnitude the buoyancy of a plume head. While this may make plume heads seem to be very important, you should bear in mind that the total length of oceanic trenches is over 30 000 km. Thus, while the buoyancy of a plume head is impressive, it is still small compared to the total buoyancy of subducted lithosphere.

The crustal component of subducted lithosphere undergoes a different sequence of pressure-induced phase transformations than the mantle component, and as a result it is sometimes less dense and sometimes denser than the surrounding mantle, with the difference usually no more than about 200 kg/m^3 (Section 5.3.4). Even if it had the same density difference, say −100 kg/m^3, extending throughout the mantle, its thickness is only about 7 km and its total contribution to slab buoyancy would be only about 20 TN/m, compared with the slab thermal buoyancy of −200 TN/m. This suggests that normally the crustal component of subducted lithosphere does not substantially affect the slab buoyancy. However, if the subducted lithosphere is young, so that its negative thermal buoyancy is small, the crustal buoyancy may be more important. This may have been more commonly true at earlier times in earth history. These possibilities will be taken up again in Chapter 14.

The very large range of the magnitudes of buoyancies of the various objects just considered serves to emphasise that we must consider the volume occupied by anomalous density, not just the magnitude of the density anomaly itself.

8.2 A simple quantitative convection model

We are now ready to consider a convection model that is simple in concept but goes to the heart of plate tectonics and its relationship with mantle convection. The approach was first used by Turcotte and Oxburgh in 1967 [1]. At that time plate tectonics was only just beginning to gain acceptance amongst geophysicists. I give a simplified version here. A more detailed version is given by Turcotte and Schubert [2], p. 279. I also acknowledge that it is only within the last five years or so that numerical models have become substantially superior to Turcotte and Oxburgh's approximate analytical model. Such is the power of capturing the simple essence of a problem.

Consider plates on a viscous mantle, as sketched in Figure 8.1a. The plates comprise a thermal boundary layer, within which the temperature changes from the surface temperature to the temperature within the interior of the mantle. Because the plates are cold, they are denser and prone to sink: they have negative buoyancy. In Figure 8.1a, one plate is depicted as subducting, and we presume here that it is sinking under its own weight. As the subducted part sinks, it drags along the surrounding viscous mantle with it. The motion of the plate is resisted by the viscous stresses accompanying

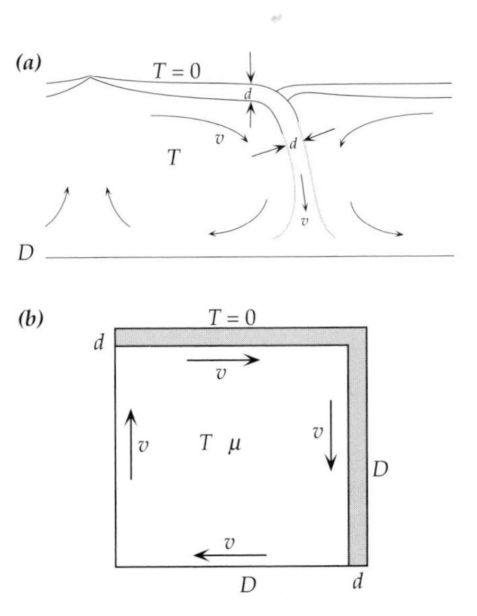

Figure 8.1. (a) Sketch of flow driven by a subducting plate. (b) Idealised form of the situation in (a).

this mantle flow. The viscous stresses are proportional to velocity. This permits an equilibrium to develop between the opposing forces: the velocity adjusts until the resistance balances the buoyancy.

Our approach is based on the same principle as that used in Chapter 6 when we considered flow down a pipe that is driven by the fluid's own weight, and the rise of a buoyant sphere. In each case, there was a balance between a buoyancy force and a viscous resistance. The system achieves balance by adjusting its velocity until the viscous resistance balances the buoyancy. This balance is stable, in the sense that a change in the velocity will induce an imbalance of the forces that will quickly return the velocity to its equilibrium value. However, we should remember that the motions are so slow in the mantle that accelerations and momenta are quite negligible, and the forces are essentially in balance at every instant, though their magnitudes may slowly change in concert.

Let us make a simple dimensional estimate of the balance between buoyancy and viscous forces, in the same way as we did for the buoyant sphere in Chapter 6. Here, because the two-dimensional sketch is assumed to be a cross-section through a structure that extends in the third dimension, the forces will be calculated per unit length in the third dimension. Let us also simplify the geometry into that depicted in Figure 8.1b.

First consider the buoyancy of the lithosphere descending down the right side of the box. Assume that this lithosphere simply turned and descended, preserving its thickness and temperature profile. From the basic formulas (8.1.1) and (8.1.2), the buoyancy is

$$B = g \cdot Dd \cdot \rho\alpha\Delta T$$

where ΔT is the average difference in temperature between the descending lithosphere and the fluid interior. This is approximately $\Delta T = -T/2$, where T is the temperature of the interior fluid. (We used the same approximation in estimating the subsidence of oceanic lithosphere in Section 7.4). Thus

$$B = -g \cdot Dd\rho\alpha T/2 \tag{8.2.1}$$

If we want to evaluate this expression, we can independently estimate the values of all quantities except the thickness, d, of the lithosphere upon subduction. This is just the thickness of the layer that has cooled by conduction of heat to the surface, as we considered in Section 7.3. It is determined by the amount of time the subducting piece of lithosphere spent at the surface. This time is

$t = D/v$. According to the discussion of thermal diffusion in Chapter 7, the thickness of the layer from which heat has diffused is approximated by

$$d = \sqrt{\kappa t} = \sqrt{\kappa D/v} \tag{8.2.2}$$

where κ is the thermal diffusivity. So we have an expression for d, but now it includes the still-unknown quantity v. We will see below how to deal with it.

Now consider the viscous resistance. As with our rough estimate for a buoyant sphere (Section 6.8.1), we estimate the viscous stresses from a characteristic velocity gradient. In this case, the velocity changes from v to $-v$ across the dimensions of the box, so a representative velocity gradient is $2v/D$. The resisting viscous stress σ acting on the side of the descending slab is then

$$\sigma = \mu \cdot 2v/D$$

This is a force per unit area. We get the force per unit length (in the third dimension) by multiplying σ by the vertical length, D, of the slab:

$$R = D\sigma = D \cdot 2\mu v/D = 2\mu v \tag{8.2.3}$$

The buoyancy and resistance are balanced when $B + R = 0$. From (8.2.1) and (8.2.3), this occurs when

$$v = -g \cdot Dd\rho\alpha T/4\mu \tag{8.2.4}$$

This expression for v also involves d. We can combine Equations (8.2.2) and (8.2.4) to solve for the two unknowns v and d. The result is

$$v = D\left(\frac{g\rho\alpha T\sqrt{\kappa}}{4\mu}\right)^{2/3} \tag{8.2.5}$$

Using $D = 3000\,\mathrm{km}$, $\rho = 4000\ \mathrm{kg/m^3}$, $\alpha = 2 \times 10^{-5}/^\circ\mathrm{C}$, $T = 1400\,^\circ\mathrm{C}$, $\kappa = 10^{-6}\,\mathrm{m^2/s}$ and $\mu = 10^{22}\,\mathrm{Pa\,s}$, this yields $v = 2.8 \times 10^{-9}\,\mathrm{m/s} = 90\,\mathrm{mm/a}$. This is quite a good estimate of the velocity of the faster plates.

Other quantities can be estimated from these results. From Equation (8.2.2), the thickness of the lithosphere is 33 km. This is of the same order of magnitude as the observed oceanic lithosphere, though about a factor of two too small. If we had used the more accurate estimate of $d = 2\sqrt{(kt)}$ that is obtained from the error function solution for the cooling lithosphere (Equation (7.3.3)), we would have obtained 66 km. Also our estimate of the time the lithosphere spent cooling at the surface is a bit small, because we assumed implicitly in Figure 8.1b that the plate is only as wide as the mantle is deep, that is about 3000 km. At a velocity of 90 mm/a = 90 km/Ma, the plate will be only 33 Ma old when it subducts. Observed lithosphere of this age is about 60 km thick. If the box were longer, the plate would be older and thicker. This problem is left as an Exercise.

The surface heat flux, q, can also be estimated from the temperature gradient through the boundary layer: $q = KT/d$, where K is the thermal conductivity. Using $K = 3\,\mathrm{W/m\,K}$, this gives $q = 130\,\mathrm{mW/m^2}$. This compares with an observed heat flux of about $90\,\mathrm{mW/m^2}$ for lithosphere of this age, and a mean heat flux of about $100\,\mathrm{mW/m^2}$ for the whole sea floor.

The point of these estimates is not that they are not very accurate, but that they are of the right order of magnitude. In the absence of the simple theory developed above, one could not make a sensible estimate even of the orders of magnitude to be expected. Given the crudity of the approximations made, the agreement within about a factor of two is very good, perhaps better than is really justified.

The agreement of these estimates with observations suggests that we have a viable theory for mantle convection that explains why plates move at their observed velocities. Think about the significance of that statement for a moment. Plate tectonics is recognised as a fundamental mechanism driving geological processes. Within a few pages, with some simple physics and simple approximations, we have produced a theory that is consistent with some primary observations of plate tectonics (their velocities, thicknesses and heat fluxes). We thus have a candidate theory for the underlying mechanism for a very wide range of geological processes. We will be further testing the viability (and sufficiency) of this theory through much of the rest of this book.

8.3 Scaling and the Rayleigh number

The approximate theory just developed yields not only reasonable numerical estimates of observed quantities, but also information on how these quantities should scale. Thus, for example, according to Equation (8.2.5), if the viscosity were a factor of 10 lower at some earlier time in earth history, the plate velocities would not be 10 times greater, but $10^{2/3} = 4.6$ times greater. Similarly, we can combine Equations (8.2.2) and (8.2.5) and deduce that

$$\left(\frac{D}{d}\right)^3 = \frac{g\rho\alpha T D^3}{4\kappa\mu} \tag{8.3.1}$$

This implies that the boundary layer thickness would have been 2.15 times less (15 km) and the heat flow 2.15 times higher (275 mW/m^2) with a viscosity 10 times lower.

Equation (8.3.1) is written in this particular form to make a more general point. The left side involves a ratio of lengths, and it is therefore dimensionless. One can work through the dimensions of the right side and confirm that it is also dimensionless, as it should be. This particular, rather arbitrary looking, collection of constants actually encapsulates the scaling properties that we have just looked at, and others besides. In fact it encapsulates many of the scaling properties of convection in a fluid layer in general, not just the mantle convection we are concerned with here. For this reason it has been recognised by fluid dynamicists as having a fundamental significance for all forms of thermal convection. It was Lord Rayleigh who first demonstrated this, and this dimensionless combination (without the numerical factor) is known as the Rayleigh number in his honour. It is usually written

$$Ra = \frac{g\rho\alpha T D^3}{\kappa\mu} \tag{8.3.2}$$

For the mantle, using values used in the last section, we can estimate that $Ra \approx 3 \times 10^6$.

We can see explicitly the way in which the Rayleigh number encapsulates the scaling properties by rewriting the above results in terms of Ra. Thus, from Equation (8.3.1),

$$d/D \sim Ra^{-1/3} \tag{8.3.3}$$

where '$\sim$' implies proportionality and 'of the order of'. The ratio d/D is obviously dimensionless also, and we can view this ratio as a way of scaling d, relative to a length scale that is characteristic of the problem, namely the depth of the fluid layer, D. Similarly, from Equation (8.2.5)

$$v(D/\kappa) \equiv v/V \sim Ra^{2/3} \tag{8.3.4}$$

The dimensions of κ are (length^2/time), so the ratio κ/D has the dimensions of velocity. We can thus regard $V = \kappa/D$ as a velocity scale characteristic of the problem. Then Equation (8.3.4) shows how the actual flow velocity v relates to the velocity scale V derived from the geometry of the problem and the properties of the material.

Fluid dynamicists are enamoured of these dimensionless ratios, for the very good reason that they encapsulate important scaling information, and they have named lots of them after people. Thus the combination vD/κ is called the Peclet number, written Pe:

$$Pe \equiv vD/\kappa = v/V \qquad (8.3.5)$$

Then Equation (8.2.5) reduces to $Pe \sim Ra^{2/3}$. Using values from the last section, we can estimate that for the mantle $Pe \approx 9000$.

I will not go through an exhaustive catalogue of these dimensionless numbers here, but a couple of further examples are worth noting. First, it is instructive to combine the scaling quantities V and D to define a characteristic time:

$$t_\kappa \equiv D/V = D^2/\kappa \qquad (8.3.6)$$

From Chapter 7, this can be recognised as a diffusion time scale. It is an estimate of the time it would take the fluid layer to cool significantly by thermal diffusion, that is by conduction, in the absence of convection. Compare this with a time scale that is more characteristic of the convection process: $t_v = D/v$. This is the time it takes the fluid to traverse the depth of the fluid layer at the typical convective velocity, v, so it can be called the transit time. From Equations (8.3.4) and (8.3.6),

$$t_v = D/v = t_\kappa Ra^{-2/3} \qquad (8.3.7)$$

If $Ra = 3 \times 10^6$, then $t_v = 5 \times 10^{-5} t_\kappa$. Thus if Ra is large, t_v is much smaller than t_κ, reflecting the fact that, at high Rayleigh numbers, convection is a much more efficient heat transport mechanism than conduction.

Actually Equation (8.3.7) indicates that t_κ is not a very useful time scale for convection processes, since it is a measure of thermal conduction. A better one would be that given by the second equality in Equation (8.3.7). Thus we can define a time scale characteristic of convection as

$$t_v \equiv (D^2/\kappa)Ra^{-2/3} \qquad (8.3.8)$$

To complete this discussion of scaling for now, we will return to the heat flux, estimated in the last section from $q = KT/d$. Using Equation (8.3.3), you can see that

$$q = (KT/D)Ra^{1/3} \tag{8.3.9}$$

Again you can recognise (KT/D) as a scaling quantity. In this case it is the heat that would be conducted across the fluid layer (not the boundary layer) if the base were held at the temperature T and the surface at $T = 0$. In other words, it is the heat that would be conducted in the steady state in the absence of convection. Denote this as q_K. The ratio q/q_K is known as the Nusselt number, denoted as Nu:

$$Nu \equiv q/q_K = qD/KT \tag{8.3.10}$$

Then Equation (8.3.9) reduces to

$$Nu \sim Ra^{1/3} \tag{8.3.11}$$

Thus the Nusselt number is a direct measure of the efficiency of convection as a heat transport mechanism relative to conduction. For the mantle, $Nu \approx 100$. In other words, mantle convection is about two orders of magnitude more efficient at transporting heat than conduction would be.

8.4 Marginal stability

Traditional treatments of convection often begin with an analysis of marginal stability, which is the analysis of a fluid layer just at the point when convection is about to begin. This approach reflects the historical development of the topic, and the fact that the mathematics of marginal stability has yielded analytical solutions. The mantle is far from marginal stability, as we will see, and so I began the topic of convection differently, with the more directly relevant 'finite amplitude' convection problem.

Nevertheless the marginal stability problem gives us some important physical insights into convection and the Rayleigh number. However, many treatments of it give long and intricate mathematical derivations and do not always make the physics clear. I will err in the other direction, keeping the mathematics as simple as possible and endeavouring to clarify the physics.

The marginal stability problem arises from the fact that, for a fluid layer heated uniformly on a lower horizontal boundary, there is a minimum amount of heating below which convection does not occur. If the temperature at the bottom is initially equal to the temperature at the top, then of course there will be no convection. Now if the bottom temperature is slowly increased, still there will be no convection, until some critical temperature difference is reached, at which point slow convection will begin. At this point, the fluid layer has just become unstable and begins to overturn. The transition, just at the point of instability, is called marginal stability. Lord Rayleigh [3] was the first to provide a mathematical analysis of this. He showed that marginal stability occurs at a critical value of the Rayleigh number. The critical value depends on the particular boundary conditions and other geometric details, but is usually of the order of 1000. The mathematical analysis of marginal stability is reproduced by Chandrasekhar [4] and by Turcotte and Schubert [2] (p. 274).

Consider the two layers of fluid sketched in Figure 8.2. The lower layer is less dense, and the interface between them has a bulge of height h and width w. Take h to be quite small. This bulge is buoyant relative to the overlying fluid, and its buoyancy is approximately

$$B = g\Delta\rho wh$$

per unit length in the third dimension. Its buoyancy will make it grow, so that its highest point rises with some velocity $v = \partial h/\partial t$, and its growth will be resisted by viscous stresses.

The viscous resistance will have different forms, depending on whether the width of the bulge is smaller or larger than the layer depth D. If $w \ll D$, the dominant shear resistance will be proportional to the velocity gradient v/w. The resisting force is then

$$R_s = \mu(v/w)w = \mu v = \mu\partial h/\partial t$$

where v/w is a characteristic strain rate and the subscript 's' denotes small w. Equating B and R_s to balance the forces yields

$$\frac{\partial h}{\partial t} = \frac{g\Delta\rho w}{\mu} h \tag{8.4.1}$$

which has the solution

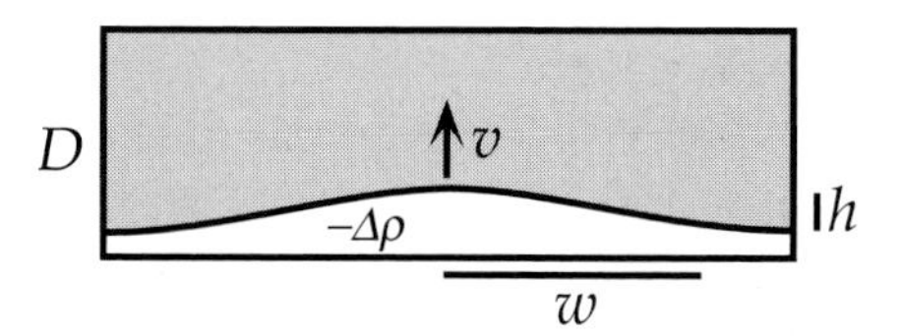

Figure 8.2. Sketch of two layers of fluid with the denser fluid above and with an undulating interface that is unstable.

$$h = h_0 \exp(t/\tau_s) \tag{8.4.2}$$

where h_0 is a constant and

$$\tau_s = \frac{\mu}{g\Delta\rho w} \tag{8.4.3}$$

In other words, the bulge grows exponentially with a time constant τ_s, because the interface is unstable: the lighter fluid wants to rise to the top. This kind of instability is called the Rayleigh–Taylor instability. It occurs regardless of the reason for the density difference between the two fluids.

Notice that τ_s gets smaller as w gets bigger. That is, broader bulges grow more quickly. However, there is a limit to this: when the width of the bulge is comparable to the depth, D, of the fluid layer, the top boundary starts to interfere with the flow and to increase the viscous resistance. If w is much larger than D, then the dominant viscous resistance comes from horizontal shear flow with velocity u along the layer. By conservation of mass, $uD = vw$. The characteristic velocity gradient of this shear flow is then $u/D = vw/D^2$. The resulting shear stress acts across the width w of the bulge, so the resisting force in this case is

$$R_l = \mu(u/D)w = \mu v w^2/D^2$$

where subscript 'l' denotes large w. Balancing R_l and B then yields

$$\frac{\partial h}{\partial t} = \frac{g\Delta\rho D^2}{\mu w} h \tag{8.4.4}$$

which has the same form as Equation (8.4.1) except for the constants. It also has the same form of exponentially growing solution (Equation (8.4.2)), but with a different time scale τ_l:

$$\tau_l = \frac{\mu w}{g\Delta\rho D^2} \tag{8.4.5}$$

Notice here that τ_l gets *bigger* for larger w, whereas τ_s gets smaller, and their values are equal when $w = D$. We have considered the two extreme cases $w \ll D$ and $w \gg D$. As w approaches D from either side, the time scale of the growth of the instability gets smaller. This implies that the time scale is a minimum near $w = D$. In other words, a bulge whose horizontal scale is $w = D$ is the fastest growing bulge, and its growth time scale is

$$\tau_{\mathrm{RT}} = \frac{\mu}{g\Delta\rho D} \tag{8.4.6}$$

where the subscript 'RT' connotes the Rayleigh–Taylor time scale. A more rigorous analysis that yields this result is given by Turcotte and Schubert [2] (p. 251). The implication of this result is that if there are random small deviations of the interface from being perfectly horizontal, deviations that have a width comparable to the layer depth will grow exponentially with the shortest time scale and will quickly come to dominate. As a result, the buoyant layer will form into a series of rising blobs with a spacing of about $2w$.

Now let us consider the particular situation in which the density difference is due to the lower layer having a higher temperature because the bottom boundary of the fluid is hot. Then the density difference would be $\Delta\rho = \rho\alpha\Delta T$, where ΔT is a measure of the average difference in temperature between the layers. Suppose first that the thermal conductivity of the fluid is high and the growth of the bulge is negligibly slow: then temperature differences would be quickly smeared out by thermal diffusion. In the process, the bulge would be smeared out. After a time the temperature would approach a uniform gradient between the bottom and top boundaries, and the bulge would have ceased to exist.

However, I showed above that the bulge grows because of its buoyancy. Evidently there is a competition between the buoyancy and the thermal diffusion. We can characterise this competition in terms of the time scales of the two processes: τ_{RT} for the buoyant growth and τ_κ for the thermal diffusion, where

$$\tau_\kappa = D^2/\kappa \tag{8.4.7}$$

We can use D as a measure of the distance that heat must diffuse in order to wipe out the fastest growing bulge. In order for the bulge to grow, τ_{RT} will need to be significantly less than τ_κ. From Equations (8.4.6) and (8.4.7), this condition is

$$\frac{g\Delta\rho D^3}{\kappa\mu} = Ra \geq c \tag{8.4.8}$$

where c is a numerical constant and you can recognise the left-hand side of Equation (8.4.8) as the Rayleigh number.

This result tells us that there is indeed a value of the Rayleigh number that must be exceeded before the thermal boundary layer can rise unstably in the presence of continuous heat loss by thermal diffusion. If it cannot, there will be no thermal convection. Thus we have derived the essence of Rayleigh's result. In this case, we do not get a very good numerical estimate of the critical value of the Rayleigh number, since a rigorous stability analysis yields $c \approx 1000$, rather than $c \approx 1$.

The quantitative value may not be very accurate, but we have been able to see that the controlling physics is the competition between the Rayleigh–Taylor instability and thermal diffusion (the Rayleigh–Taylor instability involving an ever-changing balance between buoyancy and viscous resistance). In fact, you can see now that the Rayleigh number is just the ratio of the time scales of these two processes:

$$Ra = \frac{\tau_\kappa}{\tau_{\mathrm{RT}}} \tag{8.4.9}$$

The mantle Rayleigh number is at least 3×10^6, well above the critical value of about 1000. This indicates that the mantle is well beyond the regime of marginal stability. One way to look at this, using Equation (8.4.9), is that the thermal diffusion time scale is very long, which means that heat does not diffuse very far in the time it takes the fluid to become unstable and overturn. This means that the thermal boundary layers will be thin compared with the fluid layer thickness.

Thin boundary layers were assumed without comment in the simple theory of convection given in Section 8.2. That theory actually is most appropriate with very thin boundary layers, that is at very high Rayleigh numbers. For this reason it is known as the boundary layer theory of convection. Thus the marginal stability theory applies just above the critical Rayleigh number, while the boundary layer theory applies at the other extreme of high Rayleigh number.

8.5 Flow patterns

In a series of classic experiments, Benard [5] observed that, in a liquid just above marginal stability, the convection flow formed a system of hexagonal cells, like honeycomb, when viewed from above. Considerable mathematical effort was devoted subsequently to trying to explain this. It was presumed that it must imply that hexagonal cells are the most efficient at convecting heat. It turned out that the explanation for the hexagons lay in the effect of surface tension in the experiments, and specifically on differences in surface tension accompanying differences in temperature. Surface tension was important because Benard's liquid layers were only 1 mm or less in thickness.

There is an important lesson here. If a factor like the temperature-dependence of surface tension could so strongly influence the horizontal pattern, or 'planform', of the convection, then the fluid must not have a strong preference for a particular planform; that is, different planforms must not have much influence on the efficiency of the convection. The implication is that, in other situations, other factors influencing the material properties of the fluid in the boundary layers might also have a strong influence on planform.

Pursuing this logic, if the top and bottom thermal boundary layers in a fluid layer should have material properties that are distinctly different from each other, then each may tend to drive a distinctive pattern of convection. What then will be the resulting behaviour? The possibility of the different thermal boundary layers tending to have different planforms is not made obvious in standard treatments of convection. Whether it occurs depends both on the physical properties of the fluid and on the mode of heating, which we will look at next.

In the mantle, a hot boundary layer does have distinctly different mechanical properties from a cold boundary layer, and the two seem to behave quite differently. As well, the cold boundary layer in the earth is laterally heterogeneous, containing continents and so on, and it develops other heterogeneities in response to deformation: it breaks along faults. The effects of material properties on flow patterns are major themes of the next three chapters, which focus on the particular case of the earth's mantle.

8.6 Heating modes and thermal boundary layers

Textbook examples of convection often show the case of a layer of fluid heated from below and cooled from above. In this case there is a hot thermal boundary layer at the bottom and a cool thermal boundary layer at the top (Figure 8.3a). If, as well, the Rayleigh number is not very high, the resulting pattern of flow is such that each of the thermal boundary layers reinforces the flow driven by the other one. In other words the buoyant upwellings rise between the cool downwellings, so that a series of rotating 'cells' is formed which are driven in the same sense of rotation from both sides. This cooperation between the upwellings and downwellings disguises the fact that the boundary layers are dynamically separate entities. It is possible that they might drive different flow patterns, as I intimated in the last section. It is also possible that one of the thermal boundary layers is weak or absent.

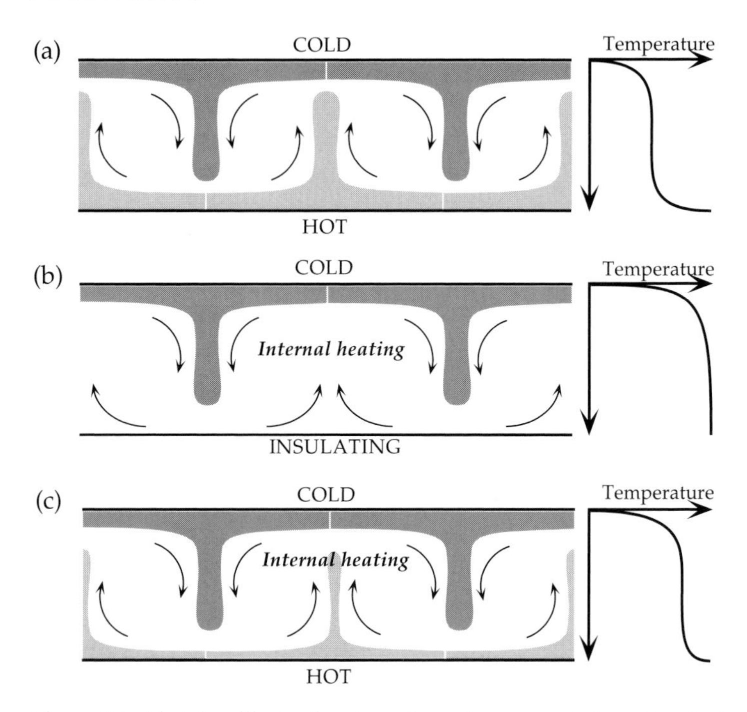

Figure 8.3. Sketches illustrating how the existence and strength of a lower thermal boundary layer depend on the way in which the fluid layer is heated.

For example, a fluid layer might be heated from within by radioactivity. If there is no heat entering the base, perhaps because it is insulating, then there will be no hot thermal boundary at the bottom. If the fluid layer is still cooled from the top, the only thermal boundary layer will be the cool one at the top (Figure 8.3b). In fact this was assumed, without comment, in the simple theory of convection presented in Section 8.2. In this case, the cool fluid sinking from the top boundary layer still drives circulation, but the *upwelling* is *passive*. By this I mean that although the fluid flows upwards between the downwellings (Figure 8.3b), it is not buoyant relative to the well-mixed interior fluid. It is merely being displaced to make way for the actively sinking cold fluid.

Although this may seem to be a trivial point here, it has been very commonly assumed, for example, that because there is clearly upwelling occurring under midocean ridges, the upwelling mantle material is hotter than normal and thus buoyant and 'actively' upwelling. We will see evidence in Chapter 10 that this is usually not true. A lot of confusion about the relationship between mantle convection and continental drift and plate tectonics can be avoided by keeping this simple point clearly in mind.

More generally, the heat input to the fluid layer might be a combination of heat entering from below and heat generated within (by radioactivity, in the case of the mantle), and states intermediate between those of Figures 8.3a and 8.3b will result (Figure 8.3c). Suppose, as implied in Figure 8.3a, that the temperature of the lower boundary is fixed. If there is no internal heating, then the temperature profile will be like that shown to the right of Figure 8.3a. If there is no heating from below, the internal temperature will be the same as the bottom boundary, as shown to the right of Figure 8.3b. If there is some internal heating, then the internal temperature will be intermediate, as in Figure 8.3c. As a result, the top thermal boundary layer will be stronger (having a larger temperature jump across it) and the lower thermal boundary layer will be correspondingly weaker. The mantle seems to be in such an intermediate state, as we will see.

The point is illustrated by numerical models in Figure 8.4. The left three panels are frames from a model with a prescribed bottom temperature and no internal heating. You can see both cool sinking columns and hot rising columns. The right three panels are from an internally heated model, and only the upper boundary layer exists. Downwellings are active, as in the bottom-heated model, but the upwellings are passive, broad and slow. Away from downwellings, isotherms are nearly horizontal, and the fluid is stably stratified. This is because the coolest fluid sinks to the bottom, and is then

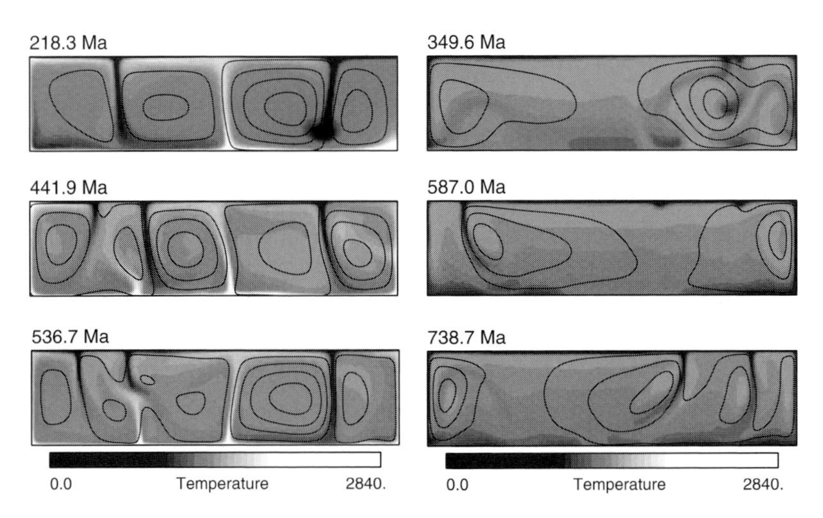

Figure 8.4. Frames from numerical models, illustrating the differences between convection in a layer heated from below (left-hand panels) and in a layer heated internally (right-hand panels). (Technical specifications of these models are given in Appendix 2.)

slowly displaced upwards by later cool fluid as it slowly warms by internal heating.

Figure 8.4 illustrates two other important points. First, the flow is unsteady. This is characteristic of convection at high Rayleigh numbers in constant-viscosity fluids. It is because the heating is so strong that the boundary layers become unstable before they have travelled a distance comparable to the depth of the fluid, which is the width of cells that allows the most vertical limbs while also minimising the viscous dissipation. Incipient instabilities in the top boundary layer are visible in the middle right panel of Figure 8.4. By the last panel they have developed into full downwellings.

Second, the two thermal boundary layers in the left sequence are behaving somewhat independently, especially on the left side of the panels. In fact in the bottom panel an upwelling and a downwelling are colliding. This illustrates the point made earlier that each boundary layer is an independent source of buoyancy, and they may interact only weakly. This becomes more pronounced at higher Rayleigh numbers.

8.6.1 Other Rayleigh numbers [*Advanced*]

We have so far specified the thermal state of the convecting fluid in terms of *temperatures* prescribed for each boundary. However, in Figures 8.3b and 8.4 (right panels) the bottom boundary is specified as *insulating*, that is as having zero heat flux through it, and the heating is specified as being internal. The temperature is not specified ahead of time. It is evident that this model is specified in terms of heat input, rather than in terms of a temperature difference between the boundaries. How then can the Rayleigh number be defined?

The philosophy of the dimensional estimates used in this chapter is that *representative* quantities are used. With appropriate choices, order-of-magnitude estimates will (usually) result. The Rayleigh number defined by Equation (8.3.2) is defined in terms of such representative quantities. This suggests that we look for *representative* and *convenient* measures in different situations.

We lack a representative temperature difference for the situation in Figure 8.3b, but we can assume that a heat flux, q, is specified. One way to proceed is to derive a quantity from q that has the dimensions of temperature; for example, we can use the temperature difference, ΔT_q, across the layer that would be required to *conduct* the specified heat flux, q:

$$\Delta T_q = qD/K$$

We can then define a new Rayleigh number as

$$R_q = \frac{g\rho\alpha D^3 \Delta T_q}{\kappa\mu} = \frac{g\rho\alpha q D^4}{K\kappa\mu} \quad (8.6.1)$$

This Rayleigh number is useful in any situation in which it is the heat input rather than a temperature difference that is specified.

It is possible in principle that some heat, say q_b, is specified at the base, and some is specified to be generated internally. If the internal heating is uniform, and generated at the rate H per unit volume of fluid, then the rate of internal heat generation per unit area of the layer surface is HD. The total heat input will then be

$$q = q_b + HD$$

Although in a laboratory setting it is not easy to prescribe a heat flux, it is easy in numerical experiments and it is useful to make the conceptual distinction between the two kinds of bottom thermal boundary layer: prescribed temperature and prescribed heat flux.

The Rayleigh numbers R_q (Equation (8.6.1)) and Ra (Equation (8.3.2)) are distinct quantities with different numerical values, as we will see, and this is why different symbols are used here for them. However they are also related. Recall that the Nusselt number, Nu, was defined as the ratio of actual heat flux, q, to the heat flux, q_K, that would be conducted with the same temperature difference across the layer (Equation (8.3.10)). In the case considered earlier, it was q_K that was specified ahead of time and q that was determined by the behaviour of the fluid layer. Here it is the reverse. However we can still use this definition of Nu. Thus, if the actual temperature difference across the layer that results from the convection process is ΔT, then $q_K = K\Delta T/D$ and

$$Nu = q/q_K = \Delta T_q/\Delta T \quad (8.6.2)$$

Thus here the Nusselt number gives the ratio of the temperature difference, ΔT_q, that would be required to conduct the heat flux q through the layer, to the actual temperature difference in the presence of convection.

Similarly, although ΔT is not known ahead of time here, it can still be used conceptually to define the Rayleigh number Ra (Equation (8.3.2)). It is then easy to see the relationship between Ra and R_q:

$$\frac{R_q}{Ra} = \frac{\Delta T_q}{\Delta T} = Nu \tag{8.6.3}$$

In the earlier discussion of scaling, we found that $Nu \sim Ra^{1/3}$, so $R_q \sim Ra^{4/3}$. Thus if Ra has the value 3×10^6 estimated earlier, for example, then R_q will be about 4.3×10^8. Thus R_q is numerically larger than Ra. Nevertheless it is a convenient way to characterise cases where it is the heat flux that is specified, rather than the temperature difference. You must of course be careful about which definition of Rayleigh number is being used in a given context, as they have different scaling properties as well as different numerical values.

This discussion illustrates the general point that different Rayleigh numbers may be defined in different contexts. There is nothing profound about this, it is merely a matter of adopting a definition that is convenient and relevant for the context, so that it encapsulates the scaling properties of the particular situation.

For the earth's mantle, however, there is a complication. An appropriate way of specifying the heat input into models of the mantle is through a combination of internal heating from radioactivity and a prescribed temperature at the base. Although the value of the temperature at the base of the mantle is not well known, the liquid core is believed to have a low viscosity, so that it would keep the temperature quite homogeneous. This means the core can be viewed as a heat bath imposing a uniform temperature on the base of the mantle. This combination of a heating rate and a prescribed uniform bottom temperature is not covered by either of the Rayleigh numbers Ra or R_q, so there is not a convenient a priori thermal prescription of mantle models. In the mantle it is the heat *output*, at the top surface that is well-constrained. This means that some trial and error may be necessary to obtain models that match the observed heat output of the mantle.

8.7 Dimensionless equations [*Advanced*]

The equations governing convection are often put into dimensionless form, that is they are expressed in terms of dimensionless variables. This is done to take advantage of the kind of scaling properties that we have been looking at, because one solution can then be scaled to a variety of contexts. There are different ways in which this can be done. We have seen an example of this already, in the different Rayleigh numbers that can be defined, depending on the way the fluid is heated. Other alternatives are more arbitrary. For example, two different time scales are commonly invoked, and others are possible. Since these alternatives are not usually presented systematically, I will do so here.

The equations governing the flow of a viscous incompressible fluid were developed in Chapter 6 (Equation (6.6.3)), and the equation governing heat flow with advection, diffusion and internal heat generation was developed in Chapter 7 (Equation (7.8.2)). The following dimensional forms of these equations are convenient here.

$$\frac{\partial \tau_{ij}}{\partial x_j} - \frac{\partial P}{\partial x_i} = -B_i = \rho g_i \tag{8.7.1}$$

$$\frac{\mathrm{D}T}{\mathrm{D}t} \equiv \frac{\partial T}{\partial t} + v_i \frac{\partial T}{\partial x_i} = \kappa \nabla^2 T + \frac{A}{\rho C_P} \tag{8.7.2}$$

In Equation (8.7.1), the buoyancy force B_i (positive upwards), is written in terms of the density and the gravity vector g_i (positive downwards). In Equation (8.7.2), the first derivative, $\mathrm{D}T/\mathrm{D}t$, is known as the total derivative, and its definition is implicit in the first identity of that equation. A is the internal heat production per unit time, per unit volume.

Three scaling quantities suffice to express these equations in dimensionless form: a length, a temperature difference and a time. For length, an appropriate choice is usually D, the depth of the convection fluid layer. Using this, we can define dimensionless position coordinates, x_i, for example, such that

$$x_i' = D x_i$$

where I have *changed notation*: the prime denotes a dimensional quantity and unprimed quantities are dimensionless, unless specifically identified as a dimensional scaling quantity, like D.

For temperature, we have seen in the last section two possible choices:

$$\Delta T = \Delta T_T = (T_\mathrm{b} - T_\mathrm{s}) \tag{8.7.3}$$

$$\Delta T = \Delta T_q = qD/K \tag{8.7.4}$$

For the moment, I will retain the general notation ΔT to cover both of these possibilities.

A time scale that is often used is the thermal diffusion time scale of Equation (8.3.6): $t_\kappa = D^2/\kappa$. Another one sometimes used is

t_κ/Ra. A third possibility emerged from the earlier discussion of scaling, namely the transit time $t_v = t_\kappa/Ra^{2/3}$ (Equations (8.3.7), (8.3.8)). Here I will carry all three possibilities by using a general time scale t_n, where

$$\begin{aligned} t_1 &= t_\kappa = D^2/\kappa \\ t_2 &= t_\kappa/Ra \\ t_3 &= t_v = t_\kappa/Ra^{2/3} \end{aligned} \tag{8.7.5}$$

Dimensional scales can be derived from D, ΔT and t_n for viscous stress, buoyancy and heat generation rate as follows. Viscous stress is viscosity times velocity gradient, so an appropriate scale is $\mu(D/t_n)/D = \mu/t_n$. Buoyancy per unit volume is $g\Delta\rho = g\rho_0\alpha\Delta T$. Using these scales in Equation (8.7.1) yields

$$\frac{\mu}{Dt_n}\left(\frac{\partial\tau_{ij}}{\partial x_j} - \frac{\partial P}{\partial x_i}\right) = g\Delta\rho(\rho g_i)$$

that is

$$\frac{\partial\tau_{ij}}{\partial x_j} - \frac{\partial P}{\partial x_i} = R_{\mathrm{F}}(\rho g_i) \tag{8.7.6}$$

where R_{F} denotes a dimensionless combination of constants in the force balance equations:

$$R_{\mathrm{F}} = \frac{g\Delta\rho D t_n}{\mu} \tag{8.7.7}$$

Similarly, for Equation (8.7.2) we need a scale for heat generation. The heat flux scale identified earlier (Equations (8.3.9) and (8.3.10)) is q_K, the heat flux that would be conducted with the same temperature difference. The heat generation rate per unit volume that corresponds to this is $q_K/D = K\Delta T/D^2$. Then Equation (8.7.2) becomes

$$\frac{\Delta T}{t_n}\left(\frac{\mathrm{D}T}{\mathrm{D}t}\right) = \frac{\kappa\Delta T}{D^2}(\nabla^2 T) + \left(\frac{K\Delta T}{\rho C_P D^2}\right)A$$

Remembering that $K/\rho C_P = \kappa$, this can be written

$$\frac{\mathrm{D}T}{\mathrm{D}t} = R_{\mathrm{H}}(\nabla^2 T + A) \tag{8.7.8}$$

where R_{H} denotes a dimensionless combination in the heat equation:

$$R_{\mathrm{H}} = \frac{\kappa t_n}{D^2} \tag{8.7.9}$$

Equations (8.7.6) and (8.7.8) are dimensionless versions of the flow and heat equations, and they involve the two dimensionless ratios R_{F} and R_{H}. The three choices of time scale proposed in Equations (8.7.5) then yield

$$t_n = t_1 : \quad R_{\mathrm{F}} = Ra \quad R_{\mathrm{H}} = 1 \tag{8.7.10a}$$

$$t_n = t_2 : \quad R_{\mathrm{F}} = 1 \quad R_{\mathrm{H}} = 1/Ra \tag{8.7.10b}$$

$$t_n = t_3 : \quad R_{\mathrm{F}} = Ra^{1/3} \quad R_{\mathrm{H}} = 1/Ra^{2/3} \tag{8.7.10c}$$

The choice of time scale is mainly a matter of convenience. With the choice t_3, one dimensionless time unit will correspond approximately with a transit time, regardless of the Rayleigh number, and it will be easier to judge the progress of a numerical calculation. On the other hand, the choice between ΔT_T and ΔT_q depends on the mode of heating of the fluid. The notation thus refers to a more substantial difference in the model than convenience, and more care must be taken to ensure the proper interpretation of results of calculations.

8.8 Topography generated by convection

The topography generated by convection is of crucial importance to understanding mantle convection, since the earth's topography provides some of the most important constraints on mantle convection. Here I present the general principle qualitatively. The particular features of topography to be expected for mantle convection, and their quantification and comparison with observations, will be given in following Chapters. We have already covered one important example in Chapter 7, the subsidence of the sea floor.

The central idea is that buoyancy does two things: it drives convective flow and it vertically deflects the horizontal surfaces of the fluid layer. Because the buoyancy is (in the thermal convection of most interest here) of thermal origin, there are intimate relationships between topography, fluid flow rates and heat transport rates.

The principle is illustrated in Figure 8.5. This shows a fluid layer with three buoyant blobs, labelled (a), (b) and (c). Blob (a) is close to the top surface and has lifted the surface. The surface uplift is required by Newton's laws of motion. If there were no force opposing the buoyancy of the blob, the blob would continuously accelerate. Of course there are viscous stresses opposing the blob locally, but these only shift the problem. The fluid adjacent to the blob opposes the blob, but then this fluid exerts a force on fluid further out. In other words, the viscous stresses *transmit* the force through the fluid, but do not result in any net opposing force. This comes from the deflected surface.

There is, in Figure 8.5, blob (a), a simple force balance: the weight of the topography balances the buoyancy of the blob. Geologists might recognise this as an *isostatic* balance. Another way to think of it is that the topography has negative buoyancy, due to its higher density than the material it has displaced (air or water, in the case of the mantle). Recalling the definition of buoyancy given earlier (Equation (8.1.1)), this implies that the excess mass of the topography equals the mass deficiency of the blob.

As I have already stressed, there is in this very viscous system an *instantaneous* force balance, even though the blob is moving. Such topography has sometimes been referred to as 'dynamic topography', but this terminology may be confusing, because it may suggest that momentum is involved. It is not. The balance is a static (strictly, a quasi-static), instantaneous balance. The 'dynamic' terminology derives from the term 'dynamic stresses', which means the stresses due to the motion, which are the viscous stresses. While this terminology may be technically correct, it is not very helpful, because it may obscure the fact that there is a simple force balance

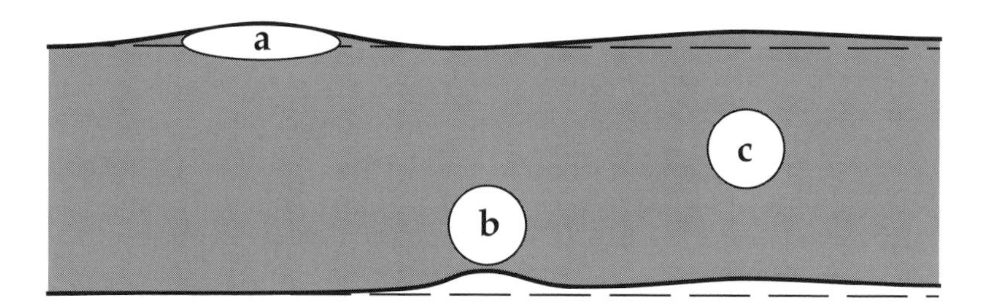

Figure 8.5. Sketch of the effects of buoyant blobs on the surfaces of a fluid layer. The layer surfaces are assumed to be free to deflect vertically, with a less dense fluid (e.g. air or water) above, and a more dense fluid (e.g. the core) below.

involved, and it may make the problem seem more complicated than it really is.

Blob (b) in Figure 8.5 is near the bottom of the fluid layer. It causes the bottom surface of the fluid to deflect. This is because the viscous stresses caused by the blob are larger close to the blob than far away, so the main effect is on the nearby bottom surface. I have implicitly assumed in Figure 8.5 that there is a denser fluid below the bottom surface, such as the core under the mantle. In this case, the topography causes denser (core) material to replace less dense (mantle) material. Thus it generates a downward compensating force, or negative buoyancy, just as does topography on the top surface. This force balances the buoyancy of blob (b).

Does blob (b) cause any deflection of the top surface? Yes, there will be a small deflection over a wide area of the surface. Blob (c) makes this point more explicitly: it is near the middle of the layer, and it deflects both the top and the bottom surfaces by similar amounts. In this case, we can see that the force balance is actually between the positive buoyancy of the blob and the *two* deflected surfaces. In fact this will always be true, even for blobs (a) and (b), but I depicted them close to one surface or the other to simplify the initial discussion, since this makes the deflection of one surface negligible.

To summarise the principle, buoyancy in a fluid layer deflects both the top and the bottom surfaces of the fluid (supposing they are deformable), and the combined weight of the topographies balances the internal buoyancy. The amount of deflection of each surface depends on the magnitude of the viscous stresses transmitted to each surface. This depends on the distance from the buoyancy to the surface. It also depends on the viscosity of the intervening fluid, a point that will be significant in following chapters.

Now apply these ideas to the thermal boundary layers we were considering above. The top thermal boundary layer is cooler and denser than the ambient interior fluid, so it is negatively buoyant and pulls the surfaces down. Because it is adjacent to the top fluid surface, it is this surface that is deflected the most. There will be, to a good approximation, an isostatic balance between the mass excess of the thermal boundary layer and the mass deficiency of the depression it causes. The result is sketched in Figure 8.6 in a form that is like that of the mantle. The topography on the left is highest where the boundary layer is thinnest. Away from this in both directions, the surface is depressed by the thicker boundary layer.

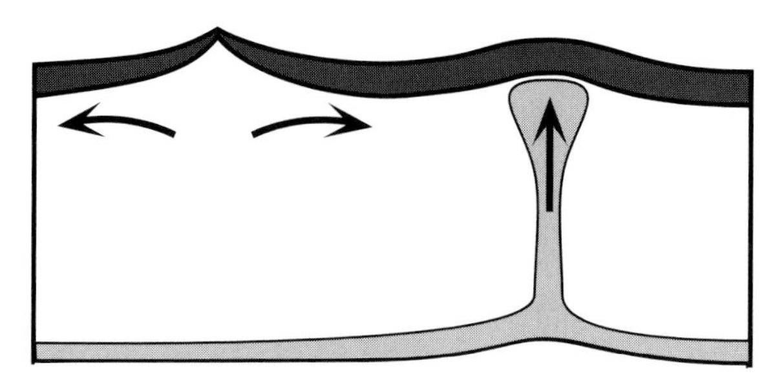

Figure 8.6. Sketch of two types of topography on the top surface of a convecting fluid layer. The top thermal boundary layer cools, thickens and subsides by thermal contraction as it moves away from the spreading centre at left, leaving a topographic high where it is thin. The bottom thermal boundary layer generates no topography on the top surface until material from it rises to the top, where it raises the top surface (upwelling on right).

On the other hand, the bottom thermal boundary layer is adjacent to the bottom surface of the fluid, and generates topography there (Figure 8.6). It does not generate significant topography on the top surface except where a buoyant column has risen to the top of the fluid layer. There the top surface is lifted. Thus it is possible for the bottom thermal boundary layer to generate topography on the top surface, but only after material from it has risen to the top.

There is an important difference between the two topographic highs in Figure 8.6. The high on the left has no 'active' upwelling beneath it: it is high because the surface on either side of it has subsided, because of the negative buoyancy of the top thermal boundary layer. In contrast, the high on the right does have an 'active', positively buoyant upwelling beneath it that has lifted it up.

You will see in the following chapters that the forms of convection driven by the two mantle boundary layers are different. As a result, the forms of topography they generate are recognisably different. Because buoyancy is directly involved both in the topography and in the convection, the observed topography of the earth contains important information about the forms of convection present in the mantle.

Even better, the topography contains *quantitative* information about the fluxes of buoyancy and heat involved. This is most readily brought out in the mantle context, where the topographic forms are distinct and lend themselves to extracting this information. However, it should by now be no surprise to you that such information is present, given the intimate involvement of buoyancy, convection and topography.

8.9 References

1. D. L. Turcotte and E. R. Oxburgh, Finite amplitude convection cells and continental drift, *J. Fluid Mech.* **28**, 29–42, 1967.
2. D. L. Turcotte and G. Schubert, *Geodynamics: Applications of Continuum Physics to Geological Problems*, 450 pp., Wiley, New York, 1982.
3. Lord Rayleigh, On convective currents in a horizontal layer of fluid when the higher temperature is on the under side, *Philos. Mag.* **32**, 529–46, 1916.
4. S. Chandrasekhar, *Hydrodynamic and Hydromagnetic Stability*, Oxford University Press, Oxford, 1961.
5. H. Benard, Les tourbillons cellulaires dans une nappe liquide transportant de la chaleur par convection en régime permanent, *Ann. Chim. Phys.* **23**, 62–144, 1901.

8.10 Exercises

1. Use Equations (8.1.1) and (8.1.2) to evaluate the buoyancy of the following. These are meant to be rough estimates, so do not calculate results to more than one or two significant figures.
(a) A ball bearing 1 cm in diameter and with density 7.7 Mg/m^3 in mantle of density 3.3 Mg/m^3.
(b) A plume head with a radius of 500 km and temperature excess of 300 °C in a mantle of density 3.3 Mg/m^3 and thermal expansion coefficient 3×10^{-5}/°C.
(c) A sheet of subducted lithosphere 100 km thick extending to a depth of (i) 600 km, (ii) 3000 km. Calculate a buoyancy per metre in the horizontal direction of the oceanic trench. Assume the slab temperature varies linearly through its thickness from 0 °C to the mantle temperature of 1300 °C; you need only consider its mean temperature deficit. Assume other parameters as above.
(d) Suppose part of the slab just considered included oceanic crust 7 km thick with a density in the mantle of 3.2 Mg/m^3. Calculate its contribution to the slab buoyancy.

2. Repeat the derivation of the approximate formula (8.2.5) for the convection velocity in the model of Figure 8.1, but this time assume that the cell length, L, is not the same as its depth, D. You will need to consider the horizontal and vertical velocities, u and v, to be different, and to relate them using conservation of mass. You will also need to include two terms in the viscous resistance, one

proportional to the velocity gradient $2u/D$ and one proportional to $2v/L$. The answer can be expressed in the form of Equation (8.2.5) with the addition of a factor involving (L/D). Using values from the text, compare the velocity when $L = D = 3000$ km and when $L = 14\,000$ km, the maximum width of the Pacific plate.

CHAPTER 9

Plates

9.1 The mechanical lithosphere

In Chapter 8 we considered convection in a fluid medium. However, the earth's mantle behaves as a fluid only in its interior, where the temperature is high. Near the surface, its viscosity is much higher, so that it is effectively rigid much of the time. This is illustrated schematically in Figure 9.1.

However, as we saw in Chapter 6, with sufficient stress the cooler mantle may yield. Close to the surface, this yielding takes the form of brittle fracture. At intermediate depths, the yielding may be more fluid-like but still result in narrow zones of deformation, which geologists call ductile shear zones. At the large scale in which we are interested here, these narrow shear zones still have the characteristics of fractures or faults, and so we may consider the lithosphere at the large scale to be a brittle solid to a first approximation. The usefulness of this approximation is illustrated, for example, by the three kinds of plate margin, which correspond to the three standard types of faults in structural geology: normal (spreading centre), reverse (subduction zone) and strike-slip (transform fault).

The implication of this 'brittle–ductile transition' is that our convecting medium changes from being effectively a viscous fluid at depth to being a brittle solid near the surface. The material of the mantle flows from one regime to the other, and so ultimately we must consider the mantle as a single medium that undergoes radical changes in properties as it flows around. We will approach this task in Chapters 10 and 11, and we will see that there are some important consequences of these changes of properties. First, however, there are some important aspects of each regime that can be understood separately. Thus in Chapter 8 we looked at convection in a

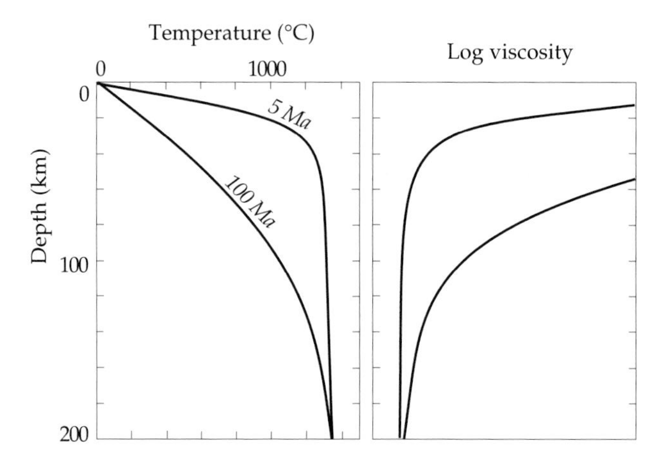

Figure 9.1. Sketch of two oceanic geotherms, ages 5 Ma and 100 Ma, and the corresponding viscosity profiles.

conventional fluid of constant viscosity. In this chapter, we look at some important specific behaviour of the lithosphere that reflects its character as a brittle solid.

We have seen already that it is important sometimes to consider the earth's surface without worrying, for the moment, about what is happening underneath. Thus Wilson's synthesis, in which he defined the plates, was done without reference to mantle convection (Section 3.4), and the description of plate motions in terms of velocity vectors and rotation vectors (Section 3.6.1) was presented in terms of the relative velocities of plates, without reference to any real or conceptual internal frame of reference.

There are two general aspects of the distinctive behaviour of the lithosphere that I want to highlight. One is that the plates, into which the lithosphere is broken, have a range of sizes and rather irregular shapes. These have been illustrated and summarised in Section 4.1. The other aspect is that the geometry of the plates changes in distinctive ways that are not like the ways fluid flow patterns change. The plates evolve steadily, following simple rules, and they may also change suddenly, if a plate breaks into two. These changes are the subject of this chapter.

What we look at in this chapter is the way plates move and change, but not the forces that cause the motions and changes. We are thus considering *kinematics*, the study of motions, as distinct from *dynamics*, the study of the way forces generate motions. Although the term dynamic is often used more loosely in popular parlance to refer to any moving or changing system, this usage is not technically correct. In Chapter 10 we will look at the way the mantle and the plates move in response to buoyancy forces, so there we will be considering dynamics. Similarly, in Chapter 11 we will look at plume dynamics.

I will use the term *plate margin*, rather than *plate boundary*, henceforth. It is useful in order to avoid confusion, since we have been considering internal boundaries in the mantle and thermal boundary layers in convection. Partly out of habit, partly for conciseness, I may use the term *ridge* interchangeably with *spreading centre*. Likewise I may interchange *trench* with *subduction zone*.

9.2 Describing plate motions

At first sight, it may seem that plates will not change much. However, it turns out that plates may grow, shrink, and even disappear without there being any major perturbations to the system, because of the different behaviour of different kinds of plate margin. It also turns out that the way the plates evolve in detail can be rather subtle. On the other hand, much of the time the plates follow a simple set of rules. It is thus possible to deduce fairly precisely how things ought to evolve, and to infer a lot about how the plates have evolved in the past. The rules are simple, but the results can be surprising, so deducing plate evolution sequences requires care in following the rules. This is aided by familiarity with a few ideas and examples, which are the subject of the next few sections.

The objective here is to understand the *kinds* of behaviour that plates exhibit, rather than to present a comprehensive reconstruction of how the plates have evolved. There are many papers on the latter topic. There are also now some lengthier treatments of plate kinematics, in both planar and spherical geometry [1, 2]. More specifically, we look here at the way the plates change their sizes and shapes even when their velocities are approximately constant and no new plate boundaries are forming by the breakup of old plates.

We do not consider in the same detail how new plate margins form, nor what might cause plate velocities to change. These are important questions, but they are not very well understood. This may be surprising, but an important reason is that these processes are not very well constrained by observations. Some important aspects can still be understood in spite of our ignorance of these processes.

The ways that certain parts of the plate system have evolved will be used later to illustrate the kinds of evolution that can be deduced from the rules of plate motion. First, those rules and some of their consequences will be presented.

9.3 Rules of plate motion on a plane

Most of the ideas I want to convey here can be illustrated in planar geometry, rather than spherical geometry. Planar geometry is much more familiar to most people, and it is easier to draw. Later I will briefly outline how plate motions work on a sphere, emphasising mainly the points that are relevant to mantle convection. Others have described the details of spherical plate kinematics [1, 2].

9.3.1 Three margins

Even when plate velocities are constant and no new plate margins are forming, the sizes and shapes of plates can change. The motions of plate margins, and the consequent evolution of plates, can be deduced from remarkably few rules. These are that the plates are rigid, and that plate margins behave as follows.

1. Spreading is symmetric at spreading centres. Equal amounts of new material attach to each of the plates that meet at a spreading centre.
2. Subduction is completely asymmetric. Material is removed from only one of the two plates that meet at a trench.
3. The relative motion of plates that meet at a transform fault is parallel to the transform fault.

The symmetry of spreading centres is an empirical rule based on the observed symmetry of magnetic stripes (Figure 3.5). It presumably comes about as follows. Suppose new oceanic crust is formed by the injection of a vertical dike of new magma (Figure 9.2). This will be hotter than its solidified surroundings, and will lose heat through its sides. If, some time later, horizontal tension has accumulated normal to the dike it will be pulled apart and new magma may intrude. If the dike has cooled symmetrically to the sides, it will be hottest and weakest at its centre. Therefore it will

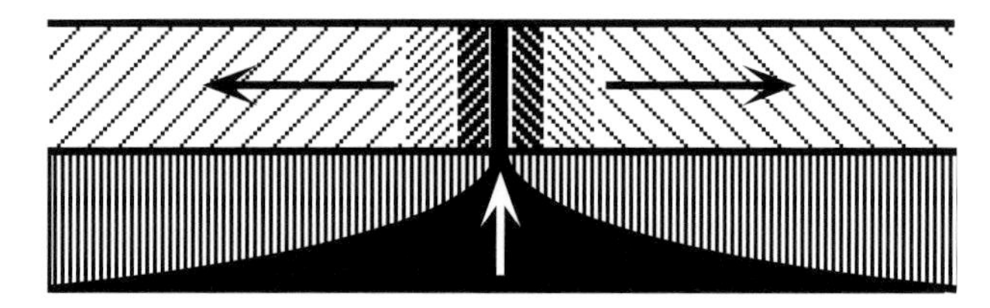

Figure 9.2. Sketch cross-section of a midocean ridge spreading centre showing the symmetric addition of crust (diagonal patterns) to each plate. Compare with the map view of Figure 3.4.

split down the centre and equal parts of it will become attached to the two plates that are pulling apart at the spreading centre.

Not all spreading is symmetric. There are some segments of spreading centres that spread asymmetrically, at least for a time, an example of which occurs on the Australian–Antarctic ridge [3]. There is evidence also that spreading centres may behave asymmetrically on short time scales. A reasonable guideline is that most spreading centres behave symmetrically most of the time at the scale resolved by the magnetic stripes. Another common feature of spreading centres is that they are oriented perpendicular to the direction of spreading. However they do sometimes deviate from this, for example south of Iceland. It is not necessary to state it as a basic rule here.

Asymmetry of subduction implies that the trench (i.e. the surface trace of the subduction zone fault) moves with the overriding (non-subducting) plate, since none of the overriding plate is removed. This rule also is to some degree empirical, and it may not always be strictly true. It is possible that some of the overriding plate is removed and carried down by the subducting plate, or that material is scraped off the subducting plate and attached to the overriding plate. This commonly happens with sediments scraped off the subducting plate. However, the resulting accretionary wedge of sediment is usually a superficial feature. Asymmetric subduction is certainly a good approximation.

9.3.2 Relative velocity vectors

Figure 9.3 depicts four different spreading centres. They are shown with different velocity vectors, but they differ only in the way the velocities are measured, each being measured from a different reference. In Figure 9.3a, the velocity of plate B is measured relative to plate A, as though you were sitting on plate A watching plate B move away from you. The others are, respectively, relative to the spreading centre (9.3b), relative to plate B (9.3c), and relative to a

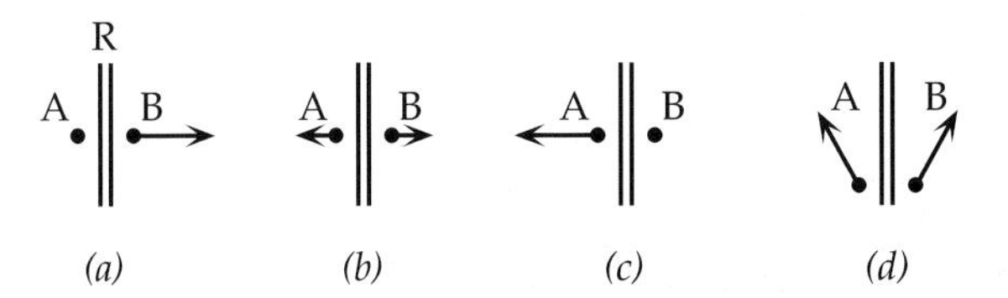

Figure 9.3. Different cases of two plates spreading from a ridge in which velocities are measured from different references. The plates have the same relative velocities in all four cases.

point moving 'south' along the ridge (9.3d) (taking north to be towards the top of the diagrams, here and subsequently).

The velocity of plate B *relative* to plate A is the velocity B would appear to have if you were moving with plate A. It is given by the vector velocity of B minus the vector velocity of A. This quantity is the same in all four cases. This is made more explicit in Figure 9.4a, which shows the velocities from Figure 9.3 plotted in terms of their components north (v_N) and east (v_E). In each case the *relative* velocity vector, represented by the line joining A and B in the velocity plot, is the same. The only difference between the four cases is the position of the line AB relative to the origin, which is determined by the frame of reference we happen to have chosen.

Since the origin is arbitrary, we can leave it out, and plot just the relative velocities of the plates. This is done in Figure 9.4b, and the result is called the *relative velocity diagram* for all of the cases shown in Figure 9.3. Included in Figure 9.4b is a point R. This represents the relative velocity of the ridge. Symmetry of spreading implies that the velocities of the two plates relative to the ridge are equal and opposite. In other words, the ridge velocity point is midway between the plate velocity points, and the ridge velocity is the vector *average* of the velocities of the plates that meet at the ridge.

Since the ridge is actually a line (presumed straight here), only ridge velocities normal to itself make sense. For an infinitely long ridge, an arbitrary velocity parallel to itself could be added without making any difference. In reality ridges often have distinguishing features along them, such as a transform offset, which removes this ambiguity. However, for limited periods and lengths, this ambiguity in ridge velocity needs to be borne in mind, as you will see later. In that case, the R point in the velocity diagram could lie anywhere

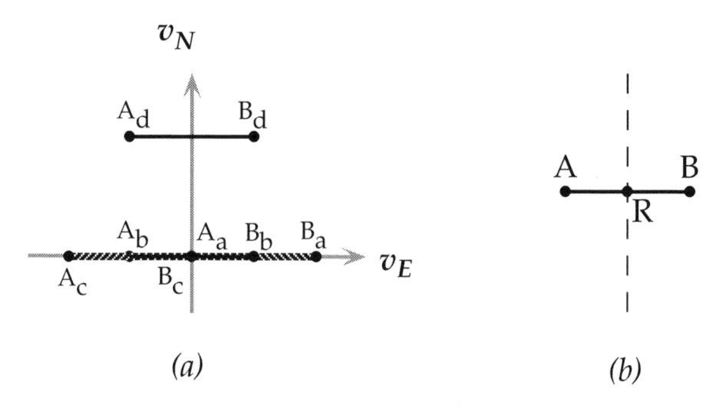

Figure 9.4. (a) Plots of the velocities of the plates for each case in Figure 9.3. (b) The same velocities referred internally to each other, rather than to an external origin.

along the dashed line, which is drawn through R and parallel to the ridge direction (usually, but not necessarily, perpendicular to the spreading direction).

Examples of trenches and their corresponding velocity diagrams are shown in Figure 9.5. The standard map symbol for a reverse fault is used to denote a trench, the 'teeth' being on the side of the overriding plate. Trenches are usually not straight, either being island arcs or taking the shape of a continental margin. The trenches in Figure 9.5 are drawn as though they are island arcs, with the appropriate sense of curvature.

Although plate B is located to the east of plate A, its velocity point is to the west of A's point in the velocity diagram, because it is moving west relative to A. According to rule 2, above, the trench moves with the overriding plate, so the trench velocity can also be represented on the velocity diagram. However, it is different in the two cases shown in Figure 9.5: it moves with plate A in case (a), and with plate B in case (b).

These simple ideas can be extended to include more than two plates, and velocities in any direction in the plane. You will see that the velocity diagram, which may look trivially simple so far, is a powerful way to keep track of plate evolutions.

9.3.3 Plate margin migration

Even with constant plate velocities, plate configurations can change. This is because only in special cases will ridges and trenches be stationary relative to each other. The reason is that spreading is symmetric and subduction is asymmetric. This means that inter-

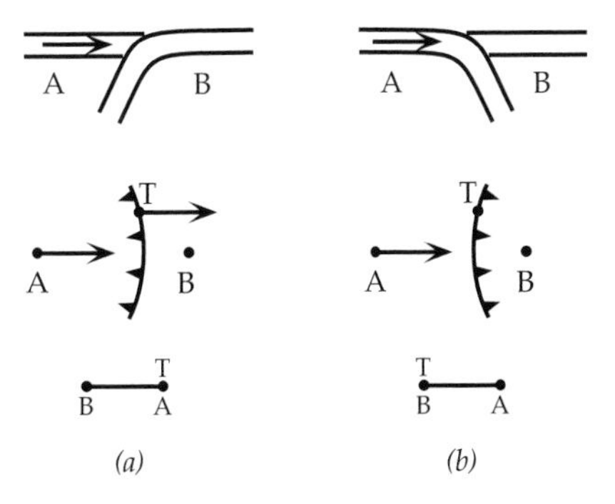

Figure 9.5. Relative velocities at a trench. The two possible trench polarities are shown (a, b), depending on which plate is being consumed. In each case, the top panel shows a cross-section, the middle panel shows a map view, and the bottom panel shows the velocity diagram.

vening plates will usually grow and shrink, and shrinking plates can disappear.

This can be illustrated most simply with three plates whose velocities have no northerly component. Figure 9.6 shows several situations in which plates have the same instantaneous (snapshot) configuration, but different velocities. The different velocities give rise to different evolution. In all cases the velocities are shown relative to plate C (and the trench). A velocity diagram is included with each case. Comparing the first three, you can see that in case (a) the ridge is moving west relative to C and so plate B is growing, in case (b) the ridge is stationary and the size of plate B is not changing, whereas in case (c) the ridge is moving east, towards the trench, and plate B is shrinking. In each case the plates are moving in the same *directions*, all that is different is the *magnitudes* of the velocities. In fact if you study the velocity diagrams you can see that the difference can just as well be regarded as a difference in the velocity of plate C relative to the others.

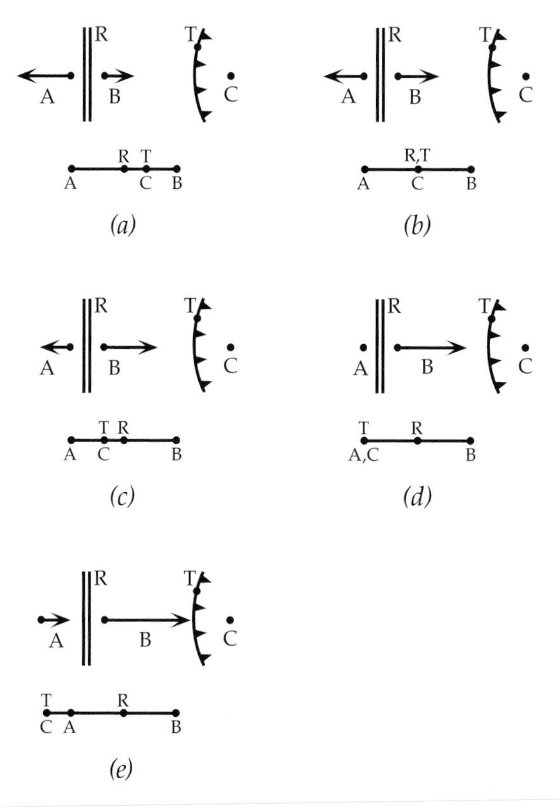

Figure 9.6. Different relative motions of ridges and trenches.

Now compare cases (c–e). In each of these cases, plate B is shrinking. The differences are in the direction of plate A relative to C. The difference does not become important until plate B shrinks to zero. At that point, plates A and C come into contact, forming a new plate margin between them. The nature of the new margin and the subsequent evolution of the system then depends on the relative velocities of A and C. In case (c), plate A is moving away from C, in which case the new margin between them will be a ridge, and this ridge will move west, so plate C will begin to grow. In case (d), plate A is stationary relative to C, so they will form a single plate when they come into contact. In case (e), plate A is moving towards C, which means the new margin between them will be a trench. The subsequent evolution will then depend on the polarity of the new trench. If it is the same as before, then A will subduct under C, following plate B into the mantle.

Examples of several of these situations can be inferred from the record of the seafloor magnetic stripes. The Phoenix plate used to subduct under Antarctica, until it disappeared and the Pacific and Antarctic plates came into contact. Now the Pacific–Antarctic ridge migrates slowly away from Antarctica, as predicted in case (c). Case (d) resembles the former situation off western North America, where the former Farallon plate has disappeared, except that the new margin, the San Andreas fault, between the Pacific and North American plates, has a strike-slip component because of the relative northward motion of the Pacific plate. Case (e) is similar to the North Pacific, where the Kula plate used to subduct under the Aleutian Islands, but now the Pacific plate subducts after it. More examples like these will be presented later.

9.3.4 Plate evolution sequences

Although you can deduce from the velocity diagrams in Figure 9.6 that the ridge in cases (c–e) will migrate towards the trench, it is not obvious at first sight exactly how this will proceed. It is useful to draw a *sequence* of sketches in order to clarify this. A simple sequence showing the development of a spreading ridge was shown in Figure 3.4. Another sequence, that illustrates the way in which case (d) of Figure 9.6 develops, is shown in Figure 9.7.

The approach is as follows. To generate the next diagram in a sequence, draw each plate with its *old margins* in their *new positions* relative to the other plates. Thus the old margin *a* does not move, because A is not moving. The old margin *b* moves to the east. The trench does not move. This will generate gaps or overlaps with neighbouring plates. A gap should be filled by drawing a ridge in

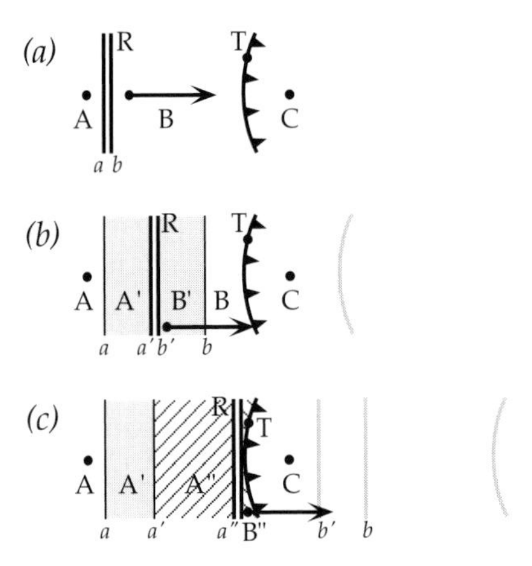

Figure 9.7. A plate evolution sequence showing the development with time of case (d) of Figure 9.6. The grey lines are former features on plate B that have been overridden by plate C.

the middle (if the spreading is symmetric). Each side of the new ridge (a' and b') represents the new margin of the plate that adjoins it. Shade the space between this plate's new margin and its old margin: this is new crust added to this plate (A$'$ and B$'$). Overlap should be eliminated by removing the overlapping area from one or other of the overlapping plates, depending on the polarity of the trench at which they meet (B is subducting under C, so part of B is removed). This procedure defines the new positions of the plate margins, according to the rules of how plate margins evolve.

In the last frame, plate B has almost disappeared. As it disappears, plate A comes in contact with plate C. Since, in this example, plate A is stationary relative to plate C, the new margin will be inactive. Of course this is a very special case: in the real world you would expect plates A and C to have some relative motion, and to form the appropriate kind of new margin between them.

This sequence assumes that there is no change in the velocity of B as it disappears. This may not happen in reality, but the point here is to illustrate the kinds of changes that can occur even without any change in plate velocities. Also it is best not to think of the ridge as being subducted. Plate B is subducted (removed), but the consequence of plate A contacting plate C is that the two old margins (ridge and trench) *coalesce to form a new margin*. Again it is better to focus on the surface features, rather than on what might be happening under the surface.

Another plate evolution sequence, in Figure 9.8, illustrates how a ridge with a transform fault offset evolves. This example is like part of the central Mid-Atlantic Ridge illustrated in Figure 3.6.

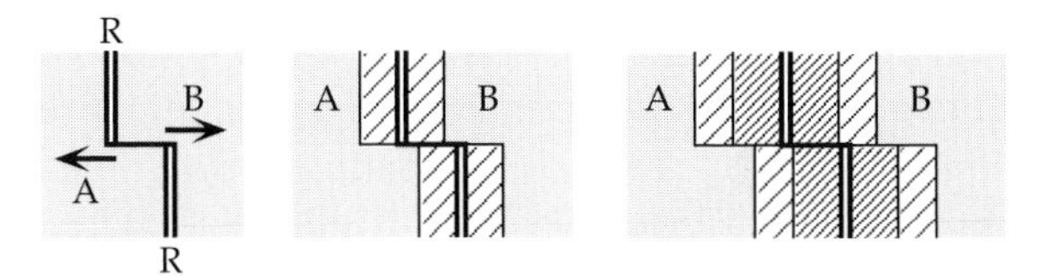

Figure 9.8. Sequence showing the evolution of a ridge with a transform fault offsetting it.

Transform fault margins translate parallel to themselves. The parts of the fault that connect ridge crest segments are shown here as heavy solid lines, indicating that they are active faults. The parts that are beyond ridge crests are shown as light lines, denoting that they are extinct faults across which there is no longer any relative motion. If the changes in shading corresponded to magnetic field reversals, then the pattern generated would represent magnetic anomaly stripes. This example shows how a transform offset of a ridge results in the magnetic anomaly pattern also being offset.

9.3.5 Triple junctions

Figure 9.9 depicts a sequence involving three plates separated by ridges. Points where three plates, and three plate margins, meet are called triple junctions. In this case the benefits of the procedure for constructing sequences just described, and of velocity diagrams, are more evident. A new feature occurs in this example, in the vicinity of the triple junction: after the old margins of B and C are displaced to their new positions, the new ridge segments need to be longer in order that they all meet again. Comparing (a) and (b), there is a triangular area (*abc*) around the triple junction that is the same

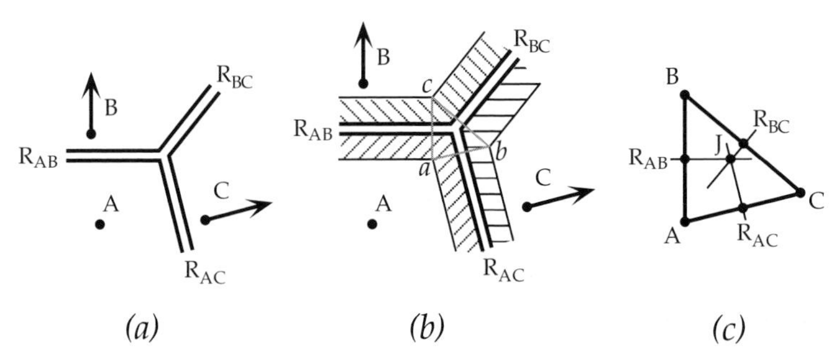

Figure 9.9. (a) and (b) Evolution of a ridge–ridge–ridge (RRR) triple junction. (c) Velocity diagram showing the three plate velocities, the three ridge velocities and the triple junction velocity (J). The ridges must lengthen into the triangle (*abc*) in (b).

shape as the velocity triangle (c), and the ridges must be extended into this region. On the velocity triangle (c), the ridge velocities are included, as light lines parallel to the corresponding ridge. They meet at a point that defines the velocity of the triple junction (J). Since these lines bisect the sides of the triangle (for symmetric spreading) the triple junction point is the *circumcentre* of the triangle (so-called because it is the centre of a circle that passes through the vertices of the triangle, that is it circumscribes the triangle). It is observed that junctions of three ridges really do work this way.

Important features can be read off the velocity diagram. For example, the triple junction point J is to the right of the line AB, which corresponds to the fact that the triple junction is moving east relative to A and B, and the ridge R_{AB} is getting longer. Since B is moving north relative to A and C is moving ENE, the relative motion of B and C is determined by vector addition. The ridge segment R_{BC} is perpendicular to this velocity vector.

If the new, shaded material on plate A is interpreted as a magnetic anomaly, you can see that it changes direction near the triple junction. 'Bent' magnetic stripes like this are observed in the Pacific, and can be seen in Figure 9.10, near the Aleutian Islands in the north-west part of the map. They are inferred to have been formed near a triple junction, but this implies that there were two additional plates that are no longer present. The eastern one, analogous to plate C in Figure 9.9, is called the Farallon plate and the northern one is called the Kula plate. A reconstructed evolutionary sequence of the plates in the north-east Pacific is shown in Figure 9.11. The inferred triple junction between the Pacific, Farallon and Kula plates can be seen at the 80 Ma, 65 Ma and 56 Ma stages.

Other types of triple junction are possible. Figure 9.12 shows a ridge–transform system that has migrated into a trench, in the manner of Figure 9.7, and created two triple junctions. At the northern triple junction, J_N, two transform faults and a trench meet, whereas at the other (J_S) a ridge, a trench and a transform meet. It is useful to denote the type of triple junction by the types of plate margin involved. Denoting a ridge by R, a trench by T and a transform fault by F, J_N can be denoted an FFT triple junction, whereas J_S is RFT. The triple junction of Figure 9.9 is RRR.

The example in Figure 9.12 is comparable to the evolution of the plates along the western margin of North America. Comparing with Figure 9.11A, we can see that plate A is analogous to the Pacific plate and plate D is analogous to the North American plate. Plate B is analogous to the small Juan de Fuca plate off Oregon and Washington states, and plate C is analogous to the Cocos plate off Central America. The transform fault contact

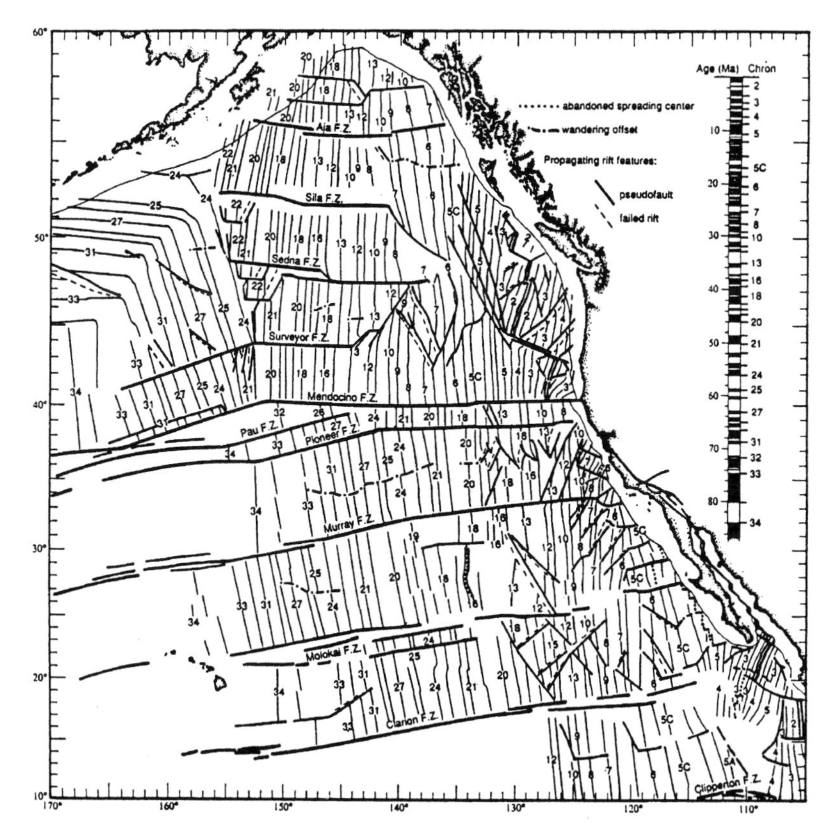

Figure 9.10. Magnetic anomalies that have been mapped in the north-east Pacific. The magnetic anomalies are the predominantly north–south lines, labelled with an identifying sequence number (which is not their age). This rather complex map also shows fracture zones and other features that interrupt the anomaly patterns. From Atwater and Severinghaus [4].

between A and D is analogous to the San Andreas fault system in California. The Juan de Fuca plate and the Cocos plate are fragments of the large Farallon plate (Figure 9.11) that used to exist between the Pacific and North American plates. The fragmentation of the Farallon plate can be seen in Figure 9.11 at the 56 Ma, 37 Ma and present stages.

It is possible to imagine all combinations of ridge, trench and transform fault meeting at a triple junction, but it turns out that some combinations can only be instantaneous juxtapositions, and they will immediately evolve into a different configuration. An example of such an 'unstable' triple junction is shown in Figure 9.13a. Because each part of the trench moves with a different plate, they are soon separated, as is illustrated in Figure 9.13b. There is then still a triple junction, and it is still of the TTF type, but its

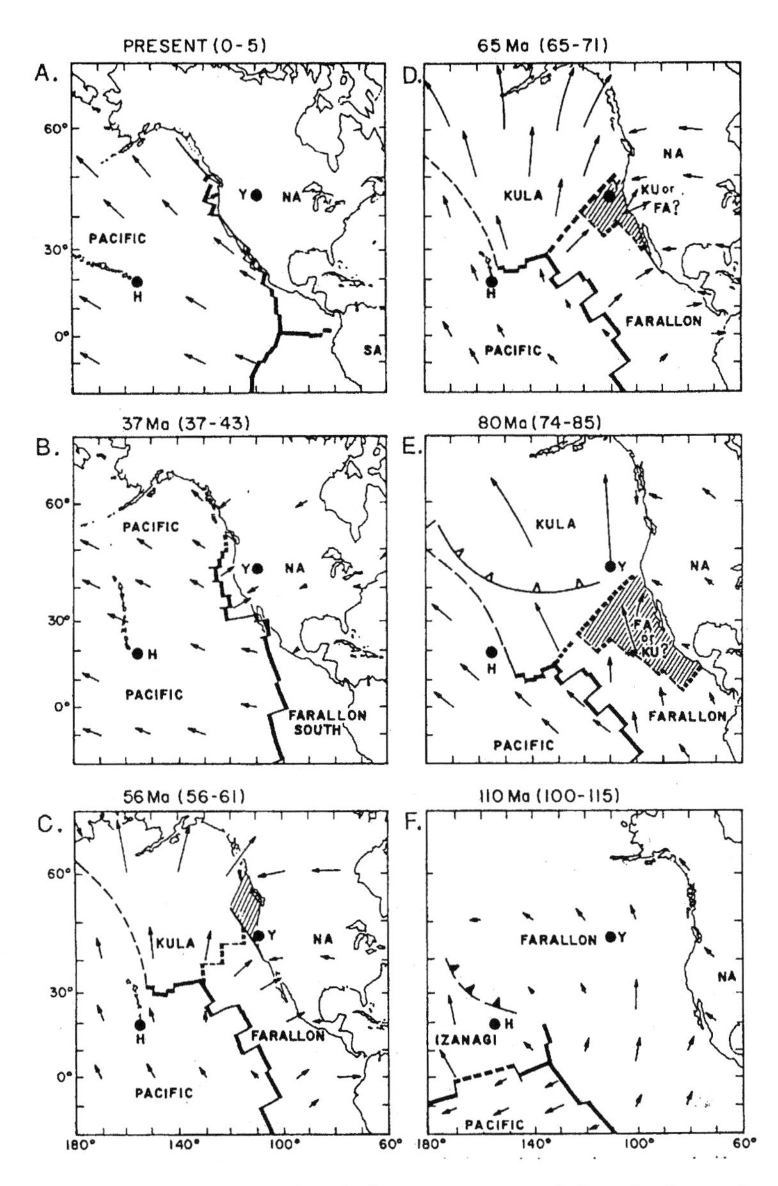

Figure 9.11. Reconstructed evolutionary sequence of plates in the north-east Pacific. From Atwater [5].

arms are now reoriented into a configuration that is 'stable', that is it can persist for a finite time. This example is taken from Central America, where the Managua fault, separating the Caribbean and North American plates, cuts through Nicaragua and joins the Central America trench.

The motions of the triple junctions in Figures 9.12 and 9.13 can also be represented in a velocity diagram using the concepts already

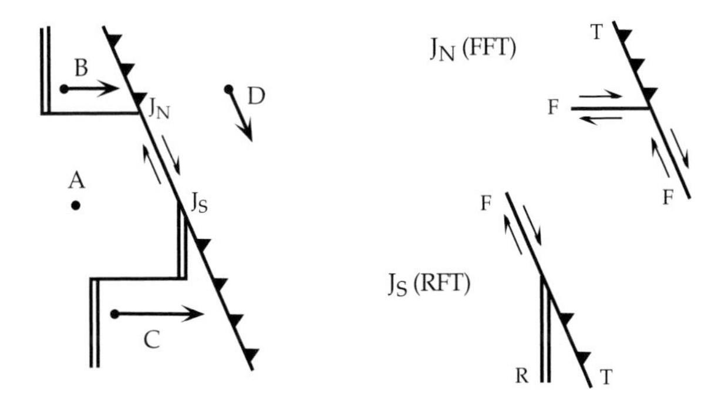

Figure 9.12. Triple junctions, J_N and J_S, created when a ridge–transform system is overridden by a trench.

outlined. However, you will have to add transform faults to your velocity diagram repertoire and bear in mind that subduction is often oblique. Subduction and transform margins can be represented on velocity diagrams by lines that are parallel to the corresponding margin, as we have already seen for ridges (Figure 9.9). A good exercise is to construct a velocity diagram including all the plates, margins and triple junctions of Figure 9.12.

9.4 Rules on a sphere

So far we have considered only plate motions on a plane, but of course the earth is not flat. The concepts we have developed so far all transfer to a spherical surface, but there are some modifications and additions for the case of a sphere. We will only note some of the important points here. A comprehensive treatment of plate tectonics on a sphere is given by Cox and Hart [1].

Euler's theorem states that any displacement of a spherical cap on a sphere can be represented as a single rotation about an axis through the centre of the sphere. Since the displacement can be taken to be relative to another spherical cap, it applies also to

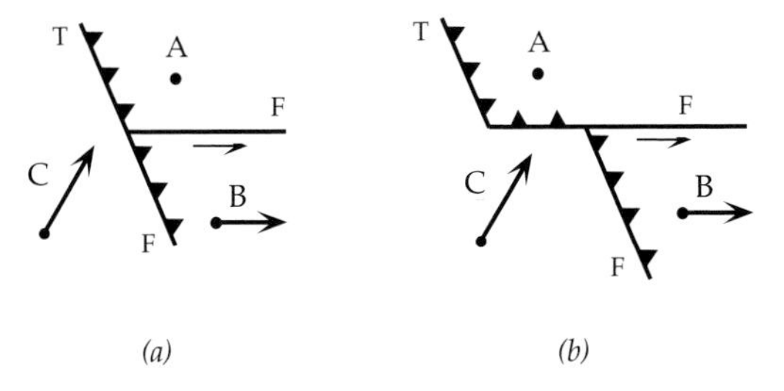

Figure 9.13. An example of an unstable triple junction (a), that immediately evolves into a different configuration (b).

the relative motions of plates. The intersection of the axis of rotation with the sphere is called the pole of (relative) rotation, or, following Menard [6], the Euler pole. The ambiguity of having two poles can be eliminated by choosing the pole for which the rotation is right-handed. The axis of rotation and the rate of rotation can be combined to define an angular velocity vector that describes the instantaneous relative motion of two plates.

There is a complication in spherical geometry that does not occur in planar geometry. Whereas infinitesimal rotation vectors add and commute, finite rotations do not. This can be seen by rotating a point from the north pole to 0 °E on the equator, followed by a rotation from 0 °E to 90 °E on the equator. Reversing the order of the two rotations does not yield the same result. Likewise taking the sum of the two rotation vectors and applying the resulting rotation does not accomplish the same result. For this reason only infinitesimal or small rotations can be treated by normal vector algebra.

A consequence of Euler's theorem is that transform faults should follow small circles centred on the Euler pole of the plates that meet at the fault. A planar version of this relationship is shown in Figure 9.14 (rotations are of course also possible in planar geometry, we just hadn't considered any until now). The fracture zones formed by transform faults will also follow small circles for as long as the Euler angular velocity vector of the two plates is constant. A consequence is that the normals to fracture zones and transform faults intersect at the Euler pole (Figure 9.14). This principle was used by Morgan [7] to locate relative rotation poles of pairs of plates.

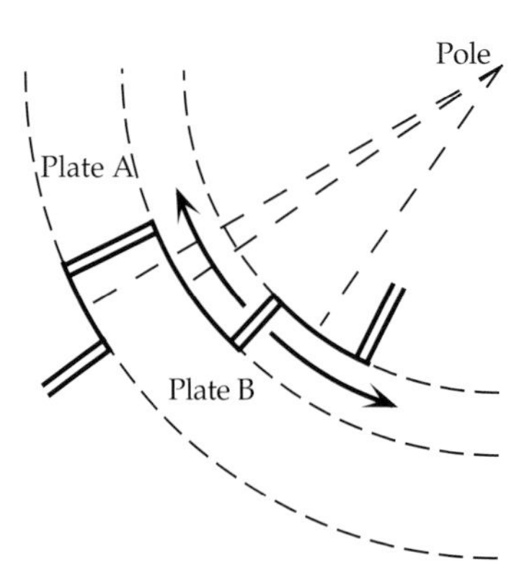

Figure 9.14. Relative rotation between two plates in the case of planar geometry. The transform faults and fracture zones form circles centred on the pole of rotation. On a sphere they form small circles.

On a sphere, the local spreading or convergence rate varies with position along plate margins, and there may even be a change in the type of margin. In Figure 9.14, the spreading rate will increase with distance from the pole. An example of this is that the spreading rate of the Mid-Atlantic Ridge is largest near the (geographic) equator and decreases towards the North America–Europe rotation pole, which is located in the Arctic. An example of a change in margin type is that the motion between the Pacific and Australian plates changes from nearly normal subduction at Tonga, north of New Zealand, to nearly strike-slip along the Macquarie Ridge, south of New Zealand, which is close to the Euler pole.

9.5 The power of the rules of plate motion

The rules of plate motion have proven to be a powerful tool for deciphering the history of the plates. We saw in Chapter 3 how the great Pacific fracture zones were extremely puzzling until it was recognised that they were formed at the Pacific–Farallon ridge, which no longer exists in this region (Figure 9.11). The 'great magnetic bight', where the magnetic stripes turn from northerly to westerly (Figure 9.10) was also puzzling. Once the unique properties of plate kinematics were discovered, it was possible to use these puzzling features to make powerful inferences, such as the former existence of two large plates in the Pacific basin (the Farallon and Kula plates).

An early and striking example of this power came from the Indian Ocean, where the sequence of events has been rather complex. The outlines of the main phases of seafloor spreading were correctly inferred by McKenzie and Sclater in 1971 [8] on the basis of a data set that was remarkably sparse for such a huge area. Given that there were four continents involved, and several distinct phases of seafloor spreading, this remains one of the more remarkable demonstrations of the power of the rules of plate motion.

Another example comes from the Pacific. In the course of teaching about this subject, I noticed that the magnetic stripes near the great magnetic bight form a peculiar 'buttress' shape. Part of it is outlined by anomaly 33 in Figure 9.10, and the shape is also evident in Figure 9.11 (stages 80 Ma, 65 Ma, 56 Ma). This shape will not extrapolate back in time without seeming to reach an impossible configuration, in which a small piece of the Pacific plate would have had to emerge separately and then merge with the main plate at its north-east corner. (This part of the plate evolution is not recorded because of a magnetic 'quiet zone', due to the cessation of magnetic reversals for a time in the late Cretaceous.) Graduate student Mark Woods pursued the idea and developed the case that the Kula plate had actually formed by breaking off the Pacific plate, in the late Cretaceous, along the Chinook fracture zone [9]. A direct implication was that a series of older Mesozoic magnetic stripes in the north-west Pacific, which had previously been attributed to Pacific–Kula separation, must have involved another plate, since the Kula plate did not then exist. This made it much easier to understand the relationship between the Mesozoic and Tertiary magnetic stripes. We named the inferred older plate Izanagi (Figure 9.11, 110 Ma stage), after one of the gods of Japanese mythology responsible for the creation of the Japanese islands. Thus the inference of a former large plate in the western Pacific resulted from noticing a small inconsistency implied by the rules of plate motion.

9.6 Sudden changes in the plate system

Plates change with time, even when no new plate margins form. There are actually three kinds of change recorded by seafloor magnetic stripes: steady growth or shrinkage of plates, changes in plate velocity, and the formation of new plate margins by plate breakup. The first kind of change is a consequence of the difference in behaviour between spreading margins and converging margins, which we have explored in some detail in this chapter. Thus plates may grow and shrink, and some plates may disappear, through the normal evolution of their margins.

A dramatic change in plate velocity occurred about 43 Ma ago when the velocity of the Pacific plate in the vicinity of Hawaii changed from north-north-west to west-north-west. This change is recorded by the 'bend' of the Hawaiian–Emperor chain of seamounts that marks the trace of the Hawaii volcanic hotspot on the Pacific plate (Figure 4.3). A number of less dramatic changes in the relative motion of the Pacific and Farallon plates is recorded by magnetic stripes on the Pacific plate (Figure 9.10). Some of these are associated with the shrinking and fragmentation of the Farallon plate.

The breakup of Pangea involved the formation of new spreading centres, and these are well recorded by magnetic anomalies in the Atlantic, Indian and Southern Oceans. Sometimes a new spreading centre has formed near an existing one, and the existing one has ceased. This has been called a 'ridge jump'. Several ridge jumps were associated with a change in the Pacific–Nazca relative motion. There was a ridge jump from one side of Greenland to the other at the time of eruption of the North Atlantic Tertiary flood basalts about 60 Ma ago.

Examples of the formation of new subduction zones are harder to find, because much of the evidence is subsequently destroyed. It is conjectured that the Mariana subduction zone began at an old fracture zone on the Pacific plate, possibly at the time of the change in Pacific motion 43 Ma ago. This relatively recent origin might help to explain the existence of sub-parallel subduction zones on either side of the Philippine plate.

Indirect evidence for episodes of subduction is recorded, in principle, in the mountain belts of island arcs and continental margins associated with subduction zones. Because the geology so recorded is complex, it is difficult to resolve detail. However it is clear, for example, that the western margin of Canada changed from being passive (like the present eastern margin) to having active subduction in the late Precambrian.

The disappearance of a number of plates from the Pacific basin can be inferred from the magnetic stripe record. The Farallon plate has not really disappeared, it has fragmented as it shrunk, into the Nazca, Cocos and Juan de Fuca plates. In the north Pacific, the Kula plate is reliably inferred to have been subducted into the Aleutian trench. The Phoenix plate (or most of it) disappeared under Antarctica. Exercise 3 illustrates a simplified version of these events. The Izanagi plate (or plates) has disappeared under Japan, as was related in Section 9.5.

9.7 Implications for mantle convection

The most important implication of plate kinematics for mantle convection is that the locations of upwellings and downwellings must be influenced, if not controlled, by the (brittle) mechanical properties of the lithosphere, rather than the (viscous) properties of the deeper mantle. This is because, by conservation of mass, there must be upwellings under spreading ridges and downwellings under subduction zones. This statement is true independently of what forces are driving the system. It is a deduction from the surface kinematics and conservation of mass. This important point will be taken up in Chapter 10.

Another implication arises from the time dependence of the configuration of plates. If plates and mantle convection are intimately related, as we will see in Chapter 10, then we should expect the pattern of mantle convection also to be unsteady. The time dependence of the plates is of a peculiar sort, being quite different from the unsteadiness of a strongly heated convecting fluid of the more familiar kind. In normal fluid convection, the flow structure can change rather randomly, and may reach a state of 'deterministic chaos'. The plate system, on the other hand, tends to evolve steadily for substantial periods, but then to suddenly shift into a different pattern of motions if a new plate boundary forms. Thus mantle convection must be consistent with the facts that plates have a range of sizes and odd, angular shapes, that plates grow and shrink, that some plates disappear, that others break up, and that plate velocities may change suddenly. Such changes are evident in Figure 9.15, which shows a selection of reconstructed plate configurations over the past 120 Ma.

The time dependence of the plates has important implications for many aspects of the interpretation of geophysical evidence, as well as for the way chemical heterogeneities will be stirred in the mantle (Chapter 13). Thus, for example, the deep expression of past subduction, as expressed in the gravity field, may not coincide with the present location of subduction zones.

The effects of spherical geometry on plate kinematics must be borne in mind, especially in relation to larger plates. This means, for example, that near a pole of rotation the plate may be rotating about a vertical axis relative to the mantle under it, and it would not be accurate to think of the mantle motion in terms of simple roll-cells of convection. In a spherical shell, the flow may connect globally in a complex way. Thus the 'return flow' from subduction under the north-west Pacific back to the East Pacific Rise may pass

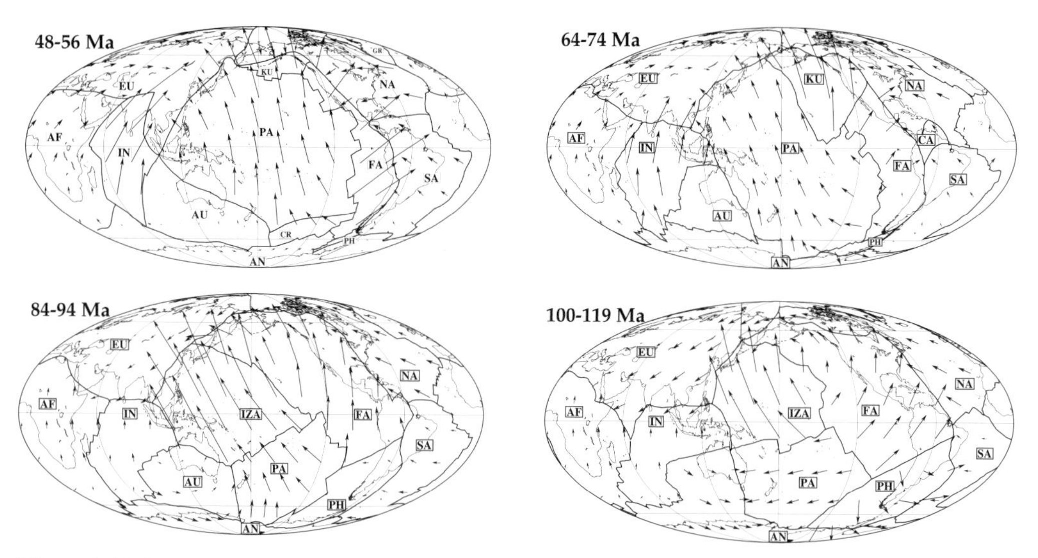

Figure 9.15. Reconstructions of plate configurations and velocities for several time intervals over the past 120 Ma. From Lithgow-Bertelloni and Richards [10]. Copyright by the American Geophysical Union.

under North America, approximating a great circle path [11], so the flow under North America may have a southerly component that would not be inferred from the local part of the plate system.

9.8 References

1. A. Cox and R. B. Hart, *Plate Tectonics: How it Works*, 392 pp., Blackwell Scientific Publications, Palo Alto, 1986.
2. C. M. R. Fowler, *The Solid Earth: An Introduction to Global Geophysics*, Cambridge University Press, Cambridge, 1990.
3. J. K. Weissel and D. E. Hayes, Asymmetric spreading south of Australia, *Nature* **231**, 518–21, 1971.
4. T. Atwater and J. Severinghaus, Tectonic maps of the northeast Pacific, in: *The Geology of North America, Vol. N, The Eastern Pacific Ocean and Hawaii*, E. L. Winterer, D. M. Hussong and R.W. Decker, eds., Geological Society of America, Boulder, CO, 1989.
5. T. Atwater, Plate tectonic history of the northeast Pacific and western North America, in: *The Geology of North America, Vol. N, The Eastern Pacific Ocean and Hawaii*, E. L. Winterer, D. M. Hussong and R. W. Decker, eds., Geological Society of America, Boulder, CO, 1989.
6. H. W. Menard, *The Ocean of Truth*, 353 pp., Princeton University Press, Princeton, NJ, 1986.
7. W. J. Morgan, Rises, trenches, great faults and crustal blocks, *J. Geophys. Res.* **73**, 1959–82, 1968.
8. D. P. McKenzie and J. G. Sclater, The evolution of the Indian Ocean since the late Cretaceous, *Geophys. J. R. Astron. Soc.* **24**, 437–528, 1971.
9. M. T. Woods and G. F. Davies, Late Cretaceous genesis of the Kula plate, *Earth. Planet. Sci. Lett.* **58**, 161–6, 1982.
10. C. Lithgow-Bertelloni and M. A. Richards, The dynamics of cenozoic and mesozoic plate motions, *Rev. Geophys.* **36**, 27–78, 1998.
11. B. H. Hager and R. J. O'Connell, Kinematic models of large-scale flow in the earth's mantle, *J. Geophys. Res.* **84**, 1031–48, 1979.

9.9 Exercises

1. Sketch an evolution sequence, in the manner of Figure 9.7, for cases (a), (c) and (e) of Figure 9.6. If the nature of a plate margin changes, continue the sequence for one stage after the change in order to show the character of the subsequent evolution.

2. (a) Construct a velocity diagram for Figure 9.12. Include the velocities of all plates, plate margins and triple

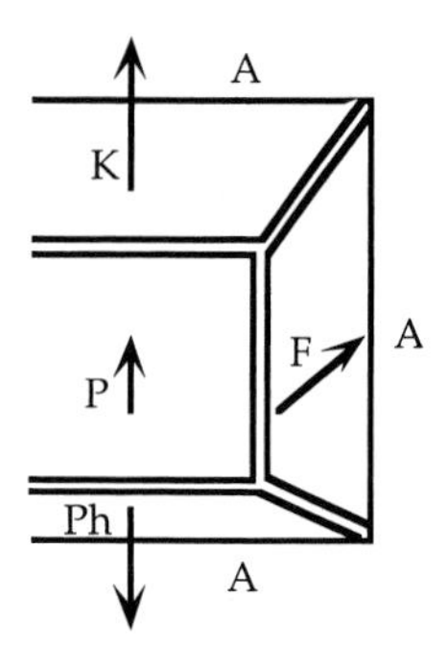

Figure 9.16 Plate configuration for Exercise 3.

junctions. (b) Sketch stages in the evolution of these plates until a steady situation is reached.

3. (a) Construct a velocity diagram for the situation in Figure 9.16. Velocities are shown relative to plate A, which surrounds the others on three sides. This is a simplification of the situation in the Pacific basin during the early Tertiary. (b) On the basis of the velocity diagram, predict the fates of plates K, F and Ph and any consequent changes in the nature of their margins with plate A. (c) Sketch an evolution sequence up to the stage where there are only two plates. (d) What would be the ultimate outcome if there are no changes in plate velocities?

CHAPTER 10

The plate mode

It is the thesis of this chapter that the plates are part of one recognisable mode of mantle convection, driven by the top thermal boundary layer of the mantle. I argued in Chapter 8 that there is no fundamental reason why the modes of flow driven by the different thermal boundary layers should be the same. You will see in Chapter 11 that mantle plumes seem to be a distinct mode of mantle convection driven by a bottom thermal boundary layer, and that in the mantle these two modes seem to behave in substantially different ways, and are not even very strongly coupled. It is therefore useful to identify explicitly two modes of mantle convection: the *plate mode* and the *plume mode*. I have previously also referred to the plate mode as the *plate-scale flow* [1, 2] because the flow seems to be quite strongly constrained to have the horizontal scale of the plates. This will be addressed in Section 10.2.

In the first half of this chapter I present a series of numerical models that illustrates the influence that various material properties have on the form of mantle convection. We look at the influence of the mechanical properties of the lithosphere, the effect of the increase of mantle viscosity with depth and the possible role of phase transformations. We also estimate the amount of heat transported by the plate–mantle system. In the second half of the chapter we look at how well the resulting conception of mantle convection matches observational constraints, especially surface topography and heat flow and the internal structure revealed by seismic tomography. The chapter concludes with a summary of the conception of the plate mode of mantle convection that is developed here.

10.1 The role of the lithosphere

We have seen, in Chapter 8, that convection involves both fluid flow and the diffusion and advection of heat. We have seen in Chapters 3 and 6 that there is good evidence that the hot interior of the mantle behaves like a fluid on geological time scales, and that it seems to be reasonably approximated as a linear viscous fluid. On the other hand, in the formulation of the theory of plate tectonics (Chapter 3), in our understanding of the ways rock rheology changes with temperature and pressure (Chapter 6), and in the detailed observations of the plates and their motions (Chapters 4 and 9) we recognise that the cool lithosphere is, to a reasonable approximation, a brittle solid.

The relationship between the moving plates and the putative convection in the underlying fluid mantle has been a puzzle. The reasons are not hard to see. A map of the plates (Figure 4.1) does not look like the surface of a convecting fluid. The ways the plates evolve (Chapter 9) are not like the ways convecting fluid motions evolve. The first plausible mechanism for continental drift to be proposed involved continents being carried passively along on top of large mantle convection cells whose origin was not clearly specified [3, 4]. This idea was carried over by some into plate tectonics, with the plates envisaged as being carried passively on convection cells. A rival school held that plates were active components and that they drive flow in the fluid mantle. Forces were identified and quantified that would drive plates, like 'slab pull' and 'ridge push' [5, 6]. Still another conception was that there was not a very direct relationship between the pattern of convection and the surface plates. Debates occurred on how many cells there might be under plates: one, several or many.

We have by now assembled the concepts that will allow us to make some sense of this puzzle. The solution involves our understanding of thermal diffusion and thermal boundary layers, the dependence of rock rheology on temperature and pressure, and the way buoyancy forces are balanced by viscous resistance in slow viscous flow.

We can start from the seemingly trivial statement at the beginning of Chapter 8 of what convection is. Thermal convection occurs when positively or negatively buoyant fluid in a thermal boundary layer becomes unstable and rises or falls into the interior of the fluid. A thermal boundary layer forms when heat diffuses into or out of the fluid through a boundary. In particular, a cool thermal boundary layer forms at the top boundary of a fluid when heat diffuses out through the top.

In our discussion of oceanic heat flow and seafloor subsidence in Chapter 7, we hypothesised that the lithosphere forms as hot mantle moves horizontally away from a spreading centre and cools by diffusion of heat to the surface. This simple model yields an excellent match to the observed decline in heat flux with seafloor age and a good first-order match to the observed subsidence of sea floor with age. We can also recognise in this picture that the oceanic lithosphere has the characteristics of a thermal boundary layer.

The lithosphere has mechanical properties that are unusual in the context of convection. Because it is cooler than the deeper interior of the mantle, it is stronger (Figures 6.18, 9.1). However, its strength is not unlimited (Figure 6.19), and it is observed to be faulted or broken into large pieces (Figure 4.1). These faults or breaks allow the pieces of the lithosphere to move relative to each other. *The distinctive feature of the earth's lithosphere is its combination of strength and mobility.* Conventional fluids lack strength in their thermal boundary layers. The lithospheres of the moon and Mars are so strong as to be unbroken and static. The earth's lithosphere is strong, but can still move.

So the thermal boundary layer at the top of the earth's mantle has unusual mechanical properties, but it is still recognisably a thermal boundary layer. We can now also recognise that it plays the key role of a thermal boundary layer. The lithosphere detaches from the earth's surface and sinks into the interior of the mantle (Figures 5.12, 5.13). It has large negative buoyancy (Section 8.1). We have calculated that this buoyancy is capable of driving flow in the mantle at velocities similar to plate velocities (Section 8.2).

The mechanics of how the lithosphere detaches and sinks are different from those of a fluid boundary layer. A fluid boundary layer becomes unstable by the Rayleigh–Taylor mechanism (Section 8.4). A lithospheric plate subducts. A portion of an oceanic plate is, evidently, usually stabilised by its own strength until such time as it arrives at a subduction fault. The existence of the fault removes the inhibition of the strength and frees the plate to sink under its accumulated weight.

The picture we have arrived at is that the plates comprise the top, cool thermal boundary layer that drives a form of mantle convection. As such, the plates are integral and active components in the system. However they are not the only active components, as we will see in Chapter 11. As well, while it is implied that the plates are active and the fluid mantle is passive, rather than vice versa, the fluid mantle strongly couples the components of the system by transmitting viscous stresses. Thus models that estimated the velocities of plates by treating them effectively as components independent of each other and driven by local forces [5, 6] were a useful step but not the whole story. It is better to think of a plate–mantle *system*.

10.2 The plate-scale flow

Here I present a series of numerical models that illustrates the influence of several factors on the form of flow in the mantle. Several of the models shown in this chapter differ in only one or two parameter values, in order to isolate the effects of changing particular parameters. Some models drawn from previous studies inevitably do not share the same parameter values. Figure 8.4 is also part of this series. So that the text is not cluttered with detail, the technical specifications of the models are collected in Appendix 2. The models are not intended to be accurate simulations of the mantle. For example, their Rayleigh number is lower than that of the mantle. They *are* intended to illustrate important effects, and for this purpose their parameters are reasonably close to those of the mantle.

10.2.1 Influence of plates on mantle flow

We consider first the effects of the top thermal boundary layer being stiff and mobile, first showing some effects separately, then in combination. It is not possible to show models that are as realistic as would ideally be desirable, because it has proven technically difficult to accurately model strong lithospheric plates separated by narrow faults. This is because the extreme gradients of material properties tend to disrupt iterative numerical methods. Nevertheless the principles can be illustrated, and a reasonable approximation to plates can be simulated.

We look first at the effect of the temperature dependence of mantle rheology, which makes the lithosphere much higher viscosity than the mantle interior. The resulting stiffness of the top thermal boundary layer tends to prevent it from dripping down under the action of its own negative buoyancy. This is illustrated on the left-hand side of Figure 10.1, which shows a numerical model of an internally heated convecting fluid layer like that in right-hand panels of Figure 8.4, but now with a viscosity that depends strongly on temperature, rather than being constant. The viscosity of the coldest fluid, at the top boundary, is 100 times the viscosity of the hottest fluid in the interior. The effect of this is to stiffen the top boundary layer so that it moves only slowly. The model in Figure 10.1 was started with a cool piece projecting down from the bound-

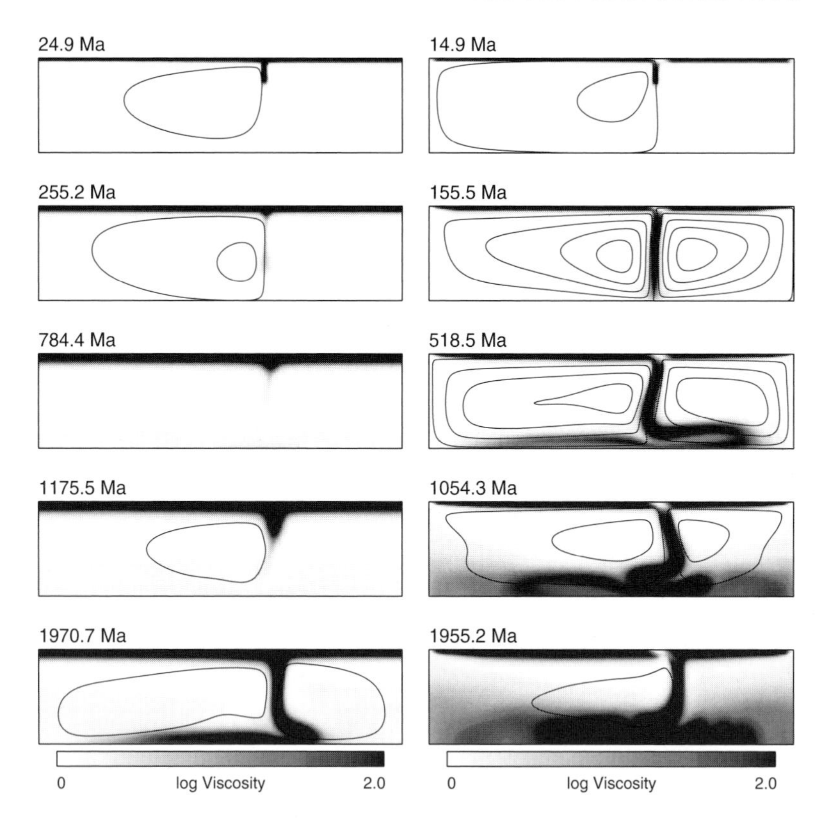

Figure 10.1. Numerical convection sequences. Left: with temperature-dependent viscosity (maximum viscosity 100 times the ambient viscosity). Right: the same with low-viscosity weak zones in the thermal boundary layer. The stiffness of the boundary layer inhibits the flow in the left sequence. The weak zones on the right allow pieces of the boundary layer to move more readily, simulating the motion of lithospheric plates. (Full technical specifications of this and subsequent numerical models are given in Appendix 2.)

ary layer, simulating a subducted lithospheric slab, in order to initiate flow. In spite of this, the top boundary layer is almost immobile for nearly a billion years. The slab pinches off and sinks, being faintly visible at times 255 Ma and 784 Ma.

The initial slab does promote an instability in the boundary layer, but it takes more than a billion years to form a drip that eventually develops into a sinking sheet driving slow internal flow (at 1970 Ma). The very slow flow of the left-hand sequence in Figure 10.1 is in contrast to the active flow of the constant-viscosity fluid in Figure 8.4 during the same interval.

Now we will look at the effect of breaks in the boundary layer. If the stiffness of the top boundary layer is interrupted by zones of weakness, then it may be more mobile. In the earth, the zones of weakness in the lithosphere are faults. In the right-hand panels of Figure 10.1, zones of weakness have been created by reducing the viscosity in three places: at each end of the box and adjacent to the initial slab. Figure 10.1 shows viscosity rather than temperature in order to reveal this difference between the two sequences. The breaks in the stiff boundary layer allow the two intervening parts of the boundary layer to move towards each other. As their material reaches the weak zones, its viscosity reduces, allowing it to turn and sink. As it sinks out of the weak zone, its viscosity rises again, and it forms a stiff sinking sheet. This sheet buckles and folds onto the bottom of the box. You can see in Figure 10.1 that the weak zones allow the fluid in the right-hand panels to be more mobile than in the left-hand panels during the first billion years.

The pieces of the 'broken' boundary layer move as nearly rigid units (with velocity that is nearly uniform, spatially). This can be seen from the fact that the streamlines are nearly parallel to the surface, velocity being inversely proportional to streamline spacing. In this respect, the broken boundary layer behaves like lithospheric plates. However, the descending flow is not like plates because it is nearly symmetrical: fluid from both sides converges and sinks.

A model showing greater asymmetry in the descending flow, like a subducting plate, is shown in Figure 10.2. This differs from the right-hand model of Figure 10.1 mainly in that there are only two weak zones, rather than three, one at the left-hand side of the box and one adjacent to the initial slab. Because there is no weak zone at the right-hand side of the box, and because the boundary conditions at the ends of the box are that the flow has mirror symmetry about the boundary, the right-hand 'plate' is effectively tied to the end of the box. As a result the right-hand plate does not move, although it does stretch to some degree. The velocity of flow under it is less than under the left-hand plate, as can be seen from the streamlines, which are more widely spaced under the right-hand plate. Especially in the last panel (966 Ma) the descending flow has considerable asymmetry, like subduction. With a higher maximum viscosity, the right-hand plate would be more rigid and subduction would be even more asymmetric. I have shown this previously in a model with a maximum viscosity 1000 times the interior viscosity [7].

The model in Figure 10.2 shows that when the top thermal boundary layer is stiff but mobile, like plates, the locations of the upwellings and downwellings are controlled by the locations of the plate boundaries. The downwelling occurs where the boundary layer detaches and sinks at a simulated subduction zone. A passive upwelling occurs at the spreading centre. Another passive upwel-

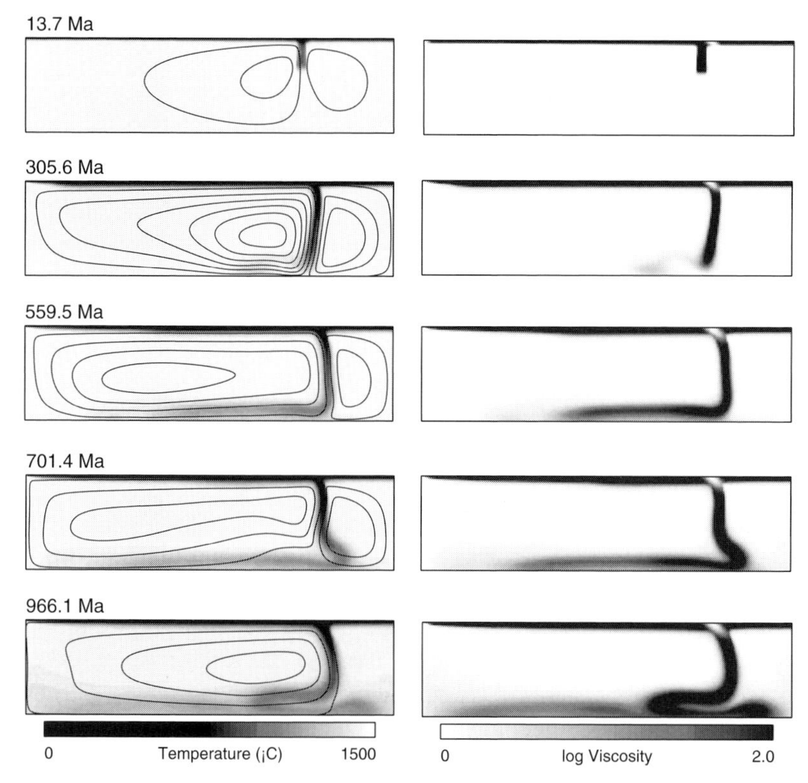

Figure 10.2. Numerical convection sequence in which the descending flow is asymmetric. This resembles subduction, which is completely asymmetric, in that only one plate subducts. The differences between this model and the right-hand model of Figure 10.1 are that here there are only two weak zones (left and middle), the interior break is closer to the end of the box, and it is displaced slightly more from the initial slab.

ling occurs under the stationary plate, but the circulation under this plate is clearly in response to the adjacent descending boundary layer. There are no downwellings other than the subduction downwelling. This is clearly because, in this model, the boundary layer is too stiff to descend anywhere other than at the subduction zone. Brad Hager has captured the essence of this relationship between the plates and the flow in the deeper mantle by saying that *the plates 'organise' the flow in the deeper mantle.*

Models like that in Figure 10.2 are sufficient to demonstrate how properties of the fluid can affect the nature of the flow, but they fall short of full simulations of lithospheric plates in a convecting mantle. Their principal remaining deficiency is that low viscosity weak zones are not the same as faults. This becomes evident if the models of Figures 10.1 and 10.2 are followed for longer periods. Often the descending flow becomes displaced from the weak zone, and the central weak zone may become an upwelling, with drips between the weak zones like that in the left-hand model of Figure 10.1. The beginning of this behaviour can be seen at 1955 Ma in the right-hand model of Figure 10.1. While it is important to appreciate this limitation of the models, it is nevertheless also true that for limited periods they do exhibit plate-like behaviour, and thus demonstrate important effects that are relevant to the mantle.

The boundary layer in the 'unbroken' model of Figure 10.1 is thicker than in the other more mobile models of Figures 10.1 and 10.2, and much thicker than in the constant viscosity models of Figure 8.4. This is because in the unbroken case the stiffer, slower fluid spends more time at the surface and cools more. This has other consequences. Because the lithosphere is thicker, the heat flux conducted through it to the surface is lower. Because heat is not escaping to the surface as quickly, the heat accumulates more in the interior and the interior becomes hotter. It is characteristic of convection with a less mobile upper boundary layer that it has higher internal temperatures. This may be relevant to some other planets, such as Mars and the moon, which seem to have had static surfaces for most of the age of the solar system.

There is a tendency in Figures 10.1 and 10.2 for the downwellings to be more widely space than in the constant viscosity case of Figure 8.4, and thus for the flow to have a larger horizontal scale. This is because the high viscosity can be partly compensated for if the strain rate can be reduced. Since strain rate is the same as velocity gradient and wider spacing lowers the velocity gradients in the boundary layer, wider spacing is favoured. We will see this effect more clearly in the next section.

10.2.2 Influence of high viscosity in the lower mantle

In Chapter 6 I described several kinds of evidence that indicate that the viscosity of the deep mantle is significantly higher than that of the shallow mantle, by a factor roughly between 10 and 100. This high viscosity layer has an effect analogous to that of a stiff upper boundary layer (Figure 10.1): it tends to increase the horizontal length scale of the flow. This is illustrated in Figure 10.3, for a model in which the viscosity increases by a factor of 100 at a depth corresponding to about 730 km in the mantle.

The initial instability of the upper thermal boundary layer comprises small downwellings whose close spacing is comparable to the depth of the low viscosity layer at the top of the box. This nicely illustrates the effect of the Rayleigh–Taylor instability, discussed in

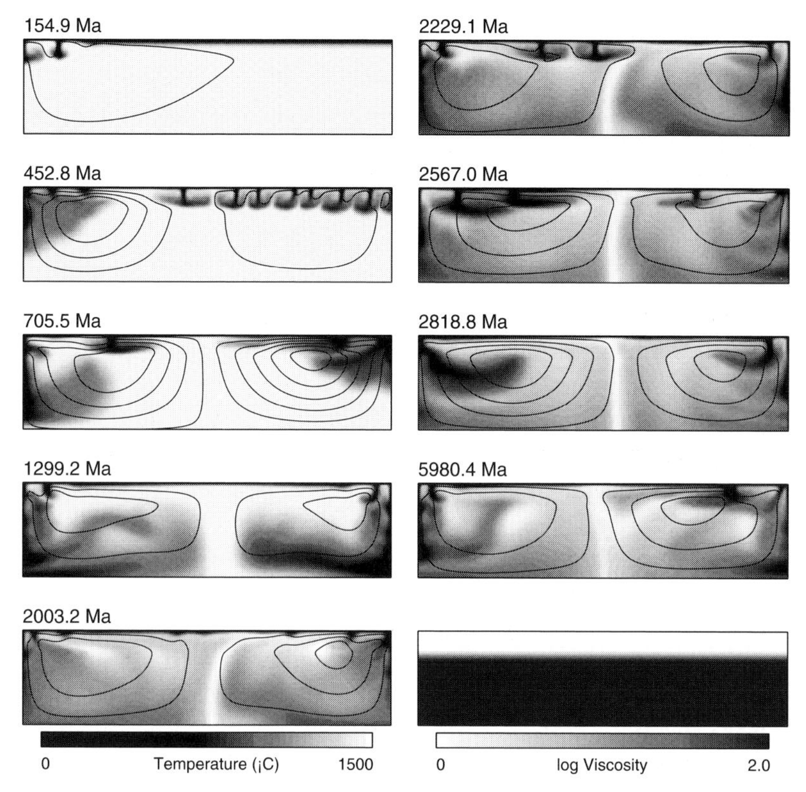

Figure 10.3. Convection sequence with layered viscosity and internal heating. The fluid in the lower part of the box has viscosity 100 times that in the upper part (lower right panel).

Section 8.4, in determining the initial scale of the convective flow. However, the small downwellings slow and accumulate as they enter the high viscosity layer, and they then generate an overturning of the lower layer which has a much larger horizontal scale (452 Ma and 705 Ma). The flow in the lower layer is slower than in the upper layer by about a factor of four, as can be seen from the spacing of the streamlines.

This larger-scale flow rises into the upper layer near the centre of the box, where it turns horizontally and sweeps the small downwellings to the sides. The large-scale, two-cell flow is persistent, in spite of the regular formation of new small-scale downwellings in the upper layer. This is particularly evident in the sequence at 2229 Ma, 2567 Ma and 2818 Ma, in which two strong downwellings form near the centre of the box. Although they seem to have the potential to break into the lower layer and generate a central downwelling, they too are swept to the side and the large cells reassert themselves.

The horizontal scale of the large cells is actually larger than it would be if the viscosity were constant through the box. This is because of two effects working together. First, the larger scale tends to reduce the velocity gradients (= strain rates) and viscous resisting stresses. In a constant viscosity box, the scale is limited by the depth of the box, as was discussed in Section 8.4. Here, however, the upper layer acts as a lubricating layer that allows the flow in the lower layer to reach a large scale before the penalty of viscous dissipation in the upper horizontal flow outweighs the benefit of lower strain rates in the lower layer. This lubrication is the second effect. The combined effect of the viscosity layering has been known for some time from two-dimensional models [8, 9, 10]. It was shown that the preferred wavelength of a marginal instability is about 4.8 times the box depth, compared with 2.8 for constant viscosity [10].

10.2.3 Influence of spherical, three-dimensional geometry

The geometry of the mantle is that of a thick spherical shell whose inner radius is only slightly more than half its outer radius: the ratio is 3482/6371 = 0.55. The surface area of the bottom of the mantle is only about 30% of the top surface area. This means that although the heat flow into the base of the mantle may be only about 10% of the heat flow out of the top (Chapter 11), the heat *flux* would be about 33% of the top heat flux. It means also that vertical flows may diverge or converge significantly, and that a structure 1000 km across at the bottom of the mantle subtends nearly twice the angle at the centre of the earth as a 1000 km structure at the top of the mantle. Other ways to state the latter point are that the deep structure corresponds to a greater fraction of the circumference of the earth, or that it corresponds to a lower-degree spherical harmonic component.

The flow in the mantle is of course three-dimensional, unlike the numerical models used for illustration so far. Although two-dimensional models can be useful, we must remember that mantle flow is not necessarily arranged into tidy rolls, especially as the surface geometry of the plates is so irregular, as we saw in Chapter 9. Thus flow that descends under the north-west Pacific might conceivably rise again in the Indian Ocean rather than in the south-east Pacific. There is a more subtle effect arising from the spherical geometry. The mantle closes upon itself. This means that the flow is really a globally connected three-dimensional flow, and its topology is possibly rather complex.

The spherical geometry has turned out to significantly magnify the effect of viscosity layering in the mantle, compared with the effect in cartesian geometry shown in Figure 10.3. Figure 10.4 illustrates this in spherical geometry by comparing cases with constant viscosity and with viscosity that increases by a factor of 30 at 670 km depth. The increase in the horizontal length scale of the flow is most obvious in the surface patterns, and it is also revealed in the shift of the dominant spherical harmonic components of the flow from around degree 15 to around degree 4–6. This is a greater change in length scale, compared with the cartesian case, although the viscosity increase is only a factor of 30 in this case. As well as an increase in the horizontal length scale, the form of the flow has been changed significantly from downwelling columns to downwelling sheets, the latter being more reminiscent of subducting plates (though the resemblance is superficial).

The effect of the higher viscosity in the lower mantle on the horizontal scale of flow is quite strong in this case. It is presumably

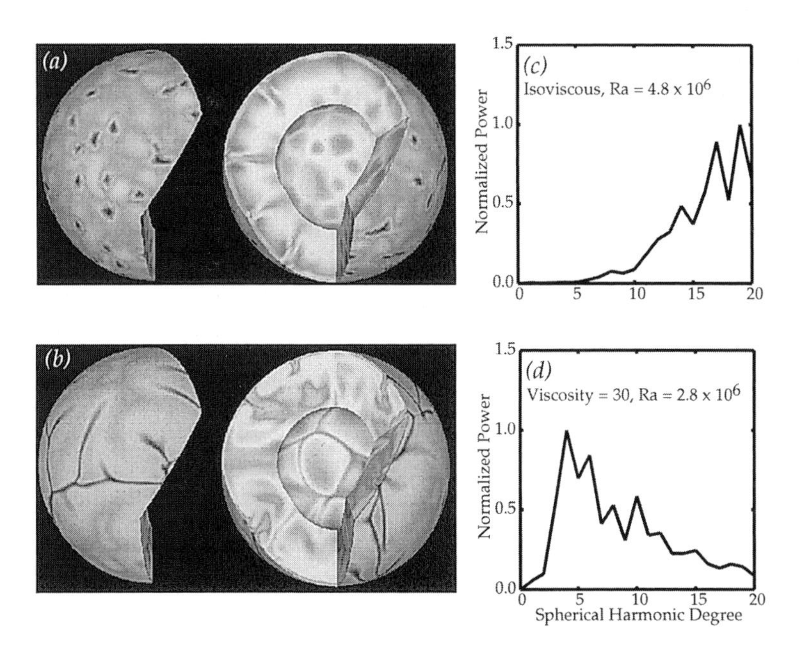

Figure 10.4. The effect in spherical geometry of viscosity that increases with depth. The top model (a) has constant viscosity, while in the bottom model (b) the viscosity increases by a factor of 30 at a depth of 670 km. The spherical harmonic power spectra (c, d) reveal the larger horizontal scale of the flow in model (b) compared with (a), which can also been seen in the surface patterns of (a) and (b). The models are incompressible with internal heating. From Bunge *et al.* [11]. Reprinted from *Nature* with permission. Copyright Macmillan Magazines Ltd.

enhanced in the spherical geometry because a horizontal scale of 5000 km at the bottom of the mantle corresponds to a spherical harmonic degree of 4, whereas 5000 km at the top of the mantle corresponds to degree 8. (The circumference of the earth is 40 000 km, and degree 8 modes divide it into about eight parts.)

Although three-dimensional and spherical effects can be significant, the main effects that we have already seen in two-dimensional flows occur also in three dimensions. This is illustrated in Figure 10.5, which shows snapshots from three models. In the first model, the viscosity is constant and the flow comprises small-scale columnar downwellings, analogous to Figure 8.4. In the second model two things have been added: the top boundary layer viscosity has been increased by a factor of 40 and the observed motions of the earth's plates have been imposed as a boundary condition. As a result the flow under fast plates has become much larger in horizontal scale, becoming more like the flow in Figure 10.2. However,

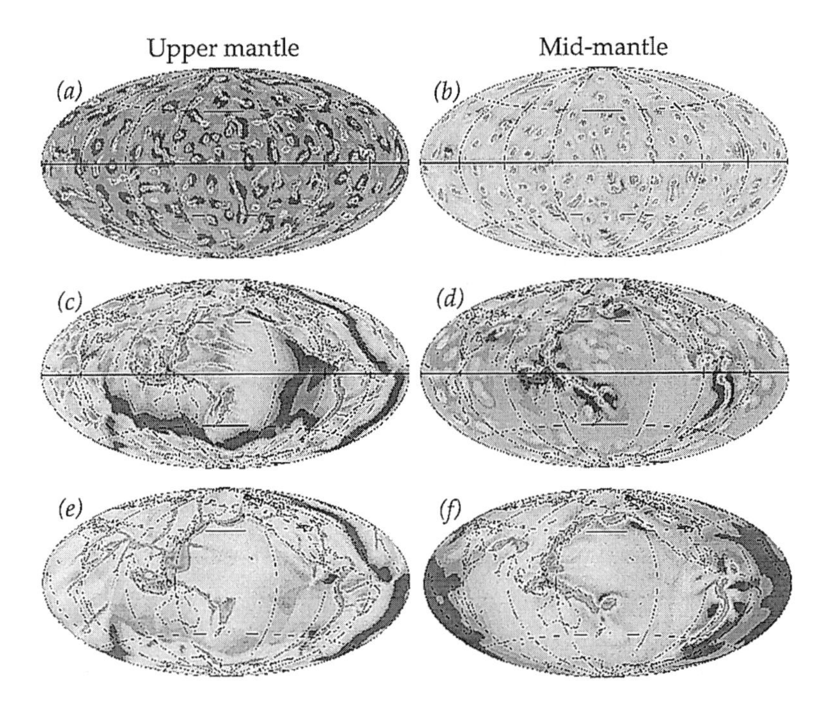

Figure 10.5. The effects of stiff plates and a high-viscosity lower mantle in three-dimensional spherical geometry. (a, b) Constant viscosity convection with a free-slip surface. (c, d) Convection with the observed plate motions imposed as a boundary condition and with the top boundary layer viscosity increased by a factor of 40. (e, f) As in the previous case but with the lower mantle viscosity also increased by a factor of 40. The heating is internal. From Bunge and Richards [12]. Copyright by the American Geophysical Union.

the flow under slow-moving plates like Africa still has many small-scale downwellings. Smaller scale rolls aligned with the plate motion can also be seen under the older, north-west part of the Pacific plate. In the third model the lower mantle viscosity has also been increased by a factor of 40. This has suppressed most of the remaining small-scale downwellings, and the flow is more plate-like. The effect of the higher viscosity in the lower mantle is to enhance the longer-wavelength components of flow, as in Figures 10.3 and 10.4, and so to allow the plates to exert greater control on the flow structure.

10.2.4 Heat transported by plate-scale flow

Convection is a heat transport mechanism and the plate-scale flow is a form of convection. We can therefore look at the rate at which the plate-scale flow transports heat and its role in the earth's thermal regime. Let us consider the life cycle of a plate. Hot mantle rises at a spreading centre and moves away horizontally, cooling by conduction. The cooled layer is stiff and strong, forming a plate. The plate thickens as cooling penetrates deeper, as described in Section 7.3. The cooled plate returns to the interior of the mantle after it subducts. There it absorbs heat from its surroundings, so cooling the interior (e.g. Figures 10.1, 10.2). It is by this cycle of plate formation, thickening, subduction and reheating that the mantle disposes of its internal heat.

The rate of heat transport by the plate cycle can be estimated readily with the help of the relationship between heat flux, q, and age, τ, described in Chapters 4 and Equation (7.3.4), which can be written

$$q = a/\sqrt{\tau}$$

where $a = K\Delta T/\sqrt{\pi\kappa}$. The rate at which new sea floor is formed is $S = 3\,\mathrm{km^2/a}$ [13]. The total area of sea floor is $A = 3.1 \times 10^8\,\mathrm{km^2}$. This means that the sea floor would be replaced within about 100 Ma at the present rate, and that the age of plates as they subduct would average $\tau_s = 100\,\mathrm{Ma}$. The total heat flow, Q, can be estimated by adding up the heat flux from each increment of age:

$$\begin{aligned} Q &= \int_0^{\tau_s} Sq\,\mathrm{d}\tau \\ &= 2Sa\tau_s^{1/2} \\ &= 2Aq_s \end{aligned} \tag{10.2.1}$$

Table 10.1. *Contributions to global heat flow.*

	Area (10^8 km^2)	Mean heat flux (mW/m^2)	Total heat flow (TW)	% of global
1. Sea floor	3.1	100	31	76
2. Continental crust	2.0	50	10	24
a. Crustal radiogenic	—	25	5	12
b. Mantle	—	25	5	12
3. Total mantle (1 + 2b)	5.1	70	36	88
4. Total global (1 + 2a + 2b)	5.1	80	41	100

where $q_s = 50\,\mathrm{mW/m^2}$ is the heat flux at age τ_s. Thus the average heat flux through the sea floor is about $Q/A = 100\,\mathrm{mW/m^2}$, and the total heat flow is 31 TW. This estimate is based implicitly on the assumption that the distribution of seafloor age by area is uniform, or in other words that the area of sea floor with ages in a particular range is independent of age. An estimate by Sclater and others [14] based on actual areas yielded a very similar result.

We can compare this estimate with the total heat loss of the earth, and the total heat loss of the mantle. The total heat flow out of the earth is 41 TW [14]. The continents have an average heat flux of about $50\,\mathrm{mW/m^2}$ (Chapter 4, [14]) and an area of $2 \times 10^8\,\mathrm{km^2}$, so the total heat flow out of continents is 10 TW. A substantial fraction of this is due to radioactivity in the upper crust, and this heat escapes directly to the surface by conduction. As well, heat conducts through the continental lithosphere from the mantle. We saw in Section 7.6 that this depends on the thickness of the continental lithosphere, and that for thicknesses ranging between 100 km and 250 km the mantle heat flux would be about 14–36 $\mathrm{mW/m^2}$. A typical value would be about $25\,\mathrm{mW/m^2}$, or about half of the total continental heat flux. Thus the heat lost from the mantle through the continents is about 5 TW.

These relationships are summarised in Table 10.1. About 75% of the earth's heat loss occurs through the sea floor as a result of plate-scale convection. This is nearly 90% of the total heat loss from the mantle, the balance being lost by conduction through the continents.

The plate-scale flow is thus the dominant means by which heat is lost from the mantle. The balance of mantle heat loss can be accounted for by conduction through continental lithosphere. We will discuss in Chapter 12 the possibility that there may be other modes of convection associated with the top thermal boundary layer of the mantle. However, we can conclude already here that by the fundamental criterion of the amount of heat removed from the mantle, the plate-scale flow is clearly the dominant mode of convection driven by the top thermal boundary layer of the mantle.

10.2.5 Summary

These examples have shown how the unusual mechanical properties of the top thermal boundary layer of the mantle may strongly influence the form of the flow in the deeper mantle. The lithosphere is strong, but broken into pieces that may move relative to each other. The stiffness or strength of the lithosphere prevents it from sinking into the mantle except where it is broken by a fault. Thus downwellings in the mantle are likely to be confined to beneath subduction zones. Passive upwellings must fit between these downwellings. They tend to be slow and broad, but some will be localised near the surface by spreading centres. Others may occur under large plates. The plates thus organise a large-scale mantle flow that we can call the *plate-scale flow*. The plates are an integral part of this mode of mantle convection, in that they comprise its driving thermal boundary layer.

Since the plates tend to be broader than the depth of the mantle, the horizontal scale of the deep mantle flow will be larger than is typical for constant-viscosity convection. However, the combination of higher viscosity in the lower mantle and spherical geometry also tends to favour larger horizonal scales at the surface. These two influences are thus mutually reinforcing.

The flow associated with the plates is the dominant flow in terms of heat and mass transport.

10.3 Effect of phase transformations

The mechanisms by which phase transformations might affect mantle dynamics were presented in Section 5.3. The potential of thermal deflection of phase boundaries to enhance or inhibit subduction had been recognised early in the plate tectonics era [15]. but it was hard to evaluate quantitatively with the limited computational resolution available at the time. Machatel and Weber [16] provoked interest by presenting a convection model that exhibited episodic layering. The model was relatively simple, in that it was two-dimensional and with constant viscosity, but it was at higher Rayleigh number than previous models. A calculation analogous to theirs is presented in Figure 10.6. With a moderately strong thermal deflection of the phase boundary, penetration of the phase

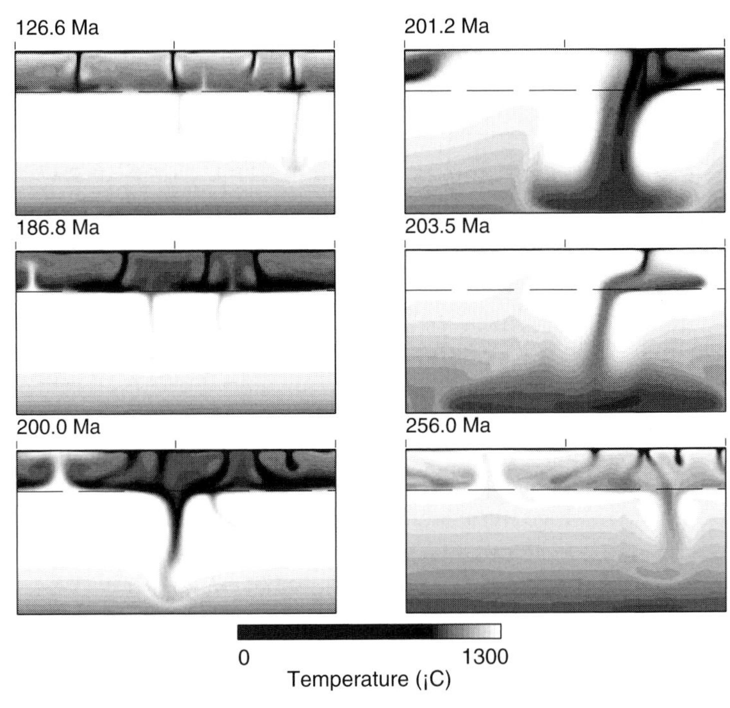

Figure 10.6. Constant viscosity convection sequence in which phase transformation buoyancy causes temporary layering of the flow. The buoyancy corresponds to a Clapeyron slope of −3 MPa/K. From [17]. Copyright by Elsevier Science. Reprinted with permission.

transformation region is blocked for an extended period, while the top layer of fluid cools by heat loss to the surface and the bottom layer warms by internal heating. Eventually the temperature difference between the layers becomes large enough that some of the fluid breaks through, triggering a complete overturn in which the cooler fluid from the top layer drains into the lower layer. The cycle may then repeat.

Although the effects of a phase transformation may be dramatic in a two-dimensional, constant viscosity model, it has been found that they are less dramatic in three dimensions and with temperature-dependent viscosity. In three dimensions, the fluid penetrates more easily. Although flow through the phase barrier is inhibited to some degree, there tend to be plume-like (columnar) breakthroughs fairly regularly, so that large differences between the temperatures of the layers do not accumulate [18]. This difference between two and three dimensions reflects a general tendency in convection that if the fluid rheology permits columns to be formed readily, they tend to predominate over sheets, which are the only form possible in two dimensions. In other words, columns seem to be more unstable and penetrative than sheets. This point will arise again in Chapter 11 on plumes.

If the viscosity is temperature-dependent, the flow can also penetrate a phase barrier more readily than in constant-viscosity convection, even in two dimensions. The reasons are different for upwellings and downwellings. For upwellings or plumes, a new plume forms a large head, as we will see in Chapter 11. Plume heads have a large buoyancy which is more capable of penetrating resistance than a narrow column [17]. For downwellings, or subducted lithosphere, the greater stiffness of the cold fluid means that more of the negative buoyancy of the stiff sheet is transmitted to the location of resistance. This is illustrated in Figure 10.7.

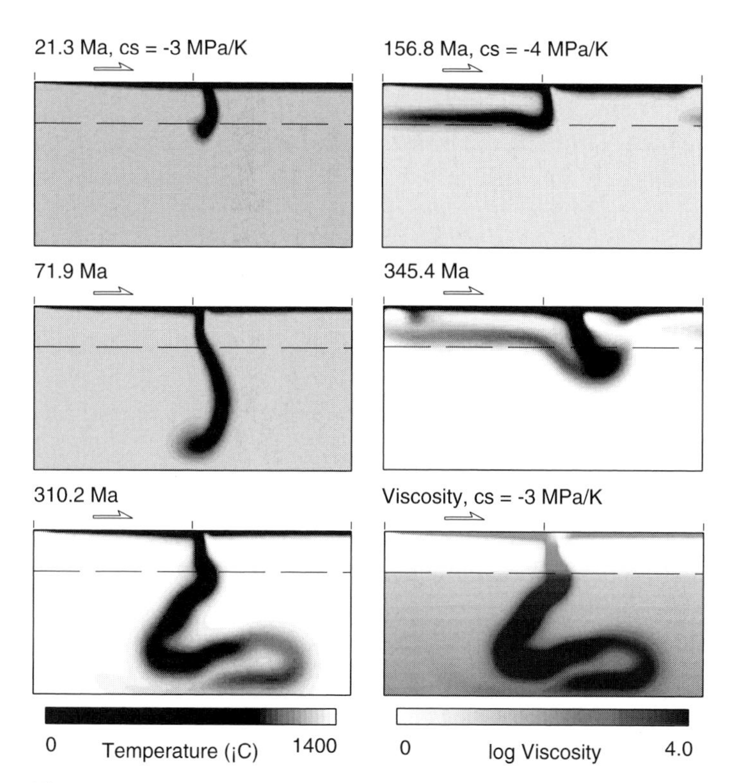

Figure 10.7. Convection sequences with a phase barrier and temperature-dependent viscosity, showing the greater ability of descending stiff sheets to penetrate a phase barrier than the constant-viscosity downwellings in Figure 10.6. Left: model with a Clapeyron slope (cs) of −3 MPa/K and reflecting end walls. The viscosity structure for the last frame is shown in the lower right panel. Right: Clapeyron slope of −4 MPa/K and periodic end walls, in which fluid that flows out at one end of the box flows back in at the other end. Penetration is delayed for a long time in the latter case. From [17]. Copyright by Elsevier Science. Reprinted with permission.

Present indications are that the phase transformation or transformations near a depth of 660 km in the mantle do not have a dramatic effect on mantle flow, although they may cause significant complications, such as temporary blockages. Support for this assessment comes from three sources. First, from the modelling studies just cited. Second, from the evidence from seismic tomography that many subducted slabs do penetrate into the lower mantle (Chapter 5). Third, from the likelihood that the net effect of the combined spinel to perovskite and garnet to perovskite transformations is equivalent to a Clapeyron slope substantially less negative than the −4 MPa/K or more that seems to be required for layering in the more realistic models just discussed. This was discussed in Section 5.3.3. It is also likely that the effects of compositional deflections of phase transformations (Section 5.3.4) are less than the effects of thermal deflections [17], although some significant uncertainty in this topic remains [19, 20].

On the other hand, the phase barrier was probably less likely to be penetrated early in earth history, when the mantle was hotter, with lower viscosity and younger, thinner subducted plates. This possibility will be addressed in Chapter 14.

10.4 Topography and heat flow

Having developed an idea of what the plate mode of mantle convection is and how it works, we can now turn to how well this conception can account for observations. In this section we look at the topography and heat flow predicted from numerical models and compare them both with observations and with the simple theory of plate cooling that was developed in Section 7.4. Given that the cooling plate model totally ignores any influence from the deeper convecting mantle, its apparent success is remarkable, but we should if possible attempt to verify that this is a reasonable assumption. The comparisons here allow us to evaluate the influence of the deeper mantle.

We also look at how the intimate relationships between topography, heat flow and convection are expressed through the plate mode. We discussed these qualitatively in Section 8.8, and here we can take advantage of the more specific context of the plate mode to quantify them. The analogous relationships for plumes will be considered in Chapter 11.

10.4.1 Topography from numerical models

The calculated topography and geoid from the last frame of the model of Figure 10.2 are shown in Figure 10.8. The plate-like character of the flow is reflected in the topography of the moving plate, which decreases from a peak of about 2.5 km at the spreading centre on the left. The trench topography is not realistic for three reasons. First and most important, the viscosity structure includes the artifice of the weakened zone simulating a subduction fault. Second, the trench depth in such models is sensitive to the deeper viscosity structure, and no attempt has been made to make it realistic. Third, the subducted lithosphere is probably thicker than is realistic. Thus the right-hand part of the plot should not be regarded as significant.

Despite the limitations of the model, the topography in Figure 10.8 reproduces the general character of seafloor topography, declining monotonically from the spreading centre on the left to the subduction zone near the right. This is demonstrated more clearly by the comparison with the topography predicted from the cooling halfspace model of Section 7.4 (Equation (7.4.2)), which predicts that the depth should increase in proportion to the square root of the cooling time. However the comparison is not rigorous, because the plate velocity in the model varies considerably with time (Figure 10.2) whereas the halfspace curve assumes a constant plate velocity.

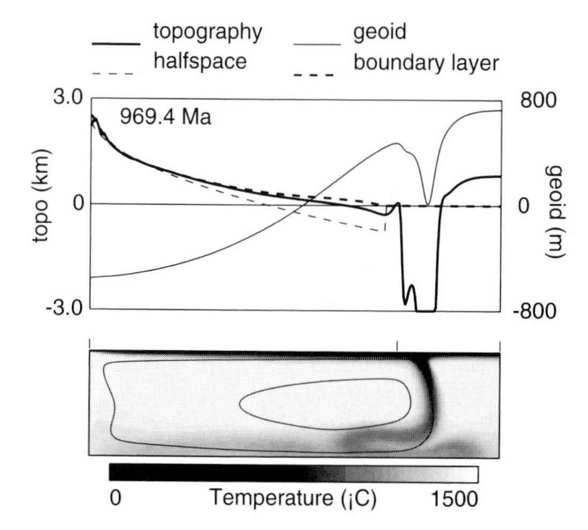

Figure 10.8. Topography and geoid calculated from the model of Figure 10.2, last panel. Results near the subduction zone are not realistic. Also shown are the topography predicted from the halfspace model (Equation (7.4.2)), using an average model plate velocity, and the topography due to the thermal contraction in the thermal boundary layer of the model.

The model topography is also compared in Figure 10.8 with the topography that would result just from thermal contraction in the actual thermal boundary layer of the model. These are quite close. In principle there would also be contributions to the topography from deep thermal structure and from pressure gradients in the fluid, so this demonstrates that in this case most of the topography is due to the thermal boundary layer. The remaining difference in this case is most likely from the deep, cool fluid, which produces some additional depression of the top surface.

A rigorous comparison with the halfspace theory is shown in Figure 10.9. For this comparison, a new model has been computed in which a horizontal velocity at the surface has been prescribed as a boundary condition, with the left-hand segment of the boundary moving to the right (see the half-arrow) and the right-hand segment stationary. Thus the top boundary is prescribed to move with steady, piecewise uniform velocity, in the manner of plates. This is analogous to the boundary condition used by Bunge and Richards [12] in Figure 10.4 (whereas in Figure 10.2 the top boundary is free-slip).

The flow is fairly similar to that in Figure 10.2, and the computed topography and geoid have similar character to those in Figure 10.8. Again the result near the trench should not be regarded as significant, because of the artificial viscosity structure there. However, in Figure 10.9 the match between the computed topography and the halfspace prediction is closer than in Figure 10.8, the computed topography tending to be slightly lower ampli-

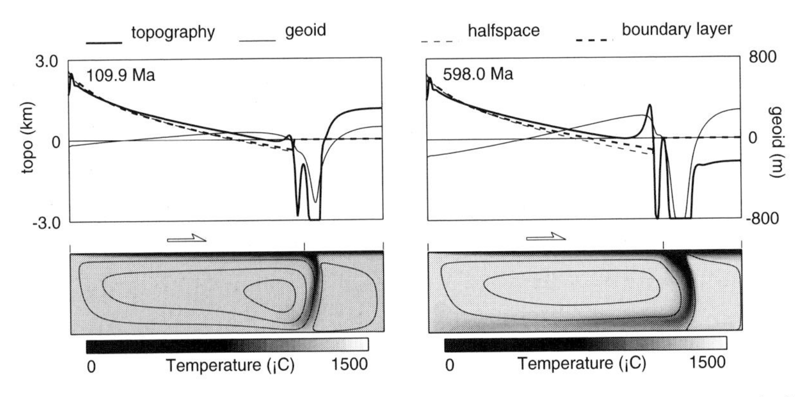

Figure 10.9. Topography and geoid calculated at two times from a numerical convection model in which more plate-like behaviour is induced by prescribing piecewise uniform, steady horizontal velocity at the surface. Results near the subduction zone are not realistic. In this case the model topography, the halfspace prediction and the contribution from the thermal boundary layer are all similar.

tude at both of the times shown. The topography due just to the thermal boundary layer of the model is quite close to the halfspace prediction, especially at the earlier time, indicating that the thickening of the boundary layer of the model closely follows the cooling halfspace assumption. The boundary layer topography is also close to the model topography, indicating that the contribution from deep thermal structure is small. The remaining difference between the model topography and the boundary layer contribution is most likely due to a pressure gradient in this case. However, other models (not shown here) suggest that deep thermal structure can have a significant effect also.

These results show that numerical convection models that include a reasonable simulation of lithospheric plates can yield topography that corresponds quite well with observed seafloor topography (Chapter 4, Figures 4.5, 4.6). They also show that the topography is due mainly to the thermal contraction in the top thermal boundary layer and not to deeper thermal structure or dynamical effects like pressure gradients. This supports the longstanding interpretation that midocean ridge topography is due to the cooling of the lithosphere. Furthermore, it is a result that depends on the models being heated internally, rather than from the base, as we will discuss in Chapter 12.

The fact that there are some differences between the numerical models and the halfspace prediction is not necessarily a deficiency of the models, because for the earth we do not know all of the parameters (such as thermal expansion and thermal diffusivity) well enough to make an accurate independent prediction of the halfspace subsidence rate of the sea floor. The key observation is that the observed seafloor topography follows approximately the *form* of the halfspace model (at least for ages younger than about 40 Ma). There seems in fact to be a tendency for the topography of convection models to have a shallower slope than the halfspace prediction. (The comparison can be made rigorously for the model, since all of the parameter values are known.) This is evident for example in some earlier results of this kind [21]. This may indicate that the halfspace subsidence rate of the mantle is greater than has been inferred from young sea floor, and this could have significant implications for the interpretation of the geoid.

10.4.2 Geoids from numerical models

Geoid perturbations calculated from the same models are included in Figures 10.8 and 10.9. These show more variation between cases than the topography, though the three examples share a general form. In each case the geoid shows a very broad high over the subduction zone, but with a narrower low superimposed close to the subduction zone. The substantial differences between the different cases arise because the geoid is a small difference between large contributions of opposite sign and as a result it is quite sensitive to details of the model, such as vertical viscosity structure. This is discussed in Section 6.9.3. This sensitivity is potentially a source of important information, but at the time of writing the extant models do not satisfactorily reconcile the geoid and topography simultaneously. Because I am confining this book to the better-established aspects of mantle convection, I will not go beyond a brief qualitative discussion here.

An important point in the discussion in Section 6.9.3 is that the net geoid becomes more positive if subducting lithosphere encounters greater resistance at depth. In the case considered there, the greater resistance comes from an increase in viscosity with depth. In the models presented here the main resistance comes from the bottom boundary, and the geoid is affected if the subducted lithosphere is sufficiently stiff to transmit some stress back to the top surface. This seems to be the explanation for the broad positive geoid at the subduction end of the models. The localised geoid low results from the depression formed by the net downward pull of the subducted lithosphere. Another effect in these models is that as the cool fluid returns along the base of the box it depresses the top surface by a few hundred metres. Both the geoid and the topography can be seen to be lowered in response to the presence of deep cool fluid in Figures 10.8 and 10.9, and the effect is magnified in the geoid.

The observed geoid shows some of the character of these calculated geoids. There tends to be a broad geoid and gravity high over subduction zones, which was remarked upon in Figures 4.9a, b. There is a more localised low over oceanic trenches, though it is not clearly resolved in Figure 4.9. However, the longest wavelength components of the geoid are not obviously related to the present plate configuration. It seems to be related to the locations of past subduction zones [22], as will be discussed in Chapter 12.

10.4.3 Heat flow from numerical models

The surface heat flux calculated from the model of Figure 10.9 (left) is shown in Figure 10.10. The heat flux predicted from the halfspace model (Equation (7.3.4)) is shown for comparison. The model heat flux is slightly higher than the halfspace heat flux near the left side of the box, but generally follows it quite closely.

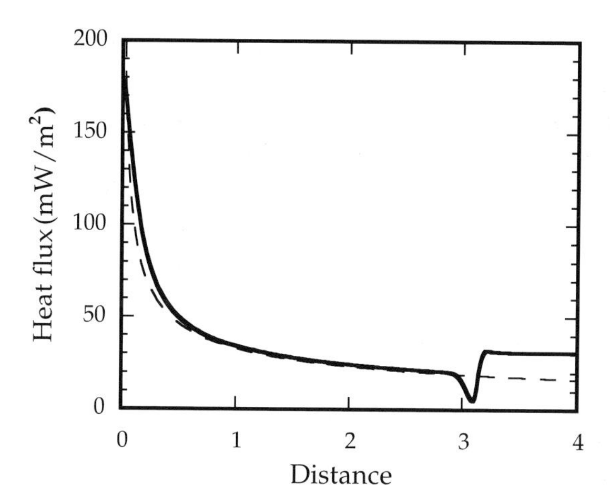

Figure 10.10. Heat flux from the model shown in Figure 10.9 (left) (——), compared with the heat flux calculated from the halfspace model (Chapter 7, Equation (7.3.4)) (– – –). Distance is the horizontal position along the box, in units of box depth.

This comparison is a test of two aspects of the model. First, that the thermal boundary layer is well resolved by the numerical grid. It seems to be, except very close to the left-hand edge, where the boundary layer thickness approaches zero. Second, that there is little deformation of the thermal boundary layer fluid as it moves to the right. This also seems to be true to a good approximation. Thus the thermal boundary layer of the model behaves like a rigid plate to a good approximation. The same conclusion could be inferred from the good match between the halfspace topography and the topography due to the boundary layer in Figure 10.9, but the test is more explicit here. This result supports the interpretations of the topography made in the last section.

10.4.4 General relationships

We can now look at the fundamental relationships that exist between topography, heat flow, buoyancy and convection, using the specific context of the plate mode of convection to derive quantitative relationships. The source of the relationships is that buoyancy and topography are both results of thermal expansion.

In Section 7.4 we looked at the topography generated by a cooling plate due to its thermal contraction, as sketched in Figure 7.5. We consider this situation again, supposing that a plate reaches thickness D before it subducts, and that it moves with velocity v. The thermal contraction of a mass column of height D is

$$h = \alpha \Delta T D \tag{10.4.1}$$

where α is the thermal expansion coefficient and ΔT is the average temperature deficit of the cooled lithosphere. Here I ignore the correction to the topography due to the surface being underwater, for simplicity.

We can estimate the rate of heat loss from this cooling plate in a different way than we did in Chapter 7. The heat lost from a mass column of unit surface area and height D is $\rho C_P \Delta T D$. If the plate is moving at velocity v, then a vertical surface within it moves a distance $v \cdot \mathrm{d}t$ in time $\mathrm{d}t$ and sweeps out a column of width $v\mathrm{d}t$. This column will have lost an amount of heat $\rho C_P \Delta T D \cdot v\mathrm{d}t$ per unit distance in the third dimension (Figure 7.5). Since cooled lithosphere that is about to subduct is continuously being replaced at the surface of the earth by hot mantle at a spreading centre, this net amount of heat has been lost from the plate in the time interval $\mathrm{d}t$. Thus the net rate of heat loss per unit area from the plate is

$$q = \rho C_P \Delta T D v \tag{10.4.2}$$

The negative buoyancy per unit area of the plate is $g \Delta\rho D = g\rho\alpha\Delta T D$. By the same logic that we just used for the heat loss, negative buoyancy is being continuously created within the plate as it cools and moves from ridge to trench at velocity v, and the rate of formation of negative buoyancy, b, is

$$b = g\rho\alpha\Delta T D v \tag{10.4.3}$$

We can now relate b and h to q:

$$\frac{b}{q} = \frac{g\alpha}{C_P} \tag{10.4.4}$$

$$\frac{h}{q} = \frac{\alpha}{\rho C_P v} \tag{10.4.5}$$

These equations give quantitative expression to the ideas of Sections 7.4 and 8.8. They express the ideas that heat loss causes thermal contraction that is manifest as topography, and that the thermal contraction generates negative buoyancy that drives flow in a viscous fluid. Notice that the ratios in these two relationships do not depend on the temperature drop, ΔT, or on the thickness of the plate, D. They depend only on material properties (α, ρ, C_P), g and the plate velocity, v. This is important because plate thickness D is not very well constrained and, in the case of plumes considered in the next chapter, ΔT is not well constrained.

A fundamental message here is that the earth's topography carries important information about mantle convection. We have seen in Section 10.2.4 that the plate-scale flow accounts for the dominant heat loss from the mantle. We have also seen, in Chapter 4, that the topography associated with the plate-scale flow, the midocean ridge system, is the dominant topography of the sea floor. The relationships presented here make it clear that this is no coincidence. The dominance of the midocean ridge topography tells us rather directly that the plate-scale flow is the dominant flow in the mantle.

10.5 Comparisons with seismic tomography

The seismic tomography images of mantle structure that are summarised in Chapter 5 provide a different kind of constraint on mantle dynamics. Some recent dynamical models have yielded some results with an encouraging resemblance to these images. Although the models are new and their robustness has not been extensively tested, they reveal a level of correspondence between theory and observations that deserves to be represented here.

10.5.1 Global structure

Bunge *et al.* [23] have presented three-dimensional spherical models of mantle convection that incorporate surface plate motions not just as they are at present (Figures 10.5), but as they have been reconstructed for the past 120 Ma. Temperatures at a depth of 1100 km from their GEMLAB 2 model are shown in Figure 10.11. The main features are a band of cool material under North and South America and cool regions under southern and eastern Asia. These features bear significant similarity to the tomographic images shown in Figure 5.14, especially under North and South America. The details in the Asian region are considerably different from the tomography, but the occurrence of cool regions in the general vicinities of eastern and southern Asia are appropriate.

While the level of agreement may not be surprising given that the prescribed surface velocities ensured that there would be convergence and downwellings in these regions, the results could have been quite different with different assumptions about other aspects

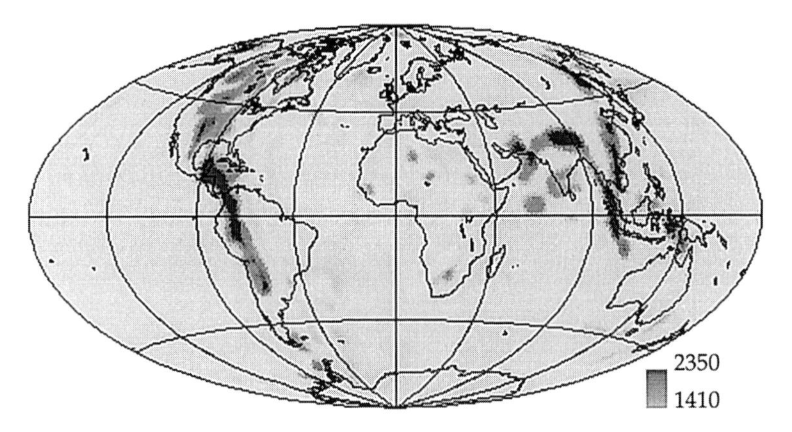

Figure 10.11. Temperatures at a depth of 1100 km from the GEMLAB 2 convection model of Bunge *et al.* [23]. The model incorporates the surface motions of the plates, reconstructed for the past 120 Ma, as a time-dependent boundary condition. It also has high viscosities in the lithosphere and the lower mantle, as in Figure 10.5(e,f). Copyright American Association for the Advancement of Science.

of the mantle. For example, if the lower mantle viscosity were less viscous than assumed, the mid-mantle downwellings would probably correspond more closely with present subduction zones. Conversely, a more viscous mantle might retain thermal structure from (unkown) plate configurations from several hundred million years ago. If subduction through the transition zone were substantially delayed by phase transformation buoyancy, then the correlation between subduction zone history and deep mantle thermal structure might be more haphazard (see Figure 10.7, right side). If, as has been vigorously advocated until recently, mantle convection occurred in two layers separated at 660 km depth, then the mid-mantle thermal structure might bear little relationship to the locations of surface subduction zones.

Thus models such as that in Figure 10.11 are an important test of the general picture of the mantle developed here. We can look forward to increasingly detailed tests of this kind, and to productive feedback into other disciplines as the dynamical models are refined. Indeed at a conference in 1997 the enthusiasm for the new tomographic images was such that people working on surface plate reconstructions began to conjecture how the tomographic images would constrain the surface reconstructions for older periods, and to anticipate useful input from dynamical models. This reflects a dramatic shift in attitudes that has occurred in the late 1990s as a result of advances in tomography (Chapter 5) and dynamical modelling (such as in Figure 10.11).

10.5.2 Subduction zones

Three-dimensional models such as that in Figure 10.11 still do not have sufficient resolution and power to fully resolve the lithospheric boundary layer, nor to incorporate such important features as temperature-dependent viscosity and weak plate margins (though current progress is rapid and such accomplishments are likely within the next few years). Two-dimensional models such as those shown in this chapter are therefore still useful. The following figures show the influence of high viscosity in the lower mantle on the trajectory and shape of descending high-viscosity sheets.

These models are similar to those of Figures 10.2 and 10.9, but with the additional artifice that there is no top thermal boundary layer on the right-hand side of the model. This is accomplished by making the temperature on this part of the top boundary equal to the internal temperature, and it is useful because it reduces the likelihood that the 'subducting' boundary layer will couple to the nonsubducting plate (as is beginning to happen in the lower right panel of Figure 10.1), so there is a cleaner separation and descent of the left-hand boundary layer. (This device was first used by U. Christensen [24].) Since the objective is to study the way the subducted 'lithosphere' behaves as it sinks to greater depths, such a device is not unreasonable.

Figure 10.12 shows a series of frames from a model in which the viscosity depends on both temperature and depth. As in Figure 10.9, the viscosity has a strong temperature dependence, truncated here to a maximum 300 times ambient value. As well in this case the viscosity increases with depth by a total factor of 100. This increase is a superposition of two variations. There is a step by a factor of 10 at a depth of 700 km, and there is a smooth exponential variation with depth by another factor of 10. The resulting viscosity variation with depth is shown in the lower right panel of Figure 10.12. Periodic boundary conditions apply to the ends of the box, which means that as fluid flows out at one end it flows back in at the other end, so there is no artificial constraint on the angle at which the cold fluid descends.

As the cool, stiff sheet in Figure 10.12 encounters the increasing resistance at depth, it buckles and forms large, open folds that occupy most of the depth of the lower mantle. Because of the high viscosity near the base of the box, the cool fluid spreads laterally only slowly, and forms a broad pile that traps some ambient fluid within the folds. Because of the presence of this pile, the sheet is not clearly differentiated from its surroundings in the lower quarter of the box.

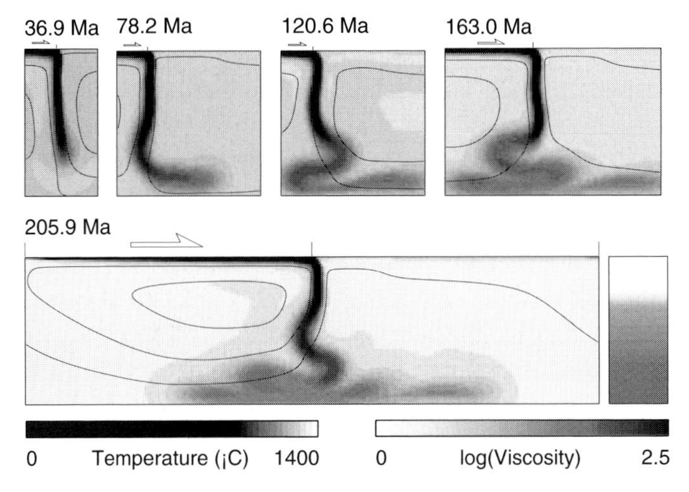

Figure 10.12. Frames from a convection model simulating subduction of an oceanic plate into mantle viscosity that increases with depth. The viscosity steps by a factor of 10 at a depth of 700 km, and it also has a smooth exponential increase with depth by a further factor of 10. Thus the bottom viscosity is 100 times the viscosity just below the lithosphere. The small panel at the lower right shows the viscosity variation with depth. The viscosity is also strongly temperature-dependent, with a maximum value 300 times the ambient value. The full computational box is shown only in the last frame (at 205.9 Ma). Only the central segment of the box is shown in the earlier frames.

Figure 10.13 demonstrates the effect of different amounts of viscosity increase with depth. The top left frame has no increase with depth (just temperature dependence), and the cool sheet falls vertically almost to the bottom before it develops tight folds and then spreads along the bottom of the box. The intermediate three models show various degrees of folding in the lower mantle. The last two models differ from each other only in that one has periodic end boundary conditions while the other has mirror end boundary conditions, which prevent flow through them. The periodic boundary conditions allow material to flow right through the box, and the result is that the descending sheet sinks obliquely to the base and then moves horizontally away to the right. In contrast, in the last model a net flow through the box is not permitted. As a result the sheet descends more vertically, but the greater increase of viscosity with depth in this case causes large open folds to occupy the whole depth of the lower mantle.

The different form of the obliquely descending sheet occurs because the higher viscosity at depth tends to inhibit buckling. The sheet has responded to the resistance at depth in two other ways. First, its speed of sinking has slowed, and it has thickened as

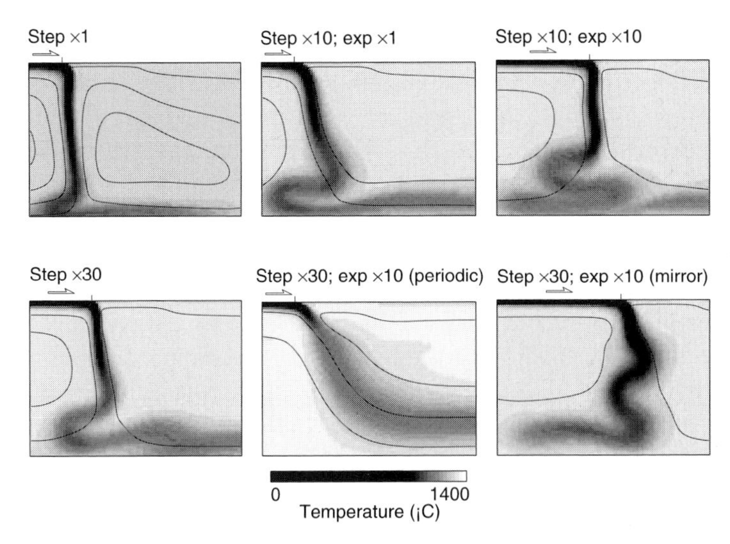

Figure 10.13. Frames from a series of convection models like that of Figure 10.12, with different variations of viscosity with depth. The cases are labelled with the magnitude of the viscosity step at 700 km and the magnitude of the superimposed smooth exponential increase, if any. The top left frame has no depth dependence. End boundary conditions are periodic, except in the lower right frame, which has no-flow ('mirror') end walls.

a result, as can be seen both in the thermal structure and in the streamlines, which diverge through the viscosity step. Second, the sheet has deflected sideways to minimise its need to bend.

What is happening in this case can be thought of in two equivalent ways. From one point of view, the entire lower mantle is flowing to the right, relative to the surface plate system. It is effectively pushed by the obliquely sinking sheet. From another point of view, the plate system is moving to the left relative to the lower mantle. These views are equivalent because there is no intrinsic frame of reference in this model. The equivalent in spherical geometry would be a relative rotation of the lithosphere and the lower mantle. Whether a rotation is induced in the mantle would depend on whether the lithosphere and the lower mantle are free to move relative to each other, which would depend on the forces induced by other subduction zones.

The first model in Figure 10.13, with no depth dependence of viscosity, bears little resemblance to the tomographic cross-sections under subduction zones shown in Figure 5.13, but the models with depth-dependent viscosity begin to reveal some of the general character of the tomographic images. Detailed comparison is not appropriate since Figure 5.13 shows cross-sections through three-dimensional structures and since the real plate system changes with time. What the models reveal is that the combination of a high-viscosity sheet sinking into a lower mantle whose viscosity increases with depth may result in large-scale buckling and folding or in oblique descent, depending on local circumstances. In these respects the models resemble the tomographic images.

The model results do not resemble the tomography if the lower mantle is no more viscous than the upper mantle (Figure 10.13, upper left panel). On the other hand a much more viscous lower mantle would induce buckling in the upper mantle. A partial or complete blockage of flow through the transition zone, by phase transformations or different composition, would yield different kinds of structures, such as that on the right of Figure 10.7. As with the three-dimensional model of Figure 10.11, the similarity in the character of the model results and the tomography is a significant test of the general picture of the mantle inferred in this chapter.

10.6 The plate mode of mantle convection

I have called this chapter *The plate mode* to emphasise that the plate-scale flow is one mode of convection out of a range of conceivable modes. At least one other mode seems to exist in the mantle – the plume mode that is the subject of the next chapter. We may accurately think of the plate mode of mantle convection as the mode driven by the negative buoyancy of subducting oceanic plates.

In this chapter we have looked in a more quantitative way at the relationship between lithospheric plates and convection in the mantle, and we have made some fairly straightforward inferences that have led us to a picture of how the plates and mantle convection are related. This picture is as follows.

The lithospheric plates are an integral part of mantle convection. They comprise the top thermal boundary layer of the mantle. A plate is 'rigid', in Wilson's terms [25], because it is cold and its viscosity is high. However, the lithosphere as a whole is mobile, rather than being frozen into immobility, because it has limited strength like a brittle solid. It has become broken into pieces (the plates) which can move relative to each other. When a plate sinks into the mantle, it does so because it has become negatively buoyant from cooling at the surface. When it sinks, it drives flow in the mantle. By conservation of mass, the descending flow must be balanced by (passive) ascending flow that ultimately emerges at midocean ridges. Trenches and ridges thereby control, to first order, the locations of downwelling and upwelling of the plate-scale flow. The cycle of formation, cooling, subduction and reheating of plates is the principal means by which the mantle loses heat. This 'plate-scale flow' is the dominant mechanism that transports heat through and out of the mantle. Its dominance is reflected in the dominance of the topography associated with this flow: the midocean ridge topography.

Although the plates seem to exert the dominant control on the structure of the plate-scale flow, we have quantitatively evaluated two other factors that may significantly influence the flow. These are the inferred increase of viscosity with depth and phase transformations in the transition zone. These are certainly not the only factors that influence the flow. For example, the continental crust is buoyant relative to the mantle and it does not readily subduct. Thus the arrival of continental crust at a subduction zone seems usually to have forced a modification of plate motions, which would then modify the mantle flow structure. It also seems that plumes have sometimes triggered the rifting of a plate [26, 27].

This general picture of the plate mode of mantle convection has been shown in this chapter to be capable of reproducing some primary observational constraints on mantle flow. Seafloor topography and heat flow can be matched to quite good accuracy. The general character of three-dimensional mantle structure revealed by seismic tomography can be reproduced, such as the belts of high wavespeed under North and South America and the Alpine–Himalaya belt, and the tendency of subducted lithosphere to buckle into open folds in the lower mantle.

10.7 References

1. G. F. Davies, Ocean bathymetry and mantle convection, 1. Large-scale flow and hotspots, *J. Geophys. Res.* **93**, 10467–80, 1988.
2. G. F. Davies and M. A. Richards, Mantle convection, *J. Geol.* **100**, 151–206, 1992.
3. A. Holmes, Continental drift: a review, *Nature* **122**, 431–3, 1928.
4. A. Holmes, Radioactivity and earth movements, *Geol. Soc. Glasgow, Trans.* **18**, 559–606, 1931.
5. D. Forsyth and S. Uyeda, On the relative importance of the driving forces of plate motion, *Geophys. J. R. Astron. Soc.* **43**, 163–200, 1975.
6. W. M. Chapple and T. E. Tullis, Evaluation of the forces that drive the plates, *J. Geophys. Res.* **82**, 1967–84, 1977.
7. G. F. Davies, Mantle convection with a dynamic plate: topography, heat flow and gravity anomalies, *Geophys. J.* **98**, 461–4, 1989.
8. C. P. McFadden and D. E. Smylie, Effect of a region of low viscosity on thermal convection in the mantle, *Nature* **220**, 468–9, 1968.
9. H. Takeuchi and S. Sakata, Convection in a mantle with variable viscosity, *J. Geophys. Res.* **75**, 921–7, 1970.

10. G. F. Davies, Whole mantle convection and plate tectonics, *Geophys. J. R. Astron. Soc.* **49**, 459–86, 1977.
11. H.-P. Bunge, M. A. Richards and J. R. Baumgartner, Effect of depth-dependent viscosity on the planform of mantle convection, *Nature* **379**, 436–8, 1996.
12. H.-P. Bunge and M. A. Richards, The origin of long-wavelength structure in mantle convection: effects of plate motions and viscosity stratification, *Geophys. Res. Lett.* **23**, 2987–90, 1996.
13. C. G. Chase, The n-plate problem of plate tectonics, *Geophys. J. R. Astron. Soc.* **29**, 117–22, 1972.
14. J. G. Sclater, C. Jaupart and D. Galson, The heat flow through the oceanic and continental crust and the heat loss of the earth, *Rev. Geophys.* **18**, 269–312, 1980.
15. G. Schubert, D. A. Yuen and D. L. Turcotte, Role of phase transitions in a dynamic mantle, *Geophys. J. R. Astron. Soc.* **42**, 705–35, 1975.
16. P. Machetel and P. Weber, Intermittent layered convection in a model mantle with an endothermic phase change at 670 km, *Nature* **350**, 55–7, 1991.
17. G. F. Davies, Penetration of plates and plumes through the mantle transition zone, *Earth Planet. Sci. Lett.* **133**, 507–16, 1995.
18. P. J. Tackley, D. J. Stevenson, G. A. Glatzmaier and G. Schubert, Effects of an endothermic phase transition at 670 km depth in a spherical model of convection in the earth's mantle, *Nature* **361**, 699–704, 1993.
19. S. E. Kesson, J. D. Fitz Gerald and J. M. G. Shelley, Mineral chemistry and density of subducted basaltic crust at lower mantle pressures, *Nature* **372**, 767–9, 1994.
20. S. E. Kesson, J. D. Fitz Gerald and J. M. Shelley, Mineralogy and dynamics of a pyrolite lower mantle, *Nature* **393**, 252–5, 1998.
21. G. F. Davies, Role of the lithosphere in mantle convection, *J. Geophys. Res.* **93**, 10451–66, 1988.
22. M. A. Richards and D. C. Engebretson, Large-scale mantle convection and the history of subduction, *Nature* **355**, 437–40, 1992.
23. H.-P. Bunge, M. A. Richards, C. Lithgow-Bertelloni, J. R. Baumgardner, S. P. Grand and B. A. Romanowicz, Time scales and heterogeneous structure in geodynamic earth models, *Science* **280**, 91–5, 1998.
24. U. R. Christensen, The influence of trench migration on slab penetration into the lower mantle, *Earth Planet. Sci. Lett.* **140**, 27–39, 1996.
25. J. T. Wilson, A new class of faults and their bearing on continental drift, *Nature* **207**, 343–7, 1965.
26. W. J. Morgan, Hotspot tracks and the opening of the Atlantic and Indian Oceans, in: *The Sea*, C. Emiliani, ed., Wiley, New York, 443–87, 1981.
27. R. I. Hill, Starting plumes and continental breakup, *Earth Planet. Sci. Lett.* **104**, 398–416, 1991.

CHAPTER 11

The plume mode

Mantle plumes are buoyant mantle upwellings that are inferred to exist under some volcanic centres. In Chapter 8 I stated the basic idea that convection is driven by thermal boundary layers that become unstable, detach from the boundary and thereby drive flow in the interior of a fluid layer. In Chapter 10 we looked at plates as a thermal boundary layer of the convecting mantle, driving a distinctive form of convection in the mantle that I called the *plate mode* of mantle convection.

Here we look at the evidence that there is a mode of mantle convection driven by a lower, hot thermal boundary layer, at the expected form of such a mode, and at the consistency of the evidence with that expectation. Since it will become clear that the form and dynamics of such upwellings, or plumes, are quite different from the downwellings of lithosphere driving the plate mode, I will call the plumes and the flow they drive the *plume mode* of mantle convection.

11.1 Volcanic hotspots and hotspot swells

In Chapter 3 I described Wilson's observation that there are, scattered about the earth's surface, about 40 isolated volcanic centres that do not seem to be associated with plates and that seem to remain fixed relative to each other as plates move around (Figure 11.1). Their fixity (or at least their slow motion relative to plate velocities) is inferred from the existence of 'hotspot tracks', that is of chains of volcanoes that are progressively older the further they are from the active volcanic centre. Wilson was building on the inferences of Darwin and Dana that a number of the island chains in the Pacific seem to age progressively along the chain.

The classic example is the Hawaiian volcanic chain of islands and seamounts, evident in the topography shown in Figure 11.2.

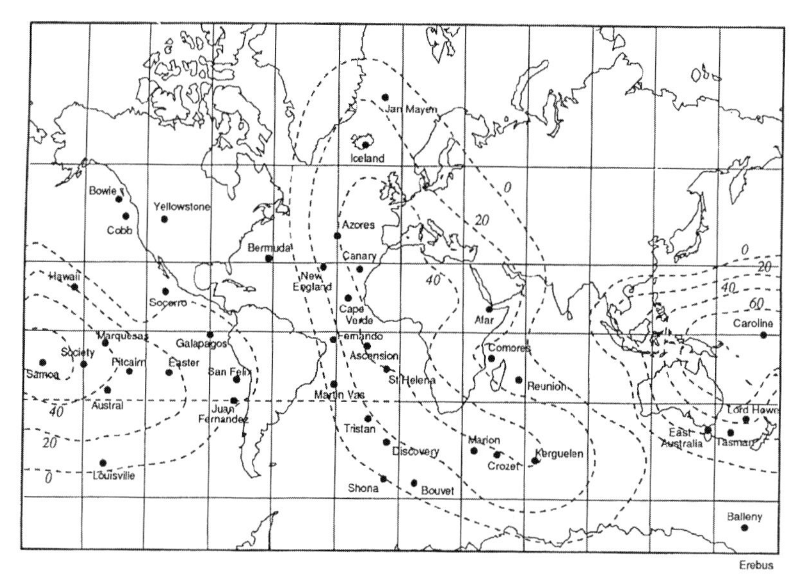

Figure 11.1. Locations of volcanic hotspots (dots). Residual geoid contours (in m) are superimposed (from Crough and Jurdy [1]). The residual geoid may reflect mainly signal from the lower mantle. Hotspots correlate with residual geoid highs but not with the present plate boundaries. From Duncan and Richards [2]. Copyright by the American Geophysical Union.

The south-eastern extremity of this chain, the island of Hawaii, is volcanically active, and the islands and seamounts to the north-west are progressively older. Wilson [3] hypothesised that the source of the eruptions was a 'mantle hotspot' located in a region of the mantle where convective velocities are small, such as the middle of a convection 'cell'. Morgan [4, 5] proposed instead that the source of the eruptions is a mantle plume, that is a column of hot, buoyant mantle rising from the core–mantle boundary.

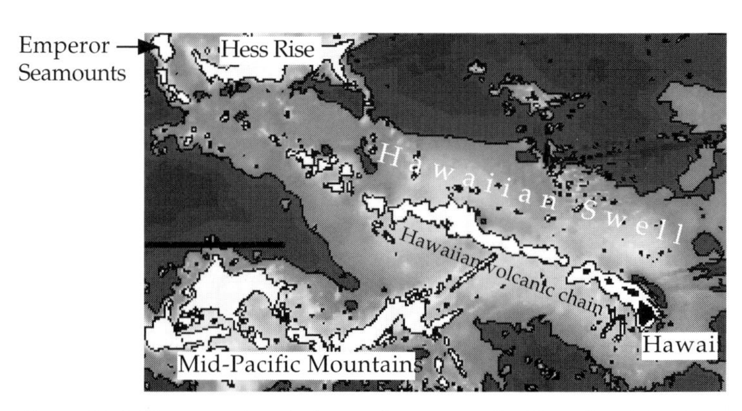

Figure 11.2. Topography of the sea floor near the Hawaiian Islands, showing the volcanic chain of islands and seamounts and the broad swell surrounding them. The contours are at depths of 3800 m and 5400 m.

Wilson's hypothesis had the disadvantages that the existence of the mantle hotspot was *ad hoc*, with no obvious reason for being there, and that it was not clear how a finite volume of warmer mantle could provide a steady supply of volcanism for tens of millions of years. Morgan's hypothesis at least implied a plausible physical source and the potential for longevity. Morgan's hypothesis immediately became the preferred one. Because of this, I proposed, in Chapter 3, dropping the concept of an internal mantle hotspot, and using the term 'volcanic hotspot' for the surface expression of the mantle phenomenon.

The number of volcanic hotspots has been variously estimated between about 40 [1, 6] and over 100 [7], but it is debatable whether many of the latter might be associated with individual mantle plumes. Figure 11.1 shows 40 hotspot locations selected by Duncan and Richards [2]. Contours of the hydrostatic geoid (i.e. relative to the shape of a rotating hydrostatic earth) are included. The suggestion is that hotspots correlate with highs in the geoid, which plausibly are due to structure in the lower mantle (Chapter 10), and specifically to regions of the deep mantle that are warmer because there has been no subduction into them in the past 200 Ma or so [8]. On the other hand, it is striking that hotspots show little correlation with the present configuration of plate boundaries.

As well as the narrow topography of the Hawaiian volcanic chain, there is evident in Figure 11.2 a broad swell in the sea floor surrounding the chain. This swell is up to about 1 km high and about 1000 km wide. Such a swell might be due to thickened oceanic crust, to a local imbalance of isostasy maintained by the strength of the lithosphere, or to buoyant material raising the lithosphere. Seismic reflection profiles show that the oceanic crust is not significantly thicker than normal [9]. Nor can such a broad swell be held up by the flexural strength of the lithosphere. The colder parts of the lithosphere behave elastically even on geological time scales, as long as their yield stress is not exceeded. For lithosphere of the age of that near Hawaii, about 90 Ma, the effective elastic thickness of the lithosphere is about 30 km thick, and it has a flexural wavelength of about 500 km [10]. However the wavelength of the swell is about 2000 km. If the swell were held out of isostatic balance by the lithosphere, the stresses would exceed the plausible yield stress of the lithosphere.

The straightforward conclusion is that the Hawaiian swell is held up by buoyant material under the lithosphere. In conjunction with the existence of the isolated volcanic centre, it is then a straightforward inference that there is a narrow column of hot mantle rising under Hawaii. Both the unusual volcanism and the supply of buoyancy to the base of the lithosphere would be explained if the column had a higher temperature than normal mantle. The volcanism occurs in a small, isolated locality far from plate boundaries, in contrast, for example, to the curvilinear volcanic island arcs near subduction zones. The isolation implies that the buoyant material is in the form of a column rather than a sheet. Since the active volcanism is confined to within an area of the order of 100 km across, it is reasonable to infer that the column diametre is of the same order. The fact that the Hawaiian hotspot track extends, through the bend into the Emperor seamounts, to ages of at least 90 Ma indicates that the mantle source is long-lived, and not due to an isolated heterogeneity within the mantle. Morgan called such a hot, narrow column a *mantle plume*.

11.2 Heat transported by plumes

Swells like that in Figure 11.2 are evident around many of the identified volcanic hotspots. Other conspicuous examples are at Iceland, which straddles the Mid-Atlantic Ridge, and at Cape Verde, off the west coast of Africa (Figure 4.3). The latter is 2 km high and even broader than the Hawaiian swell, presumably because the African plate is nearly stationary relative to the hotspot [2].

The swells can be used to estimate the rate of flow of buoyancy in the plumes. Buoyancy, as we saw in Chapter 8, is the gravitational force due to the density deficit of the buoyant material. If the plume is envisaged as a vertical cylinder with radius r and if the plume material flows upward with an average velocity u (as in Figure 7.7), then the buoyancy flow rate is

$$b = g\Delta\rho \cdot \pi r^2 u \tag{11.2.1}$$

where $\Delta\rho = (\rho_p - \rho_m)$ is the density difference between the plume and the surrounding mantle.

The way buoyancy flow rate can be inferred from hotspot swells is clearest in the case of Hawaii. The Hawaiian situation is sketched in Figure 11.3, which shows a map view and two cross-sections. As the Pacific plate moves over the rising plume column it is lifted by the plume buoyancy. There will be a close isostatic balance between the weight of the excess topography created by this uplift and the buoyancy of the plume material under the plate, as we discussed in Section 8.8. Since the plate is moving over the plume, the parts of the plate that are already elevated are being carried away from the plume. In order for the swell to persist, new

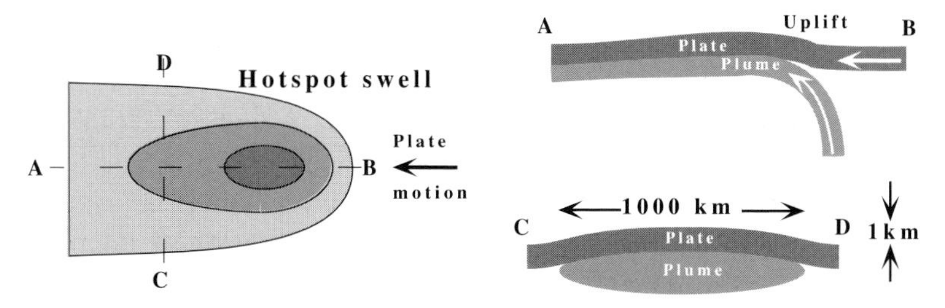

Figure 11.3. Sketch of a hotspot swell like that of Hawaii (Figure 11.2) in map view (left) and two cross-sections, showing the relationship of the swell to the plume that is inferred to be below the lithosphere. The swell is inferred to be raised by the buoyancy of the plume material. This allows the rate of flow of buoyancy and heat in the plume to be estimated.

parts of the plate have to be continuously raised as they arrive near the plume. This requires the arrival of new buoyant plume material under the plate (cross-section AB). Thus the rate at which new swell topography is generated will be a measure of the rate at which buoyant plume material arrives under the lithosphere.

The addition to swell topography each year is equivalent to elevating by a height $h = 1\,\text{km}$ a strip of sea floor with a 'width' $w = 1000\,\text{km}$ (the width of the swell) and a 'length' $v\delta t = 100\,\text{mm}$ (the distance travelled by the Pacific plate over the plume in one year at velocity $v = 100\,\text{mm/a}$). Both the sea floor and the Moho are raised, and sea water is displaced, so the effective difference in density is that between the mantle (ρ_m) and sea water (ρ_w). The rate of addition to the weight (negative buoyancy) of the new swell is then

$$W = g(\rho_\text{m} - \rho_\text{w})wvh = b \qquad (11.2.2)$$

By the argument just given, the buoyancy flow rate b in the plume is equal to W. Using the values quoted above yields $b = 7 \times 10^4\,\text{N/s}$ for Hawaii.

If the plume buoyancy is thermal, it can be related to the rate of heat transport by the plume, since both depend on the excess temperature, $\Delta T = T_\text{p} - T_\text{m}$, of the plume. Thus the difference between the plume density, ρ_p, and the mantle density is

$$\rho_\text{p} - \rho_\text{m} = \rho_\text{m}\alpha\Delta T \qquad (11.2.3)$$

while the heat flow rate is (see Section 7.7)

$$Q = \pi r^2 u\rho_\text{m}C_P\Delta T \qquad (11.2.4)$$

Taking the ratio of Q and b and using Equation (11.2.3) then yields

$$Q = C_P b/g\alpha \qquad (11.2.5)$$

Note particularly that this relationship does not depend on the excess temperature of the plume. In fact this is the same relationship as we derived in Section 10.4.4 between the buoyancy and heat flow rates of plates (Equation (10.4.4)). Thus this is another specific and quantitative example of the general relationship between convection and topography that we discussed in Section 8.8.

With $C_P = 1000\,\text{J/kg}\,^\circ\text{C}$ and $\alpha = 3 \times 10^{-5}/\,^\circ\text{C}$ this yields roughly $Q = 2 \times 10^{11}$ W, which is about 0.5% of the global heat flow. The total rate of heat transport by all known plumes has been estimated very roughly by Davies [11], and more carefully by Sleep [12], with similar results. Although there are 40 or more identified hotspots, all of them are weaker than Hawaii and many of them are substantially weaker. The total heat flow rate of plumes is about 2.3×10^{12} W (2.3 TW), which is about 6% of the global heat flow (41 TW, Table 10.1).

This value is comparable to estimates of the heat flow out of the core. Stacey [13] estimated this from the thermal conductivity of the core and its adiabatic temperature gradient, obtaining 3.7 TW for the heat that would be conducted down this gradient. Convective heat transport in the core would add to this, but compositional convection, due to continuing solidification of the inner core, might subtract from it. Another estimate can be made from thermal history calculations (Chapter 14), in which the core cools by several hundred degrees through earth history. Taking the present cooling rate to be about 70 °C/Ga, the core mass to be 1.94×10^{20} kg and the specific heat to be 500 J/kg °C yields a rate of heat loss of about 2.3 TW.

These estimates carry substantial uncertainty. As well, the estimate of plume heat flow rate should include the heat carried by plume heads (Sections 11.4, 11.5). Hill *et al.* [14] used the frequency of flood basalt eruptions in the geological record of the past 250 Ma to estimate that plume heads carry approximately 50% of the heat carried by plume tails. Thus the total heat flow rate in plumes would be approximately 3.5 TW, less than 10% of the global heat flow rate.

The approximate correspondence of the estimate of the heat transported by plumes with the rate of heat loss from the core supports Morgan's proposal that plumes come from a thermal boundary layer at the base of the mantle. According to our general discussion of convection in Chapter 8, a bottom thermal boundary layer is formed when heat enters through the bottom boundary of a fluid layer.

Stacey and Loper [15] were apparently the first to appreciate that this implies that plumes are cooling the core, in the sense that they are the agent by which heat from the core is mixed into the mantle. In this interpretation, the role of plumes is primarily to transfer heat from the core *through* the mantle, but not *out of* the mantle. Plumes bring heat to the base of the lithosphere, which is mostly quite thick and conducts heat only very slowly to the surface. For example, no excess heat flux has been consistently detected over the Hawaiian swell [16]. While in some cases, like Iceland, the lithosphere is thin and a substantial part of the excess plume heat may be lost to the surface, more commonly much of the plume heat would remain in the mantle, presumably to be mixed into the mantle after the overlying lithosphere subducts.

11.3 Volume flow rates and eruption rates of plumes

It was stressed above that the buoyancy flow rate of a plume can be estimated from the swell size without knowing the plume temperature. However, if we do have an estimate of plume temperature it is then possible to estimate the volumetric flow rate of the plume. It is instructive to compare this with the rate of volcanic eruption.

From the petrology of erupted lavas, plumes are estimated to have a peak temperature of 250–300 °C above that of normal mantle [17]. The volumetric flow rate up the plume is $\Phi_p = \pi r^2 u$, where u is the average velocity in the conduit and r is its radius. From Equations (11.2.1) and (11.2.3), this is related to the buoyancy flow rate, b, by

$$\Phi_p = b/g\rho_m \alpha \Delta T \qquad (11.3.1)$$

b was also related to the rate at which the swell volume is created, $\Phi_s = wvh$, through the weight of topography, W, in Equation (11.2.2):

$$\Phi_s = wvh = W/g(\rho_m - \rho_w) = b/g(\rho_m - \rho_w) \qquad (11.3.2)$$

so the plume volumetric flow rate is related to the swell volumetric rate of creation through

$$\Phi_p = \Phi_s(\rho_m - \rho_w)/\rho_m \alpha \Delta T \qquad (11.3.3)$$

For example, for Hawaii $\Phi_s = 0.1\,\text{km}^3/\text{a}$. If $\rho_m = 3300\,\text{kg/m}^3$, $\rho_w = 1000\,\text{kg/m}^3$, $\alpha = 3 \times 10^{-5}\,/^\circ\text{C}$ and $\Delta T = 300\,^\circ\text{C}$, then $(\rho_m - \rho_w)/\rho_m \alpha \Delta T = 75$. In other words the plume volumetric flow rate is about 75 times the rate of uplift of the swell. Thus for Hawaii $\Phi_p = 7.5\,\text{km}^3/\text{a}$.

The Hawaiian eruption rate, that is the rate at which the volcanic chain has been constructed, has been about $\Phi_e = 0.03\,\text{km}^3/\text{a}$ over the past 25 Ma [18, 19]. It is immediately evident that this is very much less than the plume volumetric flow rate. It implies that only about 0.4% of the volume of the plume material is erupted as magma at the surface. Even if there is substantially more magma emplaced below the surface, such as at the base of the crust under Hawaii [9, 20], the average melt fraction of the plume is unlikely to be much more than 1%.

Since the magmas show evidence of being derived from perhaps 5–10% partial melting of the source [17, 21], this presumably means that about 80–90% of the plume material does not melt at all, and the remainder undergoes about 5–10% partial melt. This result is important for the geochemical interpretation of plume-derived magmas and it is also useful for evaluating an alternative hypothesis for the existence of hotspot swells (Section 11.6.3).

11.4 The dynamics and form of mantle plumes

Having looked at the observational evidence for the existence of mantle plumes, and having derived some important measures of them, we now turn to the fluid dynamics of buoyant upwellings. Our understanding of the physics of such upwellings is quite well-developed, and there are some inferences and predictions that can be made with considerable confidence. This means that the hypothesis of mantle plumes can potentially be subjected to a number of quantitative observational tests.

This understanding of plume dynamics has arisen from some mathematical results, some long-standing and some more recent, and from some elegant laboratory experiments supplemented by physical scaling analyses and some numerical modelling. Plume dynamics is more tractable than plate dynamics largely because plumes are entirely fluid.

11.4.1 Experimental forms

The buoyant upwellings from a hot thermal boundary layer might have the form of sheets or columns. The downwellings driven by sinking plates clearly have the form of sheets, at least in the upper part of the mantle, since plates are stiff sheets at the surface and subduct along continuous curvilinear trenches. The stiffness of the plate would be expected to preserve this form to some depth, and recent results of seismic tomography seem to confirm this expectation (Chapter 5).

In contrast, Whitehead and Luther [22] showed experimentally and mathematically that upwellings from a buoyant fluid layer preferentially form columns rather than sheets. In experiments starting with a thin uniform fluid layer underlying a thick layer of a more dense fluid, the less dense fluid formed upwellings that started as isolated domes, rather than as sheets. Whitehead and Luther supplemented this laboratory demonstration with a mathematical analysis of second-order perturbation theory that showed that the rate of growth of a columnar upwelling is greater than the rate of growth of a sheet upwelling. This is an extension of the Rayleigh–Taylor instability that we encountered in Section 8.4.

Whitehead and Luther's experiments also demonstrated that the viscosity of an upwelling relative to the viscosity of the fluid it rises through has a strong influence on the form of the upwelling. This is illustrated in Figure 11.4, which shows buoyant upwellings

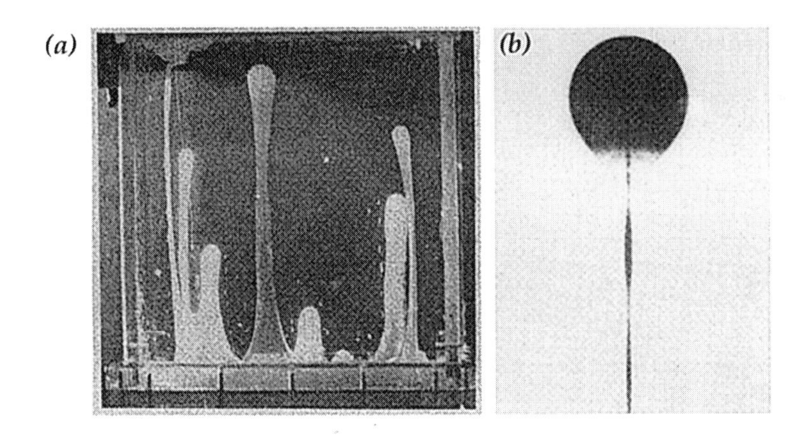

Figure 11.4. Photographs from laboratory experiments showing the effect of viscosity on the forms of buoyant upwellings. (a) The buoyant fluid is more viscous than the fluid it rises through, and the upwellings have fairly uniform diameter. In this case the buoyant fluid began as a thin uniform layer at the base of the tank. From Whitehead and Luther [22]. Copyright by the American Geophysical Union. (b) The buoyant fluid is less viscous than the fluid it rises through, and the upwelling has the form of a large spherical head and a thin columnar tail. In this case the buoyant fluid was injected through the base of the tank, and dyed to distinguish it. From Richards, Duncan and Courtillot [23]. Copyright American Association for the Advancement of Science. Reprinted with permission.

rising from the base of a tank. If the buoyant fluid is much more viscous than the ambient fluid (Figure 11.4a), the diameter of the buoyant columns is fairly uniform over its height. If the buoyant fluid is much less viscous (Figure 11.4b), then the column has a large, nearly spherical head at the top with a very thin conduit or tail connecting it to source. The reason for these different forms can be understood fairly simply, and this will be addressed in the next section.

Each of the experiments shown in Figure 11.4 involved two different fluids with different densities and viscosities. However, in the mantle we expect that the material ascending in a plume is the same material as normal mantle, but hotter. The higher temperature would make the plume less dense, and also lower its viscosity (Section 6.10.2). We might expect therefore that a mantle upwelling from a hot thermal boundary layer would form a plume, and that the plume would have a head-and-tail structure, as in Figure 11.4b. This is confirmed by the experiment illustrated in Figure 11.5a which shows a plume formed by heating a fluid whose viscosity is a strong function of temperature. The viscosity of the plume fluid is about 0.3% of the viscosity of the surrounding fluid, and the plume has a pronounced head-and-tail structure.

A striking new feature in Figure 11.5a is that the injected fluid forms a spiral inside the plume head. This is caused by thermal entrainment of surrounding, clear fluid into the head. As the head rises, heat diffuses out of it into the surrounding, cooler fluid, forming a thermal boundary layer around the head. Because this fluid is heated, it becomes buoyant, and so it tends to rise with the head. The spiral structure forms because there is a circulation within the plume head, with an upflow in the centre, where hot new fluid is arriving from the conduit, and a relative downflow around the equator, where the rise of the plume is resisted by the surrounding fluid. The fluid from the thermal boundary layer around the head is entrained into this internal circulation, flowing up next to the central conduit. This process is quantified in Section 11.4.3.

Thermal entrainment is not so important if the plume fluid is cold. Figure 11.5b shows a column of cold, dense, more viscous fluid descending into the same kind of fluid. The subdued head-and-tail structure is due to some of the surrounding fluid cooling and descending with the plume, but the resistance to the head from the surrounding lower-viscosity fluid is not sufficient to generate a significant internal circulation in the head, so there is no entrainment into it.

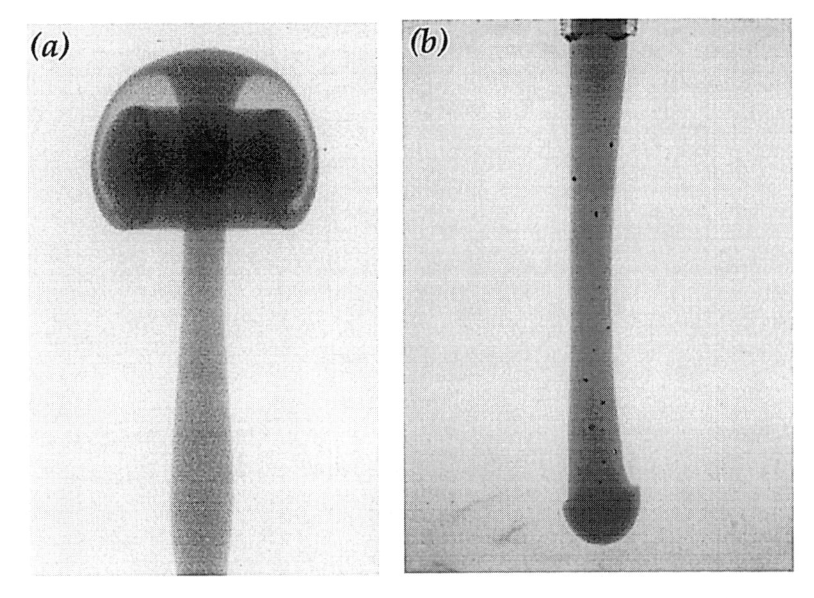

Figure 11.5. Thermal plumes in laboratory experiments, formed by injecting hot or cold dyed fluid into otherwise identical fluid. The fluid has a strong temperature dependence of viscosity. (a) The buoyant fluid is hot, and the plume viscosity is about 1/300 times that of the surrounding fluid. A spiral structure forms in the head due to thermal entrainment of ambient fluid. From Griffiths and Campbell [24]. (b) The injected fluid is cooler and hence denser and more viscous than the ambient fluid. There is little entrainment of cooled surrounding fluid, and only a very small head forms. From Campbell and Griffiths [25]. Copyright by Elsevier Science. Reprinted with permission.

Returning to the hot, low-viscosity plume of Figure 11.5a, similar structures are formed if a plume grows from a hot thermal boundary layer and the fluid viscosity is a strong function of temperature. Results of a numerical experiment scaled approximately to the mantle are shown in Figure 11.6. The panels are sections through an axisymmetric model showing the growth of a plume from an initial perturbation in the boundary layer. A line of passive tracers delineates the fluid initially within the hot boundary layer. The tracers reveal that the boundary layer fluid forms a spiral in the head due to thermal entrainment, as in Figure 11.5a. This numerical model also reveals the thermal structure within the plume. The hottest parts of the plume are the tail and the top of the head, where the tail material spreads out. Most of the head is cooler, and there are substantial thermal gradients within it. Temperatures within the head are intermediate between the plume tail temperature and the surrounding fluid.

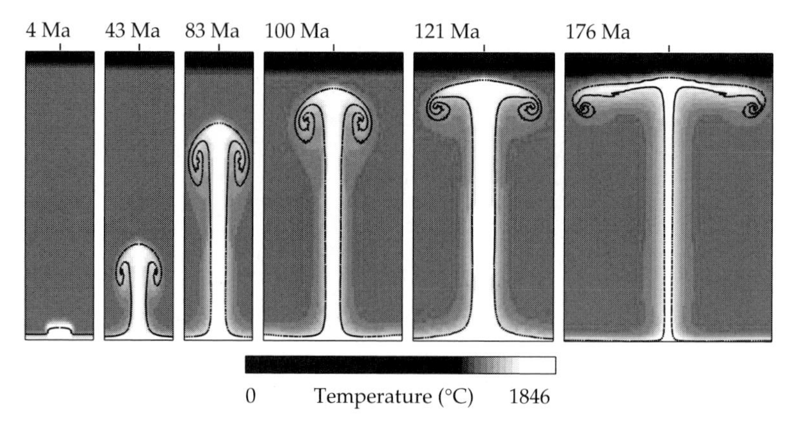

Figure 11.6. Sequence from a numerical model in which a plume grows from a thermal boundary layer. The model is axisymmetric and scaled approximately to the mantle. Viscosity is a strong function of temperature, and the ambient viscosity is 10^{22} Pa s. The bottom boundary temperature is 430 °C above the interior temperature, and the fluid viscosity there is about 1% of that of the interior fluid. A line of passive tracers delineates fluid initially within the thermal boundary layer.

11.4.2 Heads and tails

Here we look at why low-viscosity plumes form a head-and-tail structure. In the case in which the plume has a higher viscosity than the surroundings, the rise of the plume is limited mainly by the viscous resistance within the plume itself and within the boundary layer that feeds it. This means that the fluid in the plume does not rise faster than the top or head, and so it does not accumulate into a large head. The moderate variation of thickness with height is explained by the stretching of the column as the top rises faster than the stiff fluid can flow after it.

On the other hand, in the case where the plume has a lower viscosity, the plume fluid can flow readily from the boundary layer into and up the plume, and the main resistance to its rise comes from the surrounding more viscous fluid, which must be pushed out of the way. In this situation, the rise of the top of the plume is analogous to the rise of a buoyant sphere, and is regulated by the same balance of buoyancy and viscous resistance. In Chapter 6 we derived the Stokes formula for the velocity at which a buoyant sphere rises (Equation (6.8.3)). In fact you can see that the heads of the plumes in Figures 11.4b and 11.5a closely approximate a sphere. The role of this sphere is to force a path through the more viscous surroundings. Its rate of rise is initially slow, but it grows by the addition of plume fluid flowing out of the boundary layer. Once the head is large enough to force a path, the low-viscosity plume fluid can readily follow, requiring only a narrow conduit to flow through, its rate of flow being regulated by the rate at which it can flow out of the thin boundary layer. This is why the conduit trailing the head can have a much smaller radius.

The way the head-and-tail structure of plumes depends on the viscosity contrast between the plume and its surroundings is illustrated further in Figure 11.7. This shows three numerical models of plumes with different ratios of plume viscosity to surrounding viscosity: respectively 1, 1/30 and 1/200. The size of the head is similar in each case, but the conduit is thinner for the lower viscosities, reflecting the fact that the lower viscosity material requires only a thin conduit for a similar rate of flow.

11.4.3 Thermal entrainment into plumes

We will now consider the thermal structure of plumes in more detail. As the hot fluid in the conduit reaches the top of the head, it spreads radially out and around the periphery of the sphere, becoming very thin because of the greater radius of the head (Figures 11.6, 11.7). Because it is thinned, its heat diffuses out much more quickly (remember, from Chapter 7, that a diffu-

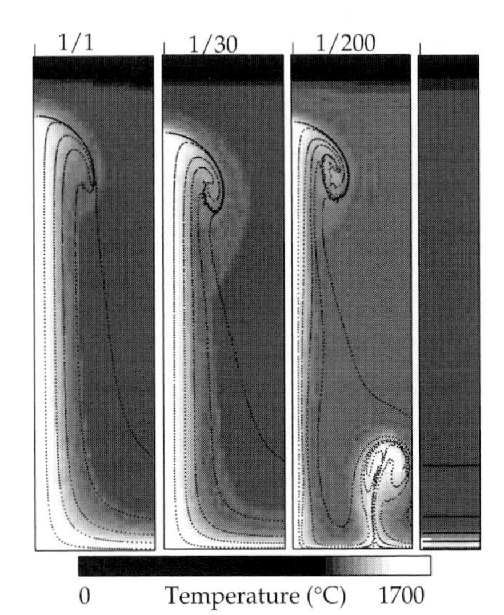

Figure 11.7. Plumes from three numerical models with different ratios of minimum plume viscosity to ambient viscosity, respectively 1, 1/30 and 1/200, showing how the tail is thinner for lower-viscosity plumes. The models are axisymmetric about the left-hand side of each panel. Several lines of tracers in this model mark fluid from different levels in the box. The initial configuration is shown in the right-hand panel. A secondary instability has developed in the right-hand model.

sion time scale is proportional to the square of the length scale involved). This heat goes partly outwards, to form the thermal boundary layer around the head, and partly inwards, to further heat the entrained material wrapping under it. As a result, the head has a temperature intermediate between that of the conduit and the surroundings. The spiral structure of the plume fluid, which is revealed by the dye in Figure 11.5a and by the tracers in Figures 11.6 and 11.7, is not evident in the thermal structure, because it is smoothed out by thermal diffusion. There are still thermal gradients in the head, but they are subdued relative to the temperature difference between the conduit and the surroundings.

The additional lines of tracers in Figure 11.7 reveal that most of the material entrained into the head comes from the lowest 10–20% of the fluid layer. Since these numerical experiments are scaled approximately to the mantle, this conclusion will apply also to plumes in the mantle. This is important for the interpretation of the geochemistry of flood basalts (Section 11.5).

We can quantify the rate of entrainment into a plume head using our understanding of thermal diffusion (Section 7.2) and of rising buoyant spheres (Section 6.8), following the approach used by Griffiths and Campbell [24]. The situation is sketched in Figure 11.8. We take the approach of using approximations that are rough, but that scale in the appropriate way. The thickness, δ, of the thermal boundary layer adjacent to the hot plume head will depend on the time the adjacent fluid is in contact with the passing plume head. This time will be of the order of $2R/U$, where the plume head radius is R and its rise velocity is U. Then, from Section 7.2,

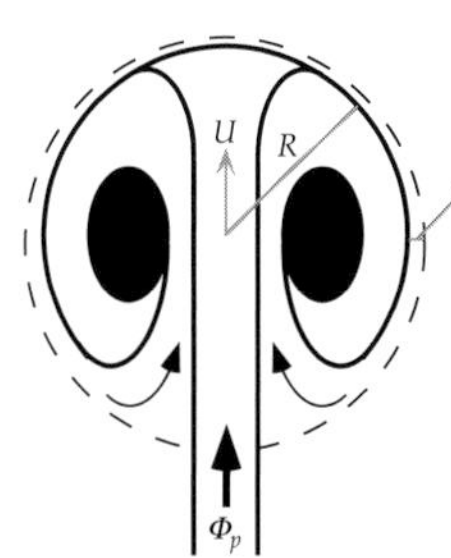

Figure 11.8. Sketch of a thermal boundary layer around a hot plume head. The fluid in the thermal boundary layer is heated by diffusion from the head. It is then buoyant and is entrained into the head. Boundary layer thickness is δ, head radius is R, head rise velocity is U and the volumetric flow rate up the plume tail is Φ_p.

$$\delta = \sqrt{\kappa t} = \sqrt{\frac{2\kappa R}{U}} \tag{11.4.1}$$

where κ is the thermal diffusivity. The horizontal cross-sectional area of the boundary layer near the head's equator is the head circumference times this thickness, $2\pi R\delta$, and the rate at which boundary layer fluid flows through this area is

$$\Phi_e = 2\pi R\delta U \tag{11.4.2}$$

We can assume that this fluid, or a constant fraction of it, becomes entrained into the head, so that Φ_e is an estimate of the volumetric rate of entrainment. The velocity, U, at which the head rises is given by the Stokes formula for a low-viscosity sphere (Section 6.8):

$$U = \frac{g\rho\alpha\Delta TR^2}{3\mu} \tag{11.4.3}$$

where ρ, α and μ are the density, thermal expansion coefficient and viscosity of the fluid respectively and ΔT is the temperature difference between the head and its surroundings.

If we take standard mantle values for these quantities (Appendix 2) with a viscosity appropriate for lower mantle, $\mu = 10^{22}$ Pa s, a temperature difference of 100 °C and a radius of 500 km, this yields a rise velocity of $U = 7 \times 10^{-10}$ m/s = 20 mm/a. The boundary layer thickness is then 40 km and the rate of entrainment is 2.7 km^3/a. This is comparable to the volume flow rate inferred for the Hawaiian plume tail of 7.5 km^3/a, which is the strongest plume tail by about a factor of 3 [11, 12]. The rate of increase of the head radius due to entrainment is

$$\frac{\partial R}{\partial t} = \frac{\Phi_e}{4\pi R^2} \tag{11.4.4}$$

With the values just derived, the rate of increase of radius is 1 mm/a = 1 km/Ma. This compares with a rise velocity of 20 mm/a.

This may suggest that entrainment is not very important, but Griffiths and Campbell integrated Equations (11.4.1–3), taking account of the influx from the tail, Φ_p, and the drop in average temperature as the entrainment proceeds. As cool fluid is entrained, the heat content of the plume is diffused through a larger volume. If the rate of inflow of fluid, Φ_p, is constant, the total heat supplied is proportional to $\Delta T_s \Phi_p (t - t_0) = \Delta T_s \Phi_p \Delta t$, where ΔT_s is the temperature excess of the source and Δt is the duration of the inflow. If the head volume at a later time is V, then conservation of energy requires that

$$\Delta T = \Delta T_s \Phi_p \Delta t / V \tag{11.4.5}$$

Combining Equations (11.4.1–3) with this yields

$$\Phi_e = 2\pi R \left[\frac{\kappa g \rho \alpha \Delta T_s \Phi_p \Delta t}{2\pi\mu}\right]^{1/2} \tag{11.4.6}$$

Then we can write an equation for the radius as a function of time as

$$\frac{\partial R}{\partial t} = \frac{\Phi_p + \Phi_e}{4\pi R^2} \tag{11.4.7}$$

Griffiths and Campbell found that plume head sizes of about 500 km radius at the top of the mantle are predicted rather consistently, independent of the tail flow rate and the temperature difference of the plume fluid source. Some of their results are shown in Figure 11.9. The initial rate of increase of the radius is much greater than it is as the head nears the top of the mantle, which explains the slow rates estimated above. Most of the curves in Figure 11.9 are for a mantle viscosity of 10^{22} Pa s, believed to be appropriate for the deep mantle where most of the head growth occurs. A lower viscosity of 10^{21} might be appropriate for the mantle in the Archean, and a smaller head is then predicted (Figure 11.9a). The plume head in the numerical experiment of Figure 11.6 approaches 1000 km in diameter near the top, consistent with their predictions. Taking the box depth to be 3000 km, the thermal halo in the fourth panel is 1000 km across and the tracers span about 800 km.

Entrainment may also occur into a plume tail. When the tail is vertical, as in Figures 11.6,7,10, this is so small that it is not evident in any obvious way. In fact Loper and Stacey have calculated that a strictly vertical plume tail with a strong viscosity contrast would entrain only a small percentage of additional material. Presumably this is because the travel time of the fluid up the conduit is short enough that diffusive heat loss to the surroundings is small. In the

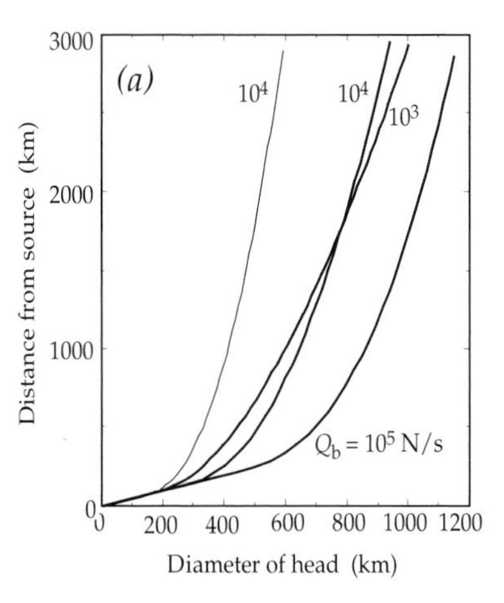

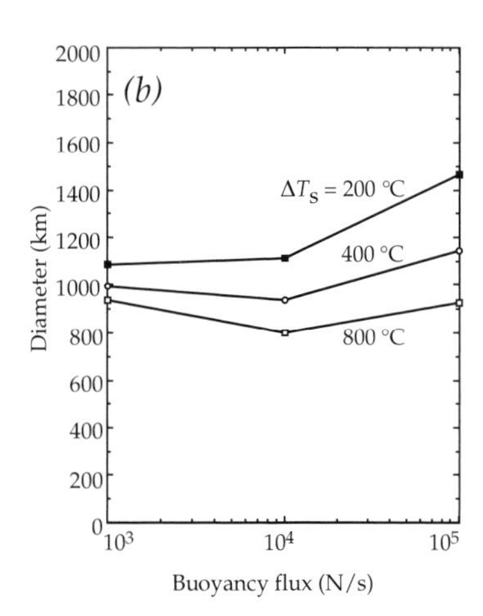

Figure 11.9. (a) Predicted plume head diameter versus height risen in a mantle of viscosity 10^{22} Pa s (heavy) and 10^{21} Pa s (light). Curves are labelled with buoyancy flow rate $Q_b = g\Delta\rho\Phi_p$. (b) Predicted plume head diameter at the top of the mantle for a mantle viscosity of 10^{22} Pa s and a range of buoyancy flow rates in the plume tail and fluid source excess temperatures, ΔT_s. From Griffiths and Campbell [24]. Copyright by Elsevier Science. Reprinted with permission.

numerical experiment depicted in the right-hand panel of Figure 11.7 the temperature in the centre of the conduit varies by only about 3% over most of its height. On the other hand, if the plume tail is inclined to the vertical, as it would be if the surrounding fluid were moving horizontally, then entrainment occurs by the same mechanism as for the plume head, and substantially larger degrees of entrainment may occur. This has been demonstrated experimentally by Richards and Griffiths [26].

11.4.4 Effects of a viscosity step and of phase changes

Figure 11.6 showed a numerical model of a thermal plume in which the viscosity depends on temperature. However, in the mantle the viscosity is also believed to vary substantially with depth, as discussed in Chapters 6 and 10. As well, phase transformations in the mantle transition zone may affect the rise of plumes, as discussed in Section 5.3, and the descent of subducted lithosphere discussed in Chapter 10.

The effects of including depth dependence of viscosity and a phase transformation are illustrated by the sequence from a numerical model shown in Figure 11.10. The viscosity increases with depth in a similar way to the models in Figure 10.12: there is a step by a factor of 20 at 700 km and an exponential increase by a factor of 10. As the plume head rises, its top feels the viscosity reducing and rises faster, stretching the plume head vertically.

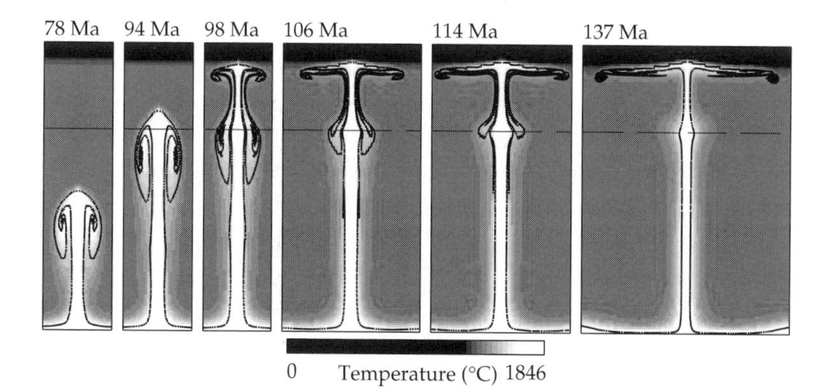

Figure 11.10. Sequence from a numerical plume model including increasing viscosity with depth and a phase transformation. The viscosity steps by a factor of 20 at 700 km depth and has an exponential increase by a factor of 10. The phase transformation at 700 km depth has a Clapeyron slope of −2 MPa/K. The plume slows and thickens through the phase transformation, but then narrows and speeds up in the low-viscosity upper layer.

This becomes pronounced as it enters the low-viscosity upper layer, where its rate of ascent increases and it necks down to a narrower diameter. As it then rises through the upper layer, it begins to form a second entrainment spiral, resulting in some convolution of the original spiral structure. The plume tail also speeds up and becomes narrower as it enters the upper layer (last frame).

This model also includes the effect of a phase transformation at 700 km depth with a moderately negative Clapeyron slope of −2 MPa/K. In this case the effect is not sufficient to block the ascent of the plume, though it does slow its rise in the vicinity of the phase transformation. This is most clearly evident in the last frame, where the plume tail bulges out as it slows, and then narrows again as it passes the phase transformation and enters the low-viscosity upper layer.

Compared with the plume in Figure 11.6, this plume reaches a shallower level. This is because it is much narrower as it rises into the upper mantle, and it does not trap as much mantle between itself and the lithosphere. Also as it spreads it is significantly thinner than in Figure 11.6, because of the lower viscosity below the lithosphere. Because it spreads faster, the high-temperature region is broader. These features are significant for the plume head model of flood basalts (Section 11.5), since they tend to promote greater melting over a broader area than in the model of Figure 11.6.

The effects of phase transformations with more negative Clapeyron slopes are illustrated by the models in Figure 11.11 [27]. As we have just seen in Figure 11.10, if the Clapeyron slope is −2 MPa/K, the plume continues through, and it is virtually unchanged except for a local bulge where its ascent is slowed by the phase transformation. If the Clapeyron slope is −3 MPa/K, then the plume is unable to penetrate. Apparently, if it does not penetrate immediately, then it spreads sufficiently rapidly that it

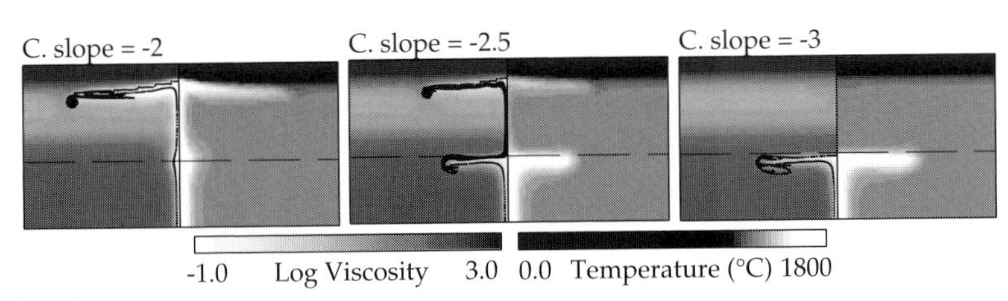

Figure 11.11. Plume models like that in Figure 11.10, but with different Clapeyron slopes (C. slope) of the phase transformation. The viscosity structure is shown on the left of these panels and the temperature on the right. From Davies [27]. Copyright by Elsevier Science. Reprinted with permission.

cannot ever penetrate. If the Clapeyron slope is −2.5 MPa/K, then the main part of the plume head penetrates but the tail is choked off and accumulates below the phase boundary. This would give rise to a tailless head in the upper mantle. (The precise value of the Clapeyron slope at which plume penetration is blocked is dependent on other details of the models, so these models should not be taken as a precise determination, but as a reasonable illustration of the process.)

11.5 Flood basalt eruptions and the plume head model

In Sections 11.1–3 we looked at observations that can be interpreted to relate to plume tails. It was the age-progressive volcanic chains that originally motivated Morgan's plume hypothesis, a model that we now identify more specifically as a plume tail. In 1981, Morgan [6] pointed out that several hotspot tracks emerged from flood basalt provinces. A notable example is the Chagos–Laccadive Ridge running south from the Deccan Traps flood basalt province in western India to Reunion Island in the Indian Ocean (Figures 4.3, 11.12).

Flood basalts are evidence of the largest volcanic eruptions identified in the geological record. They range up to 2000 km across, with accumulated thicknesses of basalt flows up to several kilometres. A map of the main identified flood basalt provinces is shown in Figure 11.12. Total volumes of extrusive eruptions range up to 10 million cubic kilometres, and evidence is accumulating that much of this volume is erupted in less than 1 million years [28]. It has been recognised within the past decade that some oceanic plateaus are oceanic equivalents of continental flood basalts. The largest flood basalt province is the Ontong–Java Plateau, a submarine plateau east of New Guinea.

Morgan [6] proposed that if flood basalts and hotspot tracks are associated, then the head-and-tail structure of a new plume, which had been demonstrated by Whitehead and Luther, would provide an explanation. Figure 11.13 illustrates the concept. The flood basalt eruption would be due to the arrival of the plume head, and the hotspot track would be formed by the tail following the head. If the overlying plate is moving, then the flood basalt and the underlying head remnant would be carried away, and the hotspot track would emerge from the flood basalt province and connect it to the currently active volcanic centre, which would be underlain by the active plume tail.

Not a lot of attention was given to Morgan's proposal until Richards, Duncan and Courtillot [23] revived and advocated the

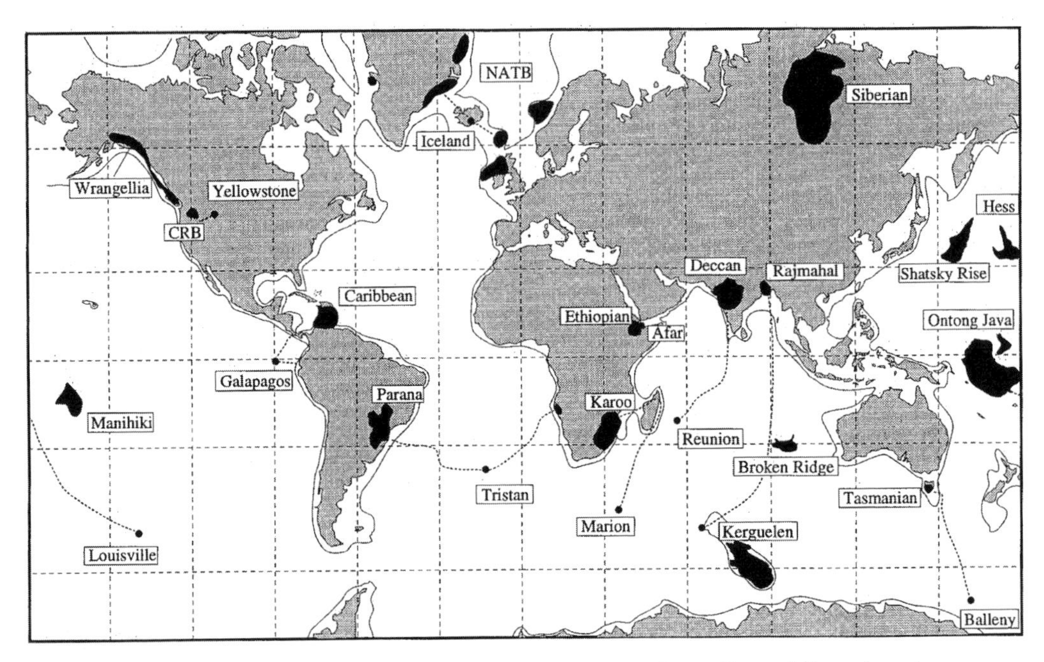

Figure 11.12. Map of continental and oceanic flood basalt provinces. Dotted lines show known or conjectured connections with active volcanic hotspots. After Duncan and Richards [2]. Copyright by the American Geophysical Union.

idea. Subsequently Griffiths and Campbell [17, 24] demonstrated the thermal entrainment process and argued in more detail for the plume head explanation of flood basalts. In particular Griffiths and Campbell argued that plume heads could reach much larger diametres, 800–1200 km, than had previously been estimated, if they rise from the bottom of the mantle, and also that they would

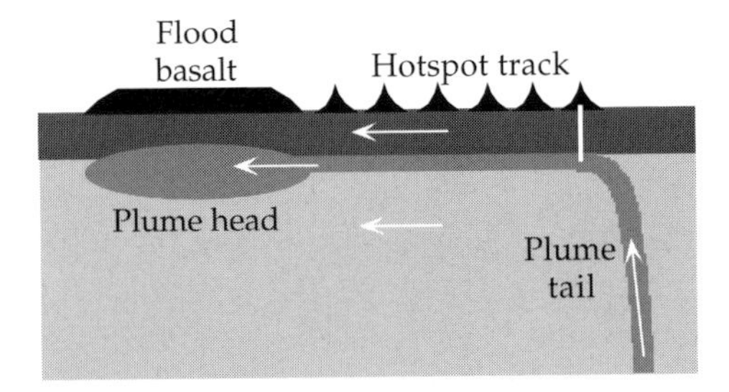

Figure 11.13. Sketch of the way a new plume with a head-and-tail structure can account for the relationship observed between some flood basalts and hotspot tracks, in which the hotspot track emerges from a flood basalt province and connects it to a currently active volcanic hotspot. It is assumed in the sketch that the plate and subjacent mantle are moving to the left relative to the plume source.

approximately double in horizontal diameter as they flattened and spread below the lithosphere (Figures 11.6, 11.10). This is in good agreement with the observed total extents of flood basalt provinces, the Karoo flood basalts being scattered over a region about 2500 km in diameter. Campbell and Griffiths argued that important aspects of the petrology and geochemistry of flood basalts could be explained by the model, in particular the concentration near the centres of provinces of picrites, which are products of higher degrees of melting than basalts. They argued that this can be explained by the temperature distribution of a plume head, which is hottest at the central conduit and cooler to the sides (Figure 11.6).

Though this model of flood basalt formation has attracted wide interest, it has not yet been fully explored quantitatively. The principal outstanding question is whether it can account quantitatively for the observed volumes of flood basalts in cases where there appears to have been little or no rifting. The perceived problem has been that normal mantle compositions do not begin to melt until they have risen to depths less than about 120 km even if they are 200 °C hotter than normal [29, 30]. Since continental lithosphere is commonly at least this thick, we would not expect plumes to melt at all under continents.

However plumes are known not to have normal mantle composition. It is widely recognised by geochemists on the basis of trace element contents that they have a larger complement of basaltic composition than normal mantle. This component of their composition is hypothesised to come from previously subducted oceanic crust that is entrained in plumes near the base of the mantle (Chapter 13; [21]). Such a composition would substantially lower the solidus temperature and enhance melt production. Some preliminary models [31] and continuing work indicate that melt volumes of the order of 1 million cubic kilometres can be produced from such a plume head. Examples of calculations of melt volume from a simplified plume head model with an enhanced basaltic component are shown in Figure 11.14. These show that it is plausible that several million cubic kilometres of magma could be erupted within about 1 Ma.

Other factors being evaluated for their influence on plume head melting are higher plume temperatures [32], the effect of mantle viscosity structure on the height to which plumes can penetrate, noted in Section 11.4.4 (Figure 11.10), or that plumes may be more effective at thinning the lithosphere and penetrating to shallow depths than has been recognised. The indications at this stage are that a satisfactory quantitative account of flood basalts will

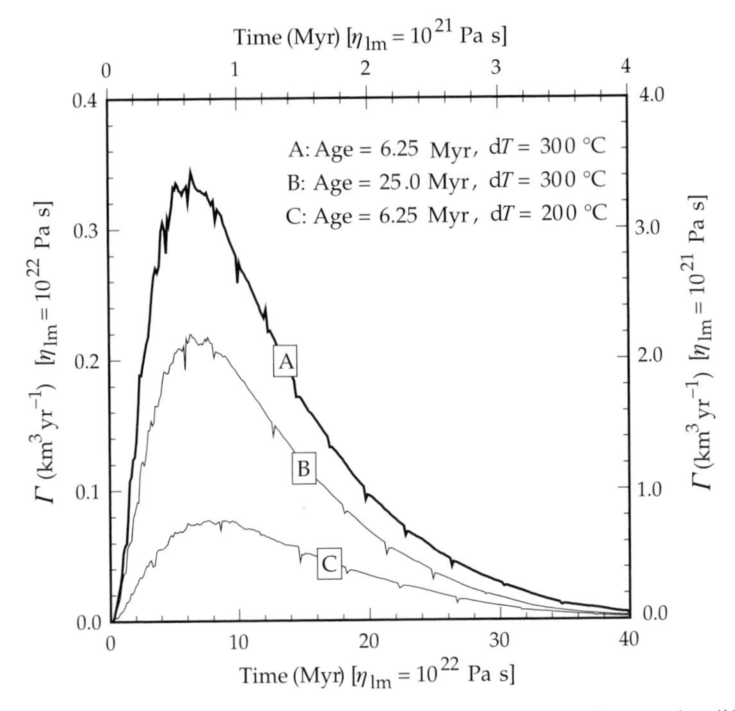

Figure 11.14. Calculated rates of magma generation, Γ, from a simplified numerical model of a plume head that includes 15% additional basaltic component. The curves assume different initial plume temperature excesses, dT, and different ages (and thus thicknesses) of lithosphere. The plume head was modelled as a sphere with initially uniform temperature. The left and bottom scales assume a mantle viscosity of 10^{22} Pa s, the right and top scales are for 10^{21} Pa s. From Cordery *et al.* [31]. Copyright by the American Geophysical Union.

emerge from the plume head model, but this has not yet been attained.

11.6 Some alternative theories

11.6.1 Rifting model of flood basalts

White and McKenzie [30] proposed a theory for the formation of very thick sequences of volcanic flows found along some continental margins and of flood basalt eruptions. The theory can usefully be separated into three parts. The first part is that the marginal volcanic provinces are produced when rifting occurs over a region of mantle that is hotter than normal because it is derived from a plume. This seems to give a very viable account of such provinces. The second part is that all flood basalts can be explained by this mechanism. The third part is that the plume material is derived mainly from a plume tail, since they assumed that plumes are part of an upper mantle convection system and that plumes therefore derive from no deeper than 670 km. In this case the plume heads would have diameters of no more than about 300 km and volumes less than about 5% of a plume head from the bottom of the mantle [24].

The second part of White and McKenzie's model encounters the difficulty that a number of flood basalt provinces are said, on the basis of field evidence, to have erupted mainly before substantial rifting occurred (e.g. Deccan Traps) or in the absence of any substantial rifting (e.g. Siberian Traps, Columbia River Basalts) [33]. It also fails to explain the very short time scale of flood basalt eruptions, less than 1 Ma in the best-constrained cases. The third part of their model implies that a sufficient volume of warm mantle would take about 50 Ma to accumulate, but at the time the Deccan Traps erupted, India was moving north at about 180 mm/yr (180 km/Ma) so it would have traversed the extent of the flood basalts in only about 10 Ma. It is difficult to see how sufficient warm mantle could accumulate from a plume tail under such a fast-moving plate.

These difficulties are avoided by the plume head model of flood basalts, since the flow rate of the plume head is much greater than the tail and much of the melting is inferred to occur from beneath the intact lithosphere upon arrival of the plume head. It is true that the volumes of the eruptions have yet to be fully explained quantitatively, but current indications are that this is not a fundamental difficulty.

11.6.2 Mantle wetspots

Green [34] has argued that volcanic hotspots can be explained by mantle 'wetspots'. From a petrological point of view, this idea has some merits, since a small amount of water (less than 0.1%) can substantially reduce the solidus temperature, at which melting first occurs. It is also true that hydrated forms of minerals are generally less dense than their dry counterparts, which could provide the buoyancy required to explain hotspot swells. The effect on density needs to be better quantified, and it would need to be shown that observed water contents of hotspot volcanics are consistent with the amounts required to explain the buoyancy. It needs also to be shown that sufficient melt can be produced to explain the observed volcanism, since although water reduces the solidus temperature, substantial degrees of melting still do not occur until the dry solidus temperature is approached.

However, a remaining difficulty would still be to explain the duration of long-lived volcanic centres like Hawaii. While a hydrated portion of the mantle, perhaps old subducted oceanic crust, might produce a burst of volcanism, there is no explanation offered for how the source might persist for 100 Ma or more. It is useful to estimate the volume of mantle required to supply the Hawaiian plume for 100 Ma. The total volume erupted into the Hawaiian and Emperor seamounts over 90 Ma is about $10^6\,km^3$. If we assume that there was about 5% melting of the source, this requires a source volume of $2 \times 10^7\,km^3$, equivalent to a sphere of diameter 340 km. If such a large and buoyant region existed as a unit in the mantle, it would rise and produce a burst of volcanism. To explain the Hawaiian volcanic chain the hydrated mantle material needs to be supplied at a small and steady rate.

The advantage of the thermal plume hypothesis is that a renewal mechanism is straightforwardly provided if the plume originates from a thermal boundary layer. It may be that the effects of water on melting and on plume buoyancy are significant, but it is far from clear that water alone could provide a sufficient explanation of the observations, while heat alone, or heat plus water, provides a straightforward and quantitatively successful account of the dynamical requirements of a theory of plumes.

11.6.3 Melt residue buoyancy under hotspot swells

J. P. Morgan and others [19] have proposed that the buoyancy supporting hotspot swells is due significantly also to the compositional buoyancy of the residue remaining after the hotspot magma has erupted. The residue will be less dense because iron partitions preferentially into the melt phase. However, the estimates made in Sections 11.2 and 11.3 indicate that the amount of melt produced is less than 1% of the volume of the plume material, in which case this will be a minor effect. Morgan and others estimate the density change of the residue as a function of mean melt fraction, f, from the formula

$$\Delta\rho = \rho_m \beta f$$

where $\beta = 0.06$ is an empirically evaluated constant. This implies that the annual volume of mantle that arrives through the plume should expand by the same fraction, βf, and this expansion is what is manifest as the plume swell. We can therefore estimate the annual contribution to the swell volume from the effect of residue buoyancy as

$$\Phi_{sr} = \Phi_p \beta f$$

Using the values $\Phi_p = 7.5\,km^3/a$ and $f = 0.01$, used earlier for Hawaii, this gives $\Phi_{sr} = 0.0045\,km^3/a$, which is only about 5% of the observed rate of swell formation of $0.1\,km^3/a$. While the residue buoyancy may be more significant locally under the volcanic chain, it seems that the direct buoyancy of the plume material is still required to account for most of the Hawaiian swell. This implies in turn that the estimates of buoyancy and heat flow rate given in Section 11.2 are reasonable.

11.7 Inevitability of mantle plumes

The earth is believed to have been strongly heated during the late stages of its formation. The heat comes from the release of gravitational energy of material falling onto the growing earth. The earth is believed to have formed from a disk of particles orbiting the sun and left over from the sun's formation. Models of the process of accumulation of material into larger bodies indicate that many bodies would grow simultaneously, but that there would be a wide distribution of sizes, with only a few large bodies and greater numbers of smaller bodies. In this situation the final stages of accumulation would involve the collision of very large bodies. A plausible and currently popular theory for the formation of the moon proposes that the moon was formed from the debris of a collision of a Mars-sized body with the earth. A collision of this magnitude would probably have melted much of the earth, and vaporised some of it. Accounts of these ideas can be found in [35, 36, 37].

Suppose that the earth was heated in this way, and that it quickly homogenised thermally, as a substantially liquid body would do. The temperature would not be uniform, but would follow an adiabatic profile with depth, due to the effect of pressure, as discussed in Chapter 7. The earth's temperature as a function of depth would therefore look like curve (a) sketched in Figure 11.15.

The earth would then lose heat through its surface. This would form an outer thermal boundary layer (a precursor to the lithosphere) and, with the mantle being very hot and possibly partially molten, rapid mantle convection could be expected. In this way the mantle would be cooled. Suppose, for the simplicity of this argument, that the entire mantle convected and cooled in this way.

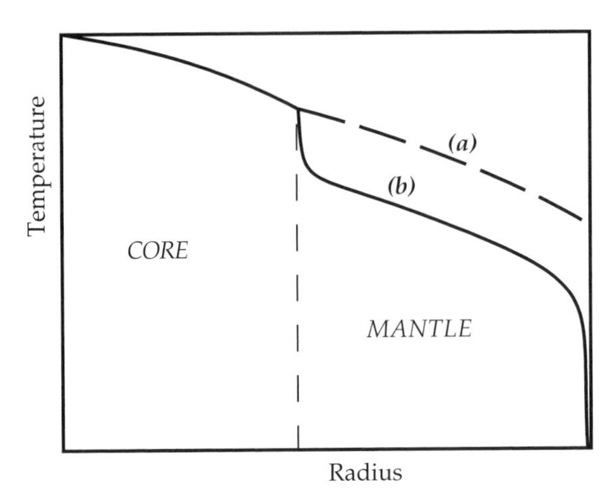

Figure 11.15. Sketch of the form of the temperature profile within the earth (a) soon after formation, and (b) later, after the mantle has cooled by heat loss to the surface. The core can only begin to lose heat after the mantle has become cooler than the core. Thereafter the heat conducting from the core into the base of the mantle forms a thermal boundary layer that can generate buoyant upwellings.

After some time, the temperature profile would have looked like curve (b) of Figure 11.15.

Initially, the core would not have been able to lose heat, because we assumed that the mantle and core had the same temperature at their interface. However, as the mantle cooled, heat would begin to conduct out of the core into the base of the mantle, and cooling of the core would commence. This heat from the core would form a thermal boundary layer at the base of the mantle, depicted in curve (b) of Figure 11.15. If the mantle viscosity were sufficiently low and the heat flow from the core sufficiently high, both of which are highly likely, this thermal boundary layer would become unstable and buoyant upwellings would rise from it. These upwellings would have a lower viscosity than the mantle they were rising through, so they would develop a head-and-tail structure, as discussed in Section 11.4.

Thus we have a general argument for the existence of thermal plumes in the mantle. The assumptions are that the core and mantle started with similar temperatures at their interface, that the mantle has been cooling, and that the conditions are such that the relevant Rayleigh number is greater than its critical value for instability and convection to occur. If the earth, now or in the past, functioned as more than two independent layers, then the argument generalises very simply: the layers would cool from the outside inwards, and plumes would be generated in each layer by heat conducting from the next deeper layer.

11.8 The plume mode of mantle convection

We have seen that the existence of volcanic island and seamount chains terminating in isolated active volcanic hotspots, such as Hawaii, and surrounded by broad topographic swells imply the existence of narrow, long-lived columns of buoyant, rising mantle material. Morgan called these mantle plumes. The buoyancy and excess melting can be explained if the plumes are 200–300 °C hotter than normal mantle, and their longevity is plausible if they derive from a hot thermal boundary layer. Their higher temperature implies that plumes would have lower viscosity than normal mantle. Fluid dynamics experiments show that the preferred form of low-viscosity buoyant upwellings is columnar, and that new plumes would start with a large, spherical head. Plume heads are calculated to reach diameters of about 1000 km near the top of the mantle, and they provide a plausible explanation for flood basalt eruptions. The association of plume heads with their following plume tails provides an explanation for hotspot tracks that emerge from flood basalt provinces.

Plumes and the flow they drive in surrounding mantle comprise a distinct mode of mantle convection, driven by a hot, lower thermal boundary layer. They therefore complement the plate mode driven by the cool, top thermal boundary layer. As with the plate mode, there will be a passive downward return flow driven by plumes that balances the upflow in plumes. The fact that hotspot locations do not correlate strongly with the current configuration of plates (Figure 11.1; [38]) indicates that the plume and plate modes are not strongly coupled. The implication is that plumes rise through the plate-scale flow without substantially disrupting it. Experiments have shown that plume tails can rise through a horizontal background flow, bending away from the vertical but retaining their narrow tubular form [39, 40, 41]. However, there is a correlation between plume locations, broad geoid highs and slower seismic wavespeeds in the deep mantle [38, 42], indicating that plumes form preferentially away from deeply subducted lithosphere.

Plumes may have been significant tectonic agents through much of earth history. They may trigger ridge jumps or occasional larger-scale rifting events [5, 43]. Plume heads have been proposed as the direct source of Archean greenstone belts and the indirect cause, through their heat, of associated granitic terrains from secondary crustal melting [44]. They may have been a significant source of continental crust, directly from continental flood basalts and through the accretion as exotic terrains of oceanic flood basalts [14, 45]. They may be the source of many dike swarms, and as a source of heat they may have been involved in some regional 'anorogenic' crustal heating and melting events and in the reworking and mineralising of a significant proportion of the continental crust [14]. The term 'plume tectonics' has been used to encapsulate their possibly substantial tectonic role [14].

A fundamental aspect of mantle convection is that the thermal boundary layers are distinct agents, as I stressed in Chapter 8. It is therefore incorrect to regard plumes and plume tectonics as a possible substitute for plate tectonics, as has been speculated not infrequently for the early earth and for Venus. Currently in the earth, plate tectonics cools the mantle. If plate tectonics did not operate, then the top boundary layer would have to operate in another way in order to remove heat from the mantle. The role of plumes is to transfer heat from the layer below (the core) into the convecting mantle. Any surface heat flow or tectonic effect from plumes is incidental, and adds to whatever tectonics are driven by the top boundary layer. This will be discussed in more detail in Chapter 14.

A further implication of this last point is that the level of activity of plumes depends on the strength of the hot thermal boundary layer at the base of the mantle. This may have varied with time, though calculations suggest that it may have been rather constant (Chapter 14). It follows also that the two thermal boundary layers need to be prescribed separately in numerical models of mantle convection. In other words, it is sensible to define separate Rayleigh numbers for each thermal boundary layer, and hence for each mode of mantle convection.

11.9 References

1. T. S. Crough and D. M. Jurdy, Subducted lithosphere, hotspots and the geoid, *Earth Planet. Sci. Lett.* **48**, 15–22, 1980.
2. R. A. Duncan and M. A. Richards, Hotspots, mantle plumes, flood basalts, and true polar wander, *Rev. Geophys.* **29**, 31–50, 1991.
3. J. T. Wilson, A possible origin of the Hawaiian islands, *Can. J. Phys.* **41**, 863–70, 1963.
4. W. J. Morgan, Convection plumes in the lower mantle, *Nature* **230**, 42–3, 1971.
5. W. J. Morgan, Plate motions and deep mantle convection, *Mem. Geol. Soc. Am.* **132**, 7–22, 1972.

6. W. J. Morgan, Hotspot tracks and the opening of the Atlantic and Indian Oceans, in: *The Sea*, C. Emiliani, ed., Wiley, New York, 443–87, 1981.
7. K. C. Burke and J. T. Wilson, Hot spots on the earth's surface, *Sci. Am.* **235**, 46–57, 1976.
8. M. A. Richards, B. H. Hager and N. H. Sleep, Dynamically supported geoid highs over hotspots: observation and theory, *J. Geophys. Res.* **93**, 7690–708, 1988.
9. A. B. Watts and U. S. ten Brink, Crustal structure, flexure and subsidence history of the Hawaiian Islands, *J. Geophys. Res.* **94**, 10 473–500, 1989.
10. D. L. Turcotte and G. Schubert, *Geodynamics: Applications of Continuum Physics to Geological Problems*, 450 pp., Wiley, New York, 1982.
11. G. F. Davies, Ocean bathymetry and mantle convection, 1. Large-scale flow and hotspots, *J. Geophys. Res.* **93**, 10 467–80, 1988.
12. N. H. Sleep, Hotspots and mantle plumes: Some phenomenology, *J. Geophys. Res.* **95**, 6715–36, 1990.
13. F. D. Stacey, *Physics of the Earth*, 513 pp., Brookfield Press, Brisbane, 1992.
14. R. I. Hill, I. H. Campbell, G. F. Davies and R. W. Griffiths, Mantle plumes and continental tectonics, *Science* **256**, 186–93, 1992.
15. F. D. Stacey and D. E. Loper, Thermal histories of the core and mantle, *Phys. Earth Planet. Inter.* **36,** 99–115, 1984.
16. R. P. Von Herzen, M. J. Cordery, R. S. Detrick and C. Fang, Heat flow and thermal origin of hotspot swells: the Hawaiian swell revisited, *J. Geophys. Res.* **94**, 13 783–99, 1989.
17. I. H. Campbell and R. W. Griffiths, Implications of mantle plume structure for the evolution of flood basalts, *Earth Planet. Sci. Lett.* **99**, 79–83, 1990.
18. D. A. Clague and G. B. Dalrymple, Tectonics, geochronology and origin of the Hawaiian-Emperor volcanic chain, in: *The Eastern Pacific Ocean and Hawaii*, E. L. Winterer, D. M. Hussong and R. W. Decker, eds., Geological Society of America, Boulder, CO, 188–217, 1989.
19. J. P. Morgan, W. J. Morgan and E. Price, Hotspot melting generates both hotspot swell volcanism and a hotspot swell?, *J. Geophys. Res.* **100**, 8045–62, 1995.
20. P. Wessel, A re-examination of the flexural deformation beneath the Hawaiian islands, *J. Geophys. Res.* **98**, 12 177–90, 1993.
21. A. W. Hofmann and W. M. White, Mantle plumes from ancient oceanic crust, *Earth Planet. Sci. Lett.* **57**, 421–36, 1982.
22. J. A. Whitehead and D. S. Luther, Dynamics of laboratory diapir and plume models, *J. Geophys. Res.* **80**, 705–17, 1975.
23. M. A. Richards, R. A. Duncan and V. E. Courtillot, Flood basalts and hot-spot tracks: plume heads and tails, *Science* **246**, 103–7, 1989.

24. R. W. Griffiths and I. H. Campbell, Stirring and structure in mantle plumes, *Earth Planet. Sci. Lett.* **99**, 66–78, 1990.
25. I. H. Campbell and R. W. Griffiths, The evolution of the mantle's chemical structure, *Lithos* **30**, 389–99, 1993.
26. M. A. Richards and R. W. Griffiths, Deflection of plumes by mantle shear flow: experimental results and a simple theory, *Geophys. J.* **94**, 367–76, 1988.
27. G. F. Davies, Penetration of plates and plumes through the mantle transition zone, *Earth Planet. Sci. Lett.* **133**, 507–16, 1995.
28. M. F. Coffin and O. Eldholm, Large igneous provinces: crustal structure, dimensions and external consequences, *Rev. Geophys.* **32**, 1–36, 1994.
29. D. P. McKenzie and M. J. Bickle, The volume and composition of melt generated by extension of the lithosphere, *J. Petrol.* **29**, 625–79, 1988.
30. R. White and D. McKenzie, Magmatism at rift zones: the generation of volcanic continental margins and flood basalts, *J. Geophys. Res.* **94**, 7685–730, 1989.
31. M. J. Cordery, G. F. Davies and I. H. Campbell, Genesis of flood basalts from eclogite-bearing mantle plumes, *J. Geophys. Res.* **102**, 20 179–97, 1997.
32. C. Farnetani and M. A. Richards, Numerical investigations of the mantle plume initiation model for flood basalt events., *J. Geophys. Res.* **99**, 13 813–33, 1994.
33. P. R. Hooper, The timing of crustal extension and the eruption of continental flood basalts, *Nature* **345**, 246–9, 1990.
34. D. H. Green and T. J. Falloon, Pyrolite: A Ringwood concept and its current expression, in: *The Earth's Mantle: Composition, Structure and Evolution*, I. N. S. Jackson, ed., Cambridge University Press, Cambridge, 311–78, 1998.
35. G. W. Wetherill, Occurrence of giant impacts during the growth of the terrestrial planets, *Science* **228**, 877–9, 1985.
36. G. W. Wetherill, Formation of the terrestrial planets, *Annu. Rev. Astron. Astrophys.* **18**, 77–113, 1980.
37. H. E. Newsom and J. H. Jones, *Origin of the Earth*, 378, Oxford University Press, New York, 1990.
38. M. Stefanick and D. M. Jurdy, The distribution of hot spots, *J. Geophys. Res.* **89**, 9919–25, 1984.
39. M. A. Richards and R. W. Griffiths, Thermal entrainment by deflected mantle plumes, *Nature* **342**, 900–2, 1989.
40. R. W. Griffiths and I. H. Campbell, On the dynamics of long-lived plume conduits in the convecting mantle, *Earth Planet. Sci. Lett.* **103**, 214–27, 1991.
41. R. W. Griffiths and M. A. Richards, The adjustment of mantle plumes to changes in plate motion, *Geophys. Res. Lett.* **16**, 437–40, 1989.
42. M. A. Richards and D. C. Engebretson, Large-scale mantle convection and the history of subduction, *Nature* **355**, 437–40, 1992.

43. R. I. Hill, Starting plumes and continental breakup, *Earth Planet. Sci. Lett.* **104**, 398–416, 1991.
44. I. H. Campbell and R. I. Hill, A two-stage model for the formation of the granite-greenstone terrains of the Kalgoorlie-Norseman area, Western Australia, *Earth Planet. Sci. Lett.* **90**, 11–25, 1988.
45. M. A. Richards, D. L. Jones, R. A. Duncan and D. J. DePaolo, A mantle plume initiation model for the Wrangellia flood basalt and other oceanic plateaus, *Science* **254**, 263–7, 1991.

CHAPTER 12

Synthesis

So far in Part 3 I have developed an approach to mantle convection based on thermal boundary layers and used this to look at the behaviour of the two boundary layers in the mantle for which there is good evidence. Here I consider how these parts assemble into a coherent picture and look at some immediate implications about how the system does and does not work.

It is also an appropriate place to discuss alternative views. Some of these are in direct opposition to the picture developed here, such as that the mantle is divided into two layers that convect separately. Others are different ways of looking at the system that have been used in the long debate. Some of these carve the total system up in a different way. Some are complementary, and useful for bringing out particular aspects, while others are unprofitable or potentially misleading.

12.1 The mantle as a dynamical system

The picture that has emerged here is of a mantle system in which two thermal boundary layers have been identified on the basis of observational evidence, one comprising the plates and the other giving rise to plumes. The boundary layers appear to transport heat at substantially different rates, and they manifest quite different geometries and flow patterns. The differences in geometrical patterns are inferred to be due to different mechanical properties of the boundary layers, and such differences can be plausibly justified on the basis of our understanding of material properties of rocks. The boundary layers seem to operate with a lot of independence, since plume locations correlate only weakly with spreading centres, the sites of (passive) upwelling in the plate-scale flow.

The combination of different heat flows and different mechanical properties gives rise to a style of convection that is sketched in

Figure 12.1 (although a three-dimensional view would be needed to depict fully the differences between the plate and plume modes of convection). This is really two styles of convection operating in the same fluid layer. One is the plate-scale mode, which is driven by the lithosphere and whose form is more like rolls, though strongly time-dependent, as was described in Chapter 9. The other is the plume mode, driven by a hot thermal boundary layer at the base and whose form is of narrow, rising columns of surprising stability, with an implicit broad, slow downwelling between.

12.1.1 Heat transport and heat generation

We saw in Chapter 10 that the plate-scale flow accounts for about 85% of the mantle heat budget, while in Chapters 10 and 11 we saw that plumes account for only about 10%. This implies that the mantle layer involved is heated mainly from within, since the heat input from below and carried up in plumes is insufficient to maintain the heat loss out of the top. Also in Chapter 11 we saw that a consistent picture emerges if the convecting mantle layer is identified with the whole mantle, since estimates of the heat flow emerging from the core are comparable to the heat inferred to be transported by plumes. (The converse implication for a layered mantle system will be taken up below.)

The source of the heating of the mantle implied by this argument is not entirely clear at present. We saw in Section 7.5 that the radioactivity of the upper mantle seems to be much too low to account for the heat emerging at the surface: if upper mantle heat production prevailed through the depth of the mantle, it would account for only about 10 mW/m^2 or 10% of the oceanic heat flux. In order to account for the observed heat flow out of the

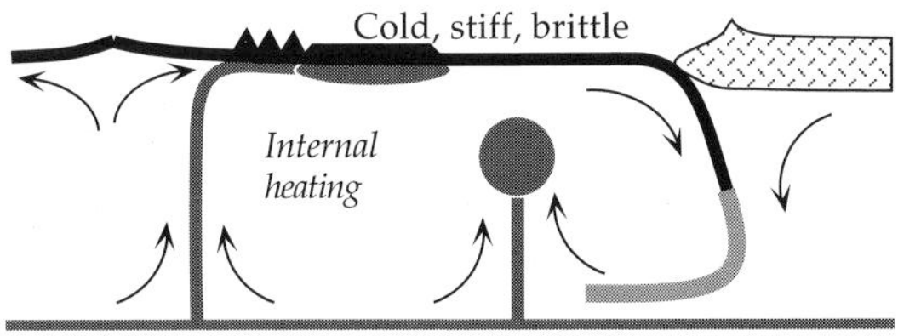

Figure 12.1. A sketch of the main components of the mantle's dynamical system so far identified. The main active component is the lithosphere, broken into plates that form, cool, subduct and reheat. The other active components are plumes which start with a head and tail structure and give rise to flood basalts and volcanic hotspot tracks. The buoyancy of the continental lithosphere (patterned, right) modulates the system.

mantle, the lower mantle would have to have heat production twice as great as chondritic meteorites and comparable to that of oceanic crust (see Tables 7.2 and 14.1). It is also possible that the mantle is not close to a steady-state balance of heat production and heat loss, but is at present losing heat at an unusually high rate that cannot be sustained in the longer term by radioactive heating. These important points will be taken up in Chapter 14 in the discussion of the thermal evolution of the mantle.

12.1.2 Role of the plates: a driving boundary layer

I observed in Chapter 3 that recognition of the nature of the mantle dynamical system was hindered by a view of convection that was too narrow. Thus it was commonly assumed that convection would have a form like the cells of Benard's convection experiments, that is it would have active upwellings in a pattern coordinated with active downwellings, and that the upwellings and downwellings would be of comparable strength, as sketched in Figure 8.3a. The picture developed in Chapters 8–11 and sketched in Figure 12.1 is of buoyant upwellings that are relatively weak and poorly correlated with the spreading centres of the plates, and of negatively buoyant downwellings that correlate strongly with subduction zones.

The plates, in this system, are thus the dominant sources of motion in the mantle. They are an integral part of mantle convection and its most active component. They are not carried passively by an unobservable or mysterious form of convection somewhere 'down there'. Convection is driven by thermal boundary layers, and it produces observable effects.

12.1.3 Passive upwelling at ridges

A clear implication of this picture, and of the observational evidence, is that upwelling at normal midocean ridges is passive. If there were active, buoyant upwelling, its buoyancy would produce extra topographic uplift, and the depth of the sea floor would not correlate simply with age, but would depend on distance from the spreading as well (Section 12.3 below).

The force of this argument is more evident if the three-dimensional nature of the flow is considered. If there were buoyant upwellings under ridges, the upwellings would have to be offset at transform faults. The fluid mantle would be expected to yield a more continuous offset of the upwelling than the sharp, faulted offsets of the brittle lithosphere. We would therefore expect either some upwelling extending beyond the end of each ridge segment or a zone of buoyant upwelling under transform faults connecting with the upwellings under spreading centre segments. There should therefore be some topographic expression of upwelling in the vicinity of transform offsets of ridges.

An illustration of how little evidence there is for such topography is provided by the topography, shown in Figure 12.2, of the sea floor near one of the longest transform offsets of any spreading centre, at the Eltanin fracture zone in the south-east Pacific. It is striking how cleanly the whole midocean ridge structure is terminated and offset by the transform fault. Where the spreading centre is terminated by the transform fault, the sea floor on the other side is at a depth normal for its age. There is no hint of a bulge, due to putative buoyant upwelling under the spreading centre, persisting across the transform fault. This topography is difficult to reconcile with a buoyant upwelling from depth, but is readily explicable if the midocean ridge topography is due to the near-surface and local process of conductive cooling, thickening and thermal contraction of the thermal boundary layer (that is, of the lithosphere).

If upwelling under normal midocean rises is passive, then neither is there any problem with spreading centres that move relative to other parts of the system: they merely pull up whatever mantle is beneath them as they move around the earth. This solves the puzzle that led Heezen (Chapter 3) to postulate an expanding

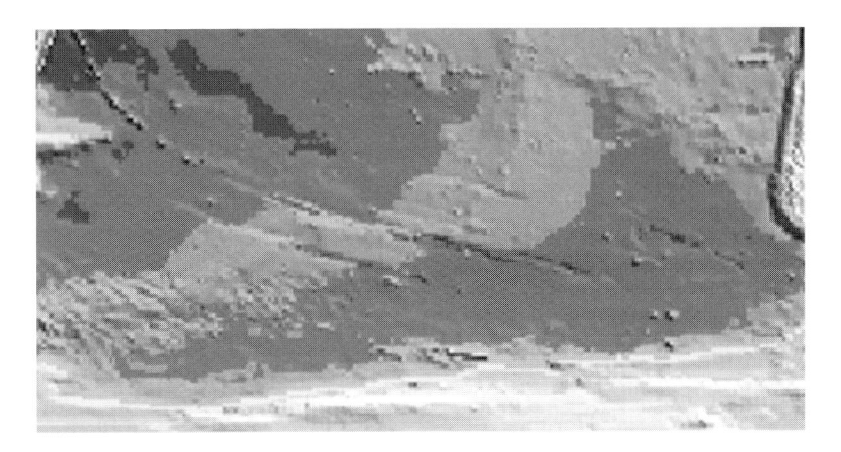

Figure 12.2. Topography of the sea floor near the Eltanin fracture zone in the south-east Pacific (compare with the global map, Figure 4.3). The East Pacific Rise is cleanly offset by the Eltanin transform fault, with no indication of rise topography extending across the fault. This is explained if the rise topography is due entirely to near-surface cooling with no contribution from a putative buoyant upwelling under the rise.

earth in order to try to explain how spreading centres could exist simultaneously on both sides of Africa.

12.1.4 Plate shapes and kinematics

I have stressed in several places in preceding chapters that the plates appear to behave like brittle solids rather than like a viscous fluid. This is the primary reason why they do not look like the top of a 'normal' convecting system. Instead the plates have odd shapes, a variety of sizes, and boundaries that are angular and segmented. Wilson recognised three kinds of plate boundaries corresponding to the three standard fault types of structural geology: normal, reverse and strike-slip (Chapter 3). The motions of plates recorded by the magnetic stripes of the sea floor show that the plate boundaries have acted like such faults much of the time, with the important proviso that seafloor spreading occurs symmetrically in the long term, though it involves normal-faulting earthquakes in the short term. The basis in rock rheology and brittle solid mechanics for the occurrence of the main fault types was presented in Chapter 6. The implications of this behaviour for the evolution of plates and for the time-dependence of the plate-scale mode of convection were explored in Chapter 9. The way in which the evolving plate system couples with flow in the mantle was illustrated in Chapter 10.

12.1.5 Forces on plates

The statement in Section 12.1.2 above that plates are active and are the main driving component of mantle convection may seem at first to contradict some common perceptions about forces that act on plates. Discussions of forces acting on plates have usually been based on the idea of conceptually separating the plates from the rest of the mantle, with the plates defined in the strict mechanical sense, that is as the surface plates plus the parts of subducted plates that are seismically active and therefore presumed to be strong. The kinds of forces considered then are slab pull, ridge push, trench resistance, basal drag and so on. This approach has yielded some useful insights, such as that slab pull is relatively large, but so is trench resistance, and that the net pull transmitted to the surface plate is smaller, and comparable to the ridge 'push' [1, 2].

In order to reconcile such conclusions with the approach of this book, we must recognise that saying that the plate-scale convection is driven by the thermal boundary layer is a more general statement than saying that slab pull is the largest plate driving force. Both slab pull and ridge push are manifestations of the thermal boundary layer. While slab pull is the obvious result of the plate subducting into the mantle, ridge 'push' is a result of the sloping density interfaces or horizontal density gradients associated with the thickening plate and its subsiding top surface [3]. ('Ridge push' is a misnomer surviving from an earlier and naive concept that plates are pushed apart by magma ascending at spreading centres. It is a gravity sliding force.)

The 'plate forces' approach has also led to some conclusions that do not survive a more complete conception of the problem. One example is the conclusion of Forsyth and Uyeda [1] that there must be a low-viscosity lubricating layer under the plates in order to explain how plates of such differing sizes as the Pacific, Nazca and Cocos plates could move with similar velocities. Another is that the net result of slab pull and trench resistance cannot be very important, otherwise the relative motions of non-subducting plates, such as Africa and Antarctica, cannot be explained.

Other conclusions become possible when it is recognised that the forces previously considered do not describe the whole system. A more complete depiction of the system of forces associated with the plate-scale flow would include the negative buoyancy of aseismic subducted lithosphere, which may persist to great depth virtually unattenuated long after the lithosphere has ceased to behave as a brittle solid. This is because the thermal anomaly of the subducted lithosphere does not actually disappear, it merely diffuses out, preserving the same total amount of buoyancy. This follows from conservation of energy. While it is on its first descent through the mantle it remains relatively concentrated, in the manner evident in Figures 10.2, 10.12 and 10.13, and retains virtually all of its driving power.

Another concept missing from the plate forces approach is that stresses propagate through the viscous fluid. The active components (surface and subducting plates) drive flow that extends for a distance comparable to the length scale of the plate (surface or descending), and this flow acts on other components. In particular, the descending plate drives a flow in the deep mantle that circulates in the same sense as the plate itself (Figures 10.2, 8, 9, 12, 13). This concept is so obvious in constant-viscosity convection as to be almost meaningless. It needs restating here because the plate is mechanically distinct and strong and, as pointed out by Elsasser [4], it can therefore act as a stress guide. However, the stress guide effect is only relevant to the negative buoyancy within the strong plate, and viscous propagation of stresses still occurs in the rest of the mantle.

With these concepts added, motion of non-subducting plates is not surprising, and it is evident that the mantle under subducting plates will tend to be moving in the same direction as the plate (Figure 10.2). Since the 'basal drag' depends on the differential motion of the plate and the underlying mantle, its magnitude can be small without requiring the viscosity under the plate to be small. Thus the requirement for a 'decoupling' low-viscosity layer is not established. Whether there is in fact a significant basal force on some plates due to a velocity gradient beneath them is a question that might be addressed most profitably by comparing cases where the relative motions are expected to be parallel, antiparallel or oblique. However, this would require global three-dimensional models of mantle flow that are more reliably precise than are available at present.

12.1.6 A decoupling layer?

The idea that plates slide on a pronounced low-viscosity layer under the lithosphere originated early in the history of plate tectonics, but the argument just presented shows that there is no necessity for such a layer arising from the need to explain plate motions. Indeed, we saw in Chapter 8 that plate velocities can be readily accounted for in general terms with a uniform viscosity mantle. In any case, it was shown early on that a low-viscosity layer of the order of 100 km thick does not decouple motions above and below it very efficiently unless its viscosity is three or more orders of magnitude lower than the adjacent mantle [5].

On the other hand, it is very plausible that there is a significant minimum in mantle viscosity just under the lithosphere. This depth range has the combination of high temperature, low pressure and close approach to melting that is most favourable to lower viscosity. The question here is not whether such a minimum exists, but whether it is so pronounced that it has a major effect on the geometrical form of mantle flow. The evidence for a viscosity minimum is good, while the evidence for substantial decoupling is doubtful.

12.1.7 Plume driving forces?

Morgan's original conception of plumes was that they were major players in driving plates [6]. This idea derived in part from the association of plumes with the initiation of continental rifting, particularly in the Atlantic. They would not be expected to be dominant players in the picture developed here, on the grounds of their secondary buoyancy fluxes. However, they may play a significant secondary role as triggers when circumstances are appropriate. Hill [7] developed the idea that new plumes with large heads (Chapter 11) may sometimes trigger rifting or a ridge jump, citing particularly the examples of the ridge jump from west of Greenland to east of Greenland close to the time at which the Iceland plume started (60 Ma) and the rifting of the North Atlantic at about 175 Ma, after the inferred arrival of a plume at 200 Ma. Continental rifting also occurred after the eruption of the Deccan Traps flood basalts in India and the Parana flood basalts of South America, but rifting did not occur in association with the Siberian or Columbia River flood basalts.

These associations are with inferred plume heads, and the proposed mechanism is gravity sliding off the uplift caused by the plume buoyancy [8], facilitated by heating and weakening of the lithosphere by the plume [7]. This mechanism accounts plausibly for the variable delay of rifting, though the specifics of such delays are not understood in any detail. The gravitational potential energy associated with plume tails is considerably smaller than that of plume heads, and there has been no active suggestion of plume tails playing a major role in driving plates since Morgan's original suggestion.

12.2 Other observable effects

In Chapters 10 and 11 we focussed on observations of midocean ridge topography, hotspot swells, seafloor heat flow and seismic tomography of the mantle in order to constrain the form of mantle convection. However, there are other observations that are consistent with the same general picture, but which have led others to propose quite different interpretations. I discuss two of them here, each involving topography related to thermal variations in the mantle that are not directly related to the surface plates. Each of these connects to a controversy regarding deviations of old sea floor from the square-root of age subsidence inferred from the simple model of a thermal boundary layer (Sections 7.4, 10.7). I will discuss the alternative hypotheses about these deviations later (Section 12.4).

12.2.1 Superswells and Cretaceous volcanism

The subsidence of the sea floor as it moves away from spreading centres fits to first order the relationship that the depth increase is proportional to the square-root of age (Chapter 4), and this relationship is predicted by the simple physical process of cooling by conduction to the earth's surface (Chapter 7). However, there are significant deviations from this relationship, particularly for older sea floor, and the deviations are usually in the sense that the sea floor is shallower than the simple model predicts (Figure 4.6).

An important observation is that the crests of the midocean rises themselves vary in depth by up to a kilometre, even away from known hotspots such as Iceland, as can be seen in Figure 4.3. That these variations are not associated with hotspots or mantle upwellings is demonstrated by the fact the one of the largest deviations is negative: this is between Australia and Antarctica, where the rise crest is about 1 km deeper than average. Nor can they be associated with any process involved in the ageing of the lithosphere, since the lithosphere at rise crests is all of zero age. Nor is it likely that they can be attributed to variations in the thickness of the oceanic crust, because these are partially compensated by isostatic balance. With densities of 2900 kg/m^3 for the crust and 3300 kg/m^3 for the mantle, it would require a reduction of crustal thickness of nearly 6 km to yield a 1 km lowering of the surface, and this is close to the total thickness of oceanic crust.

It seems that these broad (several thousand kilometres), low-amplitude (less than 1 km) variations in depths of midocean rise crests must be due to spatial variations in the sub-lithospheric mantle. An obvious possible cause is low-amplitude variations in mantle temperature, of the kind that are expected in a convecting fluid because of thermal perturbations from incompletely mixed fluid from the thermal boundary layers. For example, 1 km of topography would be generated by a temperature difference of about 40 °C over a depth of 1000 km, assuming a thermal expansion coefficient of 2.5×10^{-5} /°C.

Regardless of the cause of the variations in depth of the rise crests, if they exist at rise crests then they are very likely to exist elsewhere. To test this, we need to separate this small-amplitude signal from the larger signal of the midocean rise topography, which is due primarily to the thermal contraction of the oceanic lithosphere. A straightforward approach is to subtract the rise-type topography and see what is left; that is, we should subtract topography proportional to the square-root of age of the sea floor. In order to do this, we need to know what the amplitude should be. However, the relevant material properties (Equation (7.4.2)) are not known well enough independently to provide an accurate estimate. Since the sea floor younger than about 70 Ma seems to deviate less from the square-root of age subsidence than older sea floor, we might assume that this provides a reasonably reliable estimate of the intrinsic subsidence rate. We should note, however, that even the subsidence of young sea floor may be affected significantly by deep mantle thermal structure (Section 10.7.1), so this is not yet the ideal way to separate the effect of lithospheric thickening from the effects of the deeper mantle.

With this logic, Davies and Pribac [9] subtracted from the observed seafloor topography a depth of the form

$$d = a + bt^{1/2} \tag{12.3.1}$$

where d is seafloor depth, t is seafloor age and a and b are constants. They tried three values of the constant b: 280 m/Ma$^{\frac{1}{2}}$, 320 m/Ma$^{\frac{1}{2}}$ and 360 m/Ma$^{\frac{1}{2}}$. The results are not particularly sensitive to this choice. Figure 12.3 shows a map of residual seafloor topography resulting from using $b = 320$ m/Ma$^{\frac{1}{2}}$, and with an isostatic correction for crustal thickness also included [10]. Gaps in the map correspond to regions where the age of the sea floor is not known reliably.

The midocean rise topography of younger sea floor has been almost completely removed in Figure 12.3, confirming the appropriateness of the value of b used. As well, there is a broad swell in the central and western Pacific and a broad low between Australia and Antarctica. Swells along some parts of the Atlantic margins are not so clear at this resolution. These features are of similar character to the variations that were already evident along rise crests,

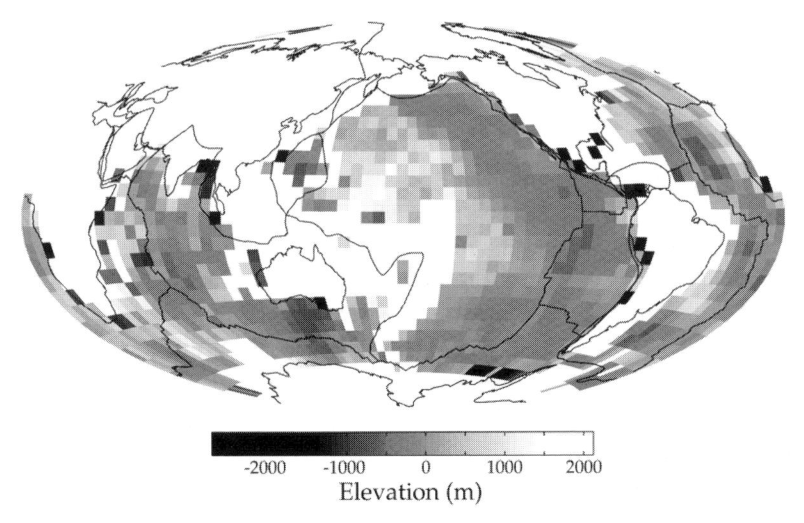

Figure 12.3. Anomalous seafloor topography. An age correction (Equation (12.3.1) with $b = 320$ m/Ma$^{1/2}$) and a correction for crustal thickness using the 5° × 5° grid of Mooney *et al.* [11] have been removed. Figure courtesy of S. V. Panasyuk, Harvard University [10].

and of course they correlate with them. They are consistent with the same hypothesis, that they are due to small-amplitude, broad-scale variations in deep mantle temperature.

McNutt and Fischer [12] and Pribac and I [13] independently coined the term superswell to describe an area several thousand kilometres in dimension that is some hundreds of metres higher than expected. The South Pacific superswell described by McNutt [12, 14] was determined using as a reference an empirical, asymptotically flattening seafloor subsidence curve (Section 12.4), which yields smaller amplitudes of positive anomalous topography in older areas. On the other hand ours was defined relative to a square-root of age subsidence curve, with the result in the Pacific that a single, much larger superswell extends from the south-east Pacific to Japan, a distance of about 9000 km, subsuming the smaller so-called South Pacific superswell.

There is independent supporting evidence for the interpretation that the topographic anomalies of Figure 12.3 are due to deep-mantle temperature variations. The residual topographic highs correspond with regions that have relatively slow seismic velocities in tomography models (Figure 5.14), away from the seismically fast bands that are interpreted as subducted lithosphere (Section 5.4). A similar pattern is evident in the geoid, which is anomalously high over the Pacific and Africa and low in between (Figure 4.9). Each of these observations is consistent with the mantle under the Pacific and Africa being relatively warm and the intervening mantle being cooler. This interpretation is consistent in turn with the observation that the locations of subduction zones in the past 100–200 Ma correlate with the low-geoid, low-topography, fast-seismic-velocity belt [15, 16]. It is obviously plausible that the mantle is generally cooler under locations of past subduction, and warmer elsewhere.

There was an era of extensive intraplate volcanism in the Pacific during the Mesozoic, including the formation of the Ontong–Java plateau, which is currently interpreted to be an oceanic flood basalt [17, 18]. Menard [19, 20] argued from evidence from guyots (wave-cut, drowned atolls) that a large area of the Pacific sea floor was shallower than expected about 100 Ma ago. He named this region the Darwin Rise, in honour of Darwin's recognition of the formation process and age progression of volcanic islands and atolls (Chapter 3). I and Pribac [9] proposed a scenario in which a region of mantle that was warming relative to surrounding regions under subduction zones slowly elevated the surface, and simultaneously promoted the formation of new plumes (Figure 12.4). Plume formation would be promoted because the upflow would sweep and thicken the bottom thermal boundary

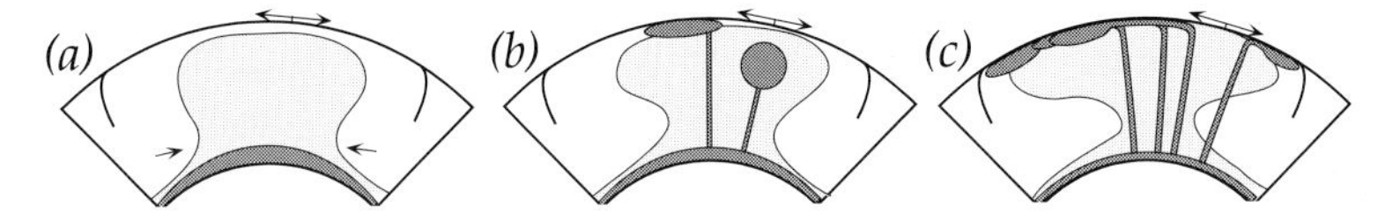

Figure 12.4. Sketches of the possible evolution of the Pacific mantle during the Cretaceous period, about 100 Ma ago. After [9]. Copyright by the American Geophysical Union.

layer, making it more unstable. We argued that the Pacific superswell is in fact the modern continuation of the Darwin Rise.

This is a straightforward and more specific explanation for the relatively high volcanic activity of the time than the so-called 'superplume' hypothesis, in which some ill-specified kind of larger plume or plumes is supposed to have risen [21]. There has been some loose use of the term superplume both as a vague explanation for observations and as a description of large upwellings seen in some numerical models whose parameters do not obviously correspond with those of the mantle.

12.2.2 Plume head topography

Plume heads ought to generate topography of the order of several hundred metres to one kilometre, and this might persist for at least tens of millions of years. An example of predicted topography from a numerical model is shown in Figure 12.5. Such topography will persist because the time it takes for heat to be lost through the lithosphere is comparable to the age of the lithosphere. If the Ontong–Java plateau is a flood basalt, and it was produced from a plume head, then this component of topography might still be present there. However, it will obviously be much smaller in amplitude than the main plateau topography, which is due to the oceanic crust being thickened up to about 30 km. There may be other examples of superswell topography that are equally hard to separate from the effects of associated volcanism.

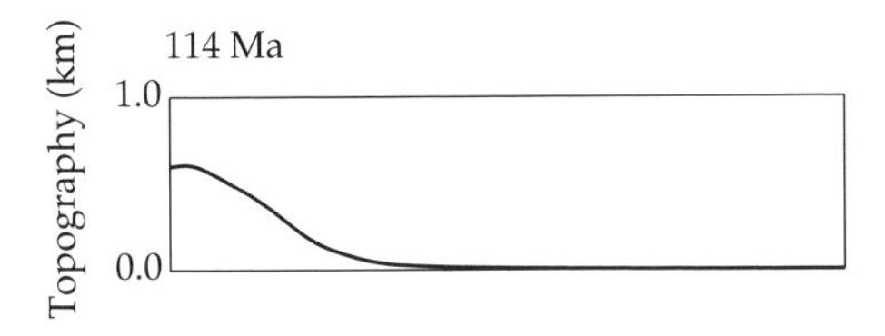

Figure 12.5. Topography generated by the plume head of Figure 11.10. The profile is from the plume axis (left) to a radius of 3000 km (right).

One circumstance where the plume head topography might be partly revealed is where the lithosphere rifts after the arrival of a plume head. This is relevant to the discussion of the elevation of old sea floor. Morgan [22] proposed that the opening of the North Atlantic was precipitated by the arrival of a plume. Hill [7] elaborated this proposal and argued that a plume head was emplaced under eastern U.S. and west Africa at about 200 Ma, and rifting began at about 175 Ma. This implies that the plume head would be rifted along with the lithosphere. A numerical model of this process is shown in Figure 12.6. As rifting proceeds, plume material is emplaced under the new sea floor. At first this happens because it takes time for the deeper plume material to rise to near the surface and turn horizontally, and new sea floor has already formed when it arrives. The amount of plume material under the new sea floor then continues to increase because it flows up from under the adjacent thicker lithosphere. Thus not only is the plume head rifted along with the lithosphere, but a lot of plume head material ends up under the new sea floor. This will cause surface uplift.

The topography due to the plume material is shown in Figure 12.7, and it is evident that significant uplift occurs for a distance of hundreds of kilometres from the old rift margin. The amount of

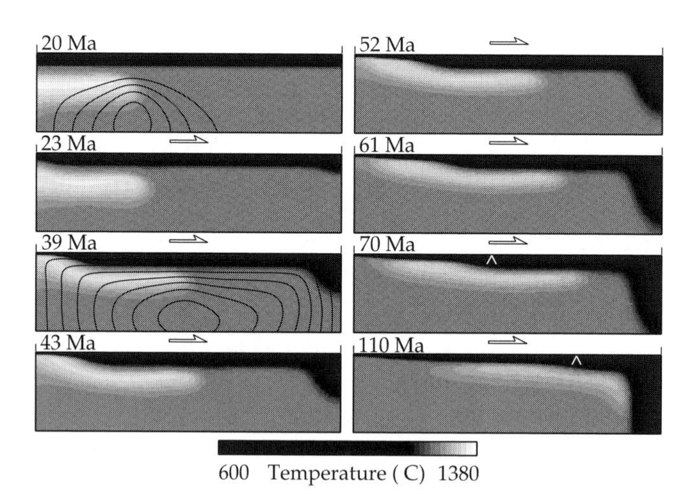

Figure 12.6. Rifting of an isolated plume head. The plume head was modelled initially as a warm blob that rose and flattened under the lithosphere (20 Ma). At that time a surface motion to the right was imposed (arrows). Mirror symmetry about the left boundary implies a spreading centre. The lithosphere and the plume under it were rifted in response. Streamlines show the flow before (20 Ma) and after (39 Ma) rifting. The location of the original rifted margin is indicated by '^' at 70 Ma and 110 Ma. The blob of lithosphere descending on the right is an artefact.

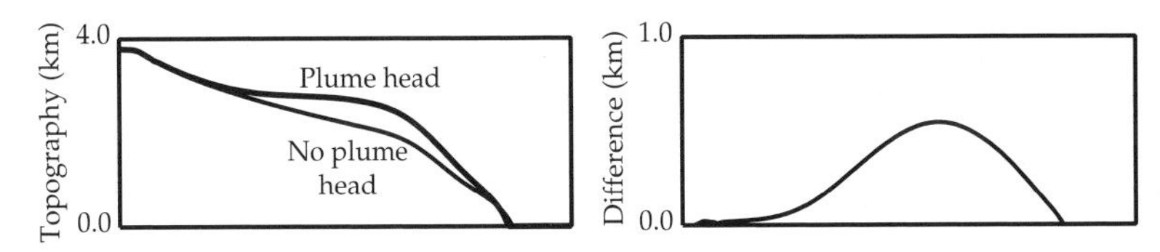

Figure 12.7. (Left) Topography from the model in Figure 12.6 (110 Ma) compared with a similar model with no plume head. (Right) The difference between the topographic curves on the left, showing the uplift due to the plume head, most of which is offshore of the old rift margin.

uplift is proportional to the temperature excess of the plume: in this case there is about 500 m of uplift for a temperature difference of 100 °C. This plume head topography is a likely contributor to the excess elevation of the old North Atlantic sea floor evident in Figure 12.3, and it may account for most of it.

We will see in Section 12.4.1 that one interpretation of the depth of old sea floor was that it asymptotically approaches a constant depth. It turns out that this was based on ship track data from the North Atlantic and the North Pacific in regions that appear here to be anomalous. According to the interpretations offered here, the older parts of these areas have been affected by anomalous mantle: deep, warm mantle in the Pacific and plume head material in the North Atlantic. Thus the early data were from areas not necessarily representative of all old sea floor. A general point originally made was valid: that the deviation from the square-root of age subsidence implied an input of additional heat. However, the hypothesis offered was unnecessarily narrow (in assuming a constant-thickness lithosphere) and poorly developed (in not specifying a clear and viable mechanism for delivering the heat).

12.3 Layered mantle convection

The question of whether the mantle convects as one or two layers has been probably the most vexed and persistent controversy about mantle convection. However I have deferred discussion of it until now for three reasons. First and most important, the fundamental concepts of mantle convection can be developed without needing to address this question. Second is my own judgement as to the relative strength of the arguments. Third, I wanted to develop the specific picture of mantle convection in what seemed a natural way, without undue interruption.

The major debate has been specifically about whether or not the mantle flows through the boundary between the upper mantle and the lower mantle at 660 km depth (as defined in Chapter 5). Whatever the outcome of this debate, there remain other possible ways in which the mantle might be layered, for example that there is a denser layer near the base of the mantle that remains largely separated, or that there is a stratification (layered or continuous) in trace element and isotopic compositions, due either to kinematic or dynamic effects, as will be discussed in Chapter 13. There is also quite strong evidence for substantial variation of viscosity with depth, as discussed in Chapter 6. Thus the present discussion is not about whether the mantle is vertically homogeneous or heterogeneous, it is about the more specific question of whether there is a barrier to flow at a depth of 660 km.

12.3.1 Review of evidence

Evidence bearing on this question has already been presented in the discussions of seismic tomography, the geoid and the effects of phase transformations on mantle dynamics. The resolving power of seismic tomography has been steadily increasing, and recent models have shown with some clarity high seismic wavespeed anomalies extending from subduction zones deep into the lower mantle, to depths of 2000 km, and greater in a couple of cases (Figures 5.13, 5.14). These images lend strong support to the idea that there is a substantial flow between the upper mantle and the lower mantle.

It is sometimes suggested that there may be separate downwellings in the lower mantle that are triggered by the impingement of subducted lithosphere at the base of the upper mantle, and the two downwellings, one beneath the other, look continuous in the seismic images. While this is possible in principle, it is an *ad hoc* argument and the phenomenon has not been demonstrated quantitatively with anything approaching realism for the mantle.

Another argument was presented by Richards and Hager [23], based on the occurrence of positive geoid and gravity anomalies over subduction zones (see Figure 4.9 and Section 6.9.3). The sign of the geoid anomalies requires that the positive mass anomalies of subducted lithosphere are compensated from below, rather than from above, by downward deflections of a deep density interface. In order to achieve the required geoid amplitude, the compensating deflections must be at great depth. Richards and Hager demonstrated that models in which flow passes through the mantle transition zone and in which the viscosity increases by a factor of about

30 through the transition zone were consistent with the observations. Models with a density barrier in the transition zone do not yield sufficient amplitude and models without the viscosity increase yield the wrong sign. A more detailed qualitative explanation is given in Section 6.9.3.

The possibility that phase transformations might block flow between the upper and lower mantles was discussed in Sections 5.3, 10.5 and 11.4. In Section 5.3 it was noted that the effect of the temperature dependence of the transformation of spinel-structured Mg_2SiO_4 to a perovskite structure plus magnesiowüstite was to resist vertical flow, but that there is substantial uncertainty in the thermodynamic parameters of this transformation. It was also noted that the transformation of the garnet structured component ($MgSiO_3$) in the same depth range would probably have the opposite effect, though its thermodynamic parameters are even less well-determined. A separate possibility is that the compositional zoning within subducted lithosphere would cause phase transformations to occur at different depths than in surrounding mantle, and again vertical flow might be inhibited (Section 5.3.4). There are large uncertainties in the details of the resulting buoyancy anomalies (Figure 5.11), and in any case the magnitude of the effects seems to be too small to be decisive [24].

On the basis of material properties it is thus far from clear at present that vertical flow would be interrupted by any of the effects of phase transformations, though the possibility remains in principle. Numerical models (Sections 10.5 and 11.4) have shown that although a strong phase transformation effect might block vertical flow, at least temporarily, both plates and plumes are likely to penetrate a phase transformation for plausible values of the thermodynamic parameters of the transformation.

12.3.2 The topographic constraint

Another constraint comes from the topography of the sea floor and measurements of radioactivity in rocks derived from the upper mantle. There is only enough radioactive heat production in the upper mantle to account for about 2% of the heat emerging at the earth's surface (Sections 7.5, 12.1.1). Some of the heat can be coming from cooling of the upper mantle, but this would be no more than about 6% (Chapter 14). This means that around 90% of the heat lost from the mantle must be coming from deeper than the upper mantle.

If the upper mantle is heated 90% from below and there is no mass flow through its base, then according to the ideas developed in

Chapter 8, there should be a well-developed hot thermal boundary layer at the base of the upper mantle, through which about 90% of the mantle's heat budget is passing. Accordingly, there should be buoyant upwellings carrying this amount of heat through the upper mantle to the base of the lithosphere. This buoyant material should cause topography by elevating the lithosphere, in the way proposed to explain the hotspot swells (Chapter 11 and Figures 11.2 and 11.3). The fundamental relationship between buoyancy and topography through thermal expansion (developed in Chapters 8–11 and especially Sections 10.4.4 and 11.2) then leads us to the conclusion that such topography should be comparable to the mid-ocean ridge topography. This is because comparable rates of heat transport are involved in the two modes of convection: the plate-scale mode of convection and the putative upper-mantle upwelling mode of convection.

Two kinds of seafloor topography have been discussed that are not explained by the thickening of oceanic lithosphere: hotspot swells and superswells. Hotspot swells are readily interpreted as being due to buoyant upwellings (Chapter 11), but these can be associated with no more than about 10% of the mantle's heat budget, rather than the 90% required here.

Although superswells might be interpreted as being due to buoyant mantle, they would not represent a large heat flow either. For example, the Pacific superswell of Figure 12.3 might be treated in the same way as the Hawaiian swell (Section 11.2) to derive a buoyancy flux and heat flow rate using Equation (11.2.2). Taking generous values of the width $w = 5000$ km and the mean elevation $h = 500$ m, we get values only 2.5 times those for the Hawaiian plume (for which $w = 1000$ km and $h = 1$ km). In other words, the buoyancy flow rate is about 1.7×10^5 N/s and the heat flow rate is about 5×10^{11} W, about 1.5% of the mantle's heat budget of 36 TW (Table 10.1). Allowing for an African superswell as well, it seems that superswells would correspond to less than 3% of the mantle's heat budget, even less than plumes and not sufficient, in combination with plumes, to approach the heat flow implied by a layered mantle.

However, the interpretation of superswells offered in Section 12.2.1 above is that they are not due to hot material rising from a thermal boundary layer, but are due to incomplete homogenisation of the mantle interior. In this interpretation, superswells do not imply *any* additional heat coming from a thermal boundary layer at the base of the upper mantle. Then the only identified evidence for buoyant upwellings is the hotspot swells, corresponding to only about 10% of the mantle heat budget. The straightforward conclusion is that the putative strong upper mantle upwellings that would be required if there were a barrier to flow at 660 km depth do not exist.

This argument does not preclude local or temporary interruptions to flow between the upper and lower mantles. It does not preclude fewer or weaker upwellings, including some plumes, originating from 660 km depth. Nor is it inconsistent with seismological evidence (Chapter 5) that some subducted lithosphere may be blocked from penetrating the lower mantle, though it would require that such blocks be only partial or temporary. However, substantial flow would still be required between the upper and lower mantles in order to transport heat out of the lower mantle at the required rate. The topographic constraint depends on there not being a substantial difference in temperature between the upper and lower mantles, and such a temperature difference would take some time to accumulate if flow between them were blocked, the order of 100 Ma or more, according to calculations presented in Chapter 14. Thus the constraint is that there should be substantial flow averaged over 100 Ma, and it does not preclude local or shorter-term blockages.

12.3.3 A numerical test

This conclusion is tested by the numerical model shown in Figure 12.8. This is a model of convection scaled to the upper mantle, with 100% of heat input through the base. There are two plates separating at a central spreading centre. Constant surface velocities (of 20 mm/a) are imposed to make it easier to compare with the predictions of the halfspace model (which predicts depth proportional to the square-root of seafloor age, Equation (7.4.2)) and to better control the numerical experiment. After 268 Ma, a velocity of 10 mm/a to the right is added to both plates so that the spreading centre migrates to the right at 10 mm/a. The viscosity structure of the last stage (405 Ma) is shown in the bottom panel. Viscosity depends on temperature and depth. Temperature dependence is capped with a maximum dimensionless value of 30. There is a low viscosity zone to 200 km below the lithosphere with a minimum dimensionless viscosity of 0.1. The dimensional reference viscosity is 10^{21} Pa s. The prescribed base heat flux is 78 mW/m^2, slightly lower than the estimated 90–100 mW/m^2 for the upper mantle.

The strong base heating generates an early flock of hot upwellings (85 Ma). As subducted lithosphere recirculates along the bottom, the upwelling becomes focussed into a single central upwelling (177 Ma), but later side upwellings develop regularly and are swept

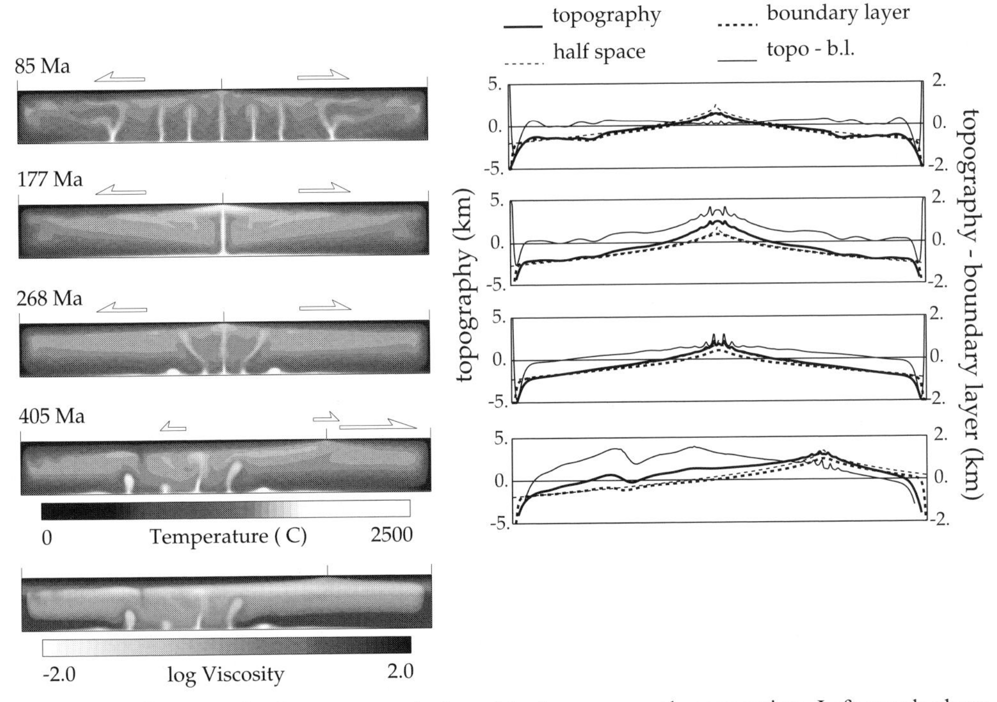

Figure 12.8. Topography from a numerical model of upper mantle convection. Left panels show temperature and viscosity structure and right panels show corresponding topography. Surface velocities of 20 mm/a are imposed (arrows). There is a stationary spreading centre until 268 Ma, at which time a velocity of 10 mm/a to the right is added to both plates, so that the spreading centre migrates. Side walls are reflecting. Model topography is compared with halfspace (square-root of age) topography and the topography due just to the top, cool boundary layer. Topography minus boundary layer represents the topography due to the deeper thermal structure.

towards the centre (268 Ma). After the migration of the spreading centre the upwellings are offset from the spreading centre (405 Ma).

The model topography is compared with the halfspace estimate and with the topography due to thermal contraction within the actual upper thermal boundary layer of the model. The latter two are generally similar, indicating that the model boundary layer behaves similarly to the square-root of age thickening of the half-space model (except for a substantial drip from the left-hand plate at 405 Ma). The plots in Figure 12.8 also show the difference between the model topography and the boundary layer contribution. This comparison is used in order to eliminate the effects of the imperfections of the model boundary layer compared with the lithosphere, and it represents the contribution from the thermal structure below the boundary layer, including that from the hot upwellings.

The early topography is close to the halfspace estimate (85 Ma), but this is a transient stage before the subducted lithosphere is fully recirculated. When upwelling is concentrated into a single column (177 Ma) there is a large topographic anomaly of 1.5 km amplitude due to the rising and spreading buoyant fluid. As the upwellings become more dispersed again (268 Ma) the anomaly broadens and its amplitude decreases to about 600 m. At this stage it is not obvious that the anomaly would be noticeable. After the spreading centre migrates, the upwellings are displaced from the spreading centre, and their associated uplift again represents an anomaly with about 1.5 km amplitude. The subsidence is quite asymmetric and the anomaly would be readily detectable.

These results are less dramatic than those I have presented in the past [25, 26] in which the midocean ridge topography was obscured by the upwelling topography. The difference seems to be due perhaps to the model being run for longer and to the subducted lithosphere being stiffer in the present model (having the maximum viscosity through much of its thickness, rather than reaching the maximum only at the surface), the main effect of which is to slow its recirculation. Nevertheless the topographic contribution from the hot upwellings is large and observable except in the special case where the upwellings are centred under the spreading centre for a long time. We should not expect this to be the norm in a globally connected spherical upper mantle circulation with irregular plate geometries and sizes. The locations of the upwellings in Figure 12.8 are controlled by the horizontal convergence of the recirculating lithosphere, and in the upper mantle this convergence would generally not occur right under a spreading centre.

12.4 Some alternative interpretations

Some quite different interpretations have been placed on some of the observations discussed so far. I will discuss two of the most prominent, the so-called 'flattening' of the old sea floor and small-scale, upper mantle convection. These have been closely related historically.

12.4.1 'Flattening' of the old sea floor

I recounted in Chapter 4 (Section 4.3.1) that early assessments of the variation of heat flow with age seemed to show that heat flow approached a constant value for sea floor older than about 70 Ma. Subsequently it was shown that when measurements affected by hydrothermal circulation were removed the heat flow followed the prediction of the conductive cooling model (Section 7.3) much more closely (Figure 4.7).

Interpretations of seafloor depth followed an analogous history. Menard [27] seems to have been the first to plot depth versus seafloor age, and he showed a rough consistency between ocean basins, with the subsidence rate decreasing with age. The implication that the same models of cooling lithosphere used to interpret heat flow also implied that depth would increase with age (regardless of spreading rate) seems to have emerged slowly, and is implicit for example in a well-known paper by Sclater and Francheteau [28]. Davis and Lister [3] showed, in a landmark paper, that a simple analysis predicts that depth would be proportional to the square-root of seafloor age (Section 7.4) and showed that available observations [29], extending to an age of about 80 Ma, followed this proportionality rather closely.

After the ages of older magnetic anomalies and sea floor were established, the depth–age data were extended to 160 Ma and it was found that they deviated from the square-root of age trend [30]. The model of a plate approaching a constant thickness, that had been developed to explain the now-revised heat flow data [31, 32], was appealed to [28, 30]. The data were interpreted as indicating an asymptotic approach to a constant depth [30]. However, the data used were from regions of the North Atlantic and North Pacific that can now be inferred to be anomalous (Figure 12.3). As I noted in Section 4.2, the modern, more complete data do not show any particular tendency to approach a constant depth (Figure 4.6). In some regions the sea floor continues to subside while in others it becomes shallower again. Only when the data are aggregated into a global average curve is there an appearance of flattening. Thus the empirical basis for an asymptotically constant seafloor depth was removed by more complete data, just as for the heat flow data.

What *is* true is that old sea floor is commonly shallower than the prediction of the conductive cooling model, though there are places where the square-root-of-age subsidence persists to ages up to as old as 150 Ma. To put these variations into perspective, recall that the amplitude of the midocean rises is about 3 km, compared with amplitudes of the regional variations of a few hundred metres, and up to a kilometre in some places (Figure 12.3). The midocean rise system is clearly the dominant topography, as a glance at Figure 4.3 confirms. The explanation offered here for the regional variations (Section 12.2) is that they are due to incompletely homogenised temperature variations through the mantle, with some contribution from plumes and possibly old plume heads.

12.4.2 Small-scale convection

The model of the lithosphere approaching a constant thickness contributed to the development of the idea of a pervasive mode of 'small-scale' convection confined to the upper mantle. It was recognised that the constant-thickness lithosphere model implied a heat input to the base of the lithosphere in order to maintain an asymptotic steady-state heat flux at the surface, and it was proposed that this was due to some form of sub-lithospheric small-scale convection. Initially this was supposed to be convection cells of the scale of the upper mantle, to which this mode of convection was assumed to be confined [32, 33, 34]. A later variation was that it is driven by instability of the lowest, softest part of the lithosphere [35], though it was assumed rather than demonstrated that the lower lithosphere had the requisite mobility, the thermal boundary layer being assumed in that study to have a stepped viscosity structure. Subsequent evaluation using a more appropriate temperature-dependent viscosity [36] showed that it is not clear that such convection would have significant amplitude, or even occur at all.

The proposal for an upper mantle scale of convection, that is for cells of about 650 km depth and a comparable width, encounters the topographic constraint already discussed for any form of upper mantle convection: there is no evidence for the substantial topographic anomalies that should accompany the upwelling and downwelling limbs of such convection (Section 12.3). The proposal for convection driven by 'dripping' lower lithosphere also encounters the topographic constraint, but in this case the model implies only that there should be depressions where lower lithosphere is detaching. The amplitude of such depressions has not been accurately estimated, though I demonstrated the principle with an example in which the amplitudes were of the order of a kilometre [37]. Since a substantial amount of heat transport is required, in order to 'flatten' the sea floor (about 40 mW/m^2 [30, 35]), a significant and detectable amount of topography is very plausible. In particular, the higher viscosity of the cooler drips would enhance such topography by coupling their negative buoyancy more strongly to the surface; an unintentional example can be seen in Figure 12.8 at 405 Ma. On the other hand, the elastic strength of the lithosphere would reduce the short-wavelength components of this signal. On balance, it is likely that there should be a network of depressions across the older sea floor, probably with amplitudes at least of the order of a few hundred metres. Such a signal should be readily observable, but it is not evident in Figures 4.3 or 12.3.

Corresponding signals in the gravity field should also be present, but are not evident (Figure 4.9).

Some such signals have been demonstrated in restricted regions, but they appear to be due to something other than small-scale convection. The best-developed signals are in the south-east Pacific, where there are undulations of the sea floor with a wavelength of about 200 km and amplitudes of less than 200 m. Associated gravity and geoid anomalies have also been detected. The gravity anomalies are of low-amplitude (5–20 mgal) and linear with wavelengths of 100–200 km and lengths of the order of 1000 km [38]. More recently narrow volcanic ridges have been discovered that coincide with the gravity lows, and Sandwell and others [39] have argued that these are incompatible with small-scale convection. They propose that the lithosphere has been stretched over a broad region and that it has developed boudins, which are thinner, necked bands oriented perpendicular to the direction of stretching.

There have in fact been claims that a pervasive system of small-scale gravity signals exists [40, 41], but serious questions have been raised about those claims. The more basic is that the geoid signal used was an artefact resulting from inappropriate filtering of low-order harmonics from the observed field. Since the low-order components have the larger amplitudes, it is necessary to use a smoothed filter in order to avoid 'aliasing' of low-order signal into the higher-order components. When this is done, much of the putative small-scale convection signal goes away, and the remaining signal correlates well with topographic features like hotspot swells and oceanic plateaus [42]. Another question is that there was no attempt to exclude the hotspot swell signals from the analysis. This is due to a difference in interpretation of hotspot swells. McKenzie and others conceived them as being due simply to part of a pervasive system of upper mantle convection. If this is taken to imply a buoyant rising sheet under the swell, then we might expect active volcanism along a line instead of at an isolated hotspot. If it is not, then the downstream swells would not tell us about the internal mantle structure and should be excluded. On the other hand most others regard the combined evidence of isolated volcanism and surrounding swells as straightforward evidence for isolated rising buoyant columns (Chapter 11) distinct in character from that expected for a pervasive system of upper mantle convection.

A further problem with the dripping lower lithosphere hypothesis is that its long-term effect would actually be to increase the rate of subsidence, not to decrease it as claimed. This is because it would enhance the rate of heat loss from the mantle, and thus would enhance the thermal contraction that is the primary reason for seafloor subsidence. This was pointed out by O'Connell and Hager [43], and the effect is evident in a numerical model [37]. Previous conclusions had only taken account of the replacement of cool lower lithosphere by warm mantle, and had overlooked the fact that the influence of the cool lithospheric material can persist after it has sunk into the mantle.

Whereas the independent evidence (seismic tomography, geoid, subduction history) for the deep, warm mantle interpretation of old seafloor topography is strongly supporting (Section 12.2), the independent evidence for the small-scale convection interpretation is absent or equivocal when it might be expected to be clear. As well, the direct evidence that the old sea floor asymptotically approaches a constant depth seems to be a misreading of the observations.

12.5 A stocktaking

There have been many ideas about mantle convection proposed in the few decades since plate tectonics became widely accepted. Some of these have become well-established, while others can be seen not to have a good empirical basis. These distinctions have been discussed in this Chapter, and here I briefly summarise my assessment.

Simple and direct arguments from well-established observations lead to the conclusion that the plates are an integral part of a large-scale circulation (the plate-scale flow) that is the dominant form of mantle convection.

While thermal mantle plumes cannot be observed as directly as plates, their existence is a straightforward inference from the occurrence of isolated volcanic hotspots with associated hotspot swells. The hotspot swells constrain the buoyancy flux of plumes, and indicate that plumes transport less than about 10% of the mantle heat budget. Plumes are thus a distinct, secondary, but well-established mode of mantle convection. By simple inference, they arise from a hot thermal boundary layer at depth.

No other mode of mantle convection has been demonstrated. In particular, there is little evidence for a pervasive system of 'small-scale' convection, and none that is not equivocal.

Upwelling under normal midocean rises is passive. If it were not, the young sea floor would not subside in proportion to the square-root of its age. Midocean rise segments where plumes exist, such as Iceland, are obvious exceptions to this.

Evidence from seismic tomography and the gravity field increasingly supports the possibility that there is a large mass flow through the mantle transition zone, and that mantle convection occurs as a single layer rather than two.

The topography of the sea floor constrains the possibility that the mantle convects as two separate layers. If this were true, there ought to be topography due to upwellings from 660 km depth that is comparable to the midocean rise topography in scale and amplitude. Such topography is not evident.

It is likely nevertheless that the mantle is stratified in viscosity and, as a consequence, in trace element and isotopic composition (Chapter 13). It is plausible that there is some density stratification near the bottom of the mantle.

There is no requirement for a 'decoupling' low-viscosity layer under the lithosphere. There is very likely to be a viscosity minimum there, but this probably does not greatly perturb the plate-scale flow. The mantle under this minimum commonly will still be driven in the same direction as the overlying plates by the sinking lithosphere.

The negative buoyancy of subducted lithosphere will persist long after it becomes aseismic, because thermal diffusion merely smears out the thermal anomaly, it does not remove it. This is a consequence of conservation of energy.

It is not very useful to try to isolate the plates from the rest of the mantle in order to determine the details of 'plate driving forces' because the plates are an integral part of a convection system: stresses are transmitted through the viscous mantle as well as through the elastic plates.

The irregular and changing shapes of the plates, particularly of ridge-transform systems, is compatible with them being part of a convection system because upwelling under spreading centres is normally passive. An isolated or migrating ridge segment will pull up whatever mantle underlies it at any given time.

Plumes probably do not influence plate motions very much, though they may trigger some changes in favourable circumstances, such as ridge jumps or even major rifting.

12.6 References

1. D. Forsyth and S. Uyeda, On the relative importance of the driving forces of plate motion, *Geophys. J. R. Astron. Soc.* **43**, 163–200, 1975.
2. W. M. Chapple and T. E. Tullis, Evaluation of the forces that drive the plates, *J. Geophys. Res.* **82**, 1867–84, 1977.

3. E. E. Davis and C. R. B. Lister, Fundamentals of ridge crest topography, *Earth Planet. Sci. Lett.* **21**, 405–13, 1974.
4. W. M. Elsasser, Convection and stress propagation in the upper mantle, in: *The Application of Modern Physics to the Earth and Planetary Interiors*, S. K. Runcorn, ed., Wiley-Interscience, New York, 223–46, 1969.
5. G. F. Davies, Whole mantle convection and plate tectonics, *Geophys. J. R. Astron. Soc.* **49**, 459–86, 1977.
6. W. J. Morgan, Plate motions and deep mantle convection, *Mem. Geol. Soc. Am.* **132**, 7–22, 1972.
7. R. I. Hill, Starting plumes and continental breakup, *Earth Planet. Sci. Lett.* **104**, 398–416, 1991.
8. G. Houseman and P. England, A dynamical model of lithosphere extension and sedimentary basin formation, *J. Geophys. Res.* **91**, 719–29, 1986.
9. G. F. Davies and F. Pribac, Mesozoic seafloor subsidence and the Darwin Rise, past and present, in: *The Mesozoic Pacific*, M. Pringle, W. Sager, W. Sliter and S. Stein, eds., American Geophysical Union, Washington, D.C., 39–52, 1993.
10. L. Panasyuk, Residual topography of the earth, pers. comm., 1998.
11. W. D. Mooney, G. Laske and T. G. Masters, CRUST 5.1: a global crustal model at $5^\circ \times 5^\circ$, *J. Geophys. Res.* **103**, 727–47, 1998.
12. M. K. McNutt and K. M. Fischer, The south Pacific superswell, in: *Seamounts, Islands, and Atolls*, B. H. Keating, P. Fryer, R. Batiza and G. Boehlert, eds., AGU, Washington, D.C., 25–34, 1987.
13. F. Pribac and G. F. Davies, Mantle superswells: regressions and rifts?, *Eos, Trans. Amer. Geophys. Union* **68**, 1451, 1987.
14. M. K. McNutt, Superswells, *Rev. Geophys.* **36**, 311–44, 1998.
15. M. A. Richards and D. C. Engebretson, Large-scale mantle convection and the history of subduction, *Nature* **355**, 437–40, 1992.
16. Y. Ricard, M. Richards, C. Lithgow-Bertelloni and Y. Le Stunff, A geodynamic model of mantle density heterogeneity, *J. Geophys. Res.* **98**, 21,895–909, 1993.
17. R. A. Duncan and M. A. Richards, Hotspots, mantle plumes, flood basalts, and true polar wander, *Rev. Geophys.* **29**, 31–50, 1991.
18. M. F. Coffin and O. Eldholm, Large igneous provinces: crustal structure, dimensions and external consequences, *Rev. Geophys.* **32**, 1–36, 1994.
19. H. W. Menard, *Marine Geology of the Pacific*, McGraw-Hill, New York, 1964.
20. H. W. Menard, Darwin reprise, *J. Geophys. Res.* **89**, 9960–68, 1984.
21. R. L. Larson, Latest pulse of the earth: evidence for a mid-Cretaceous superplume, *Geology* **19**, 547–50, 1991.
22. W. J. Morgan, Hotspot tracks and the opening of the Atlantic and Indian Oceans, in: *The Sea*, C. Emiliani, ed., Wiley, New York, 443–87, 1981.

23. M. A. Richards and B. H. Hager, The earth's geoid and the large-scale structure of mantle convection, in: *Physics of the Planets*, S. K. Runcorn, ed., Wiley, New York, 247–72, 1988.
24. G. F. Davies, Penetration of plates and plumes through the mantle transition zone, *Earth Planet. Sci. Lett.* **133**, 507–16, 1995.
25. G. F. Davies, Ocean bathymetry and mantle convection, 1. Large-scale flow and hotspots, *J. Geophys. Res.* **93**, 10 467–80, 1988.
26. G. F. Davies, Effect of a low viscosity layer on long-wavelength topography, upper mantle case, *Geophys. Res. Lett.* **16**, 625–8, 1989.
27. H. W. Menard, Elevation and subsidence of oceanic crust, *Earth Planet. Sci. Lett.* **6**, 275–84, 1969.
28. J. G. Sclater and J. Francheteau, The implications of terrestrial heat flow observations on current tectonic and geochemical models of the crust and upper mantle of the Earth, *Geophys. J. R. Astron. Soc.* **20**, 509–42, 1970.
29. J. G. Sclater, R. N. Anderson and M. L. Bell, Elevation of ridges and evolution of the central eastern Pacific, *J. Geophys. Res.* **76**, 7888–915, 1971.
30. B. Parsons and J. G. Sclater, An analysis of the variation of ocean floor bathymetry and heat flow with age, *J. Geophys. Res.* **82**, 803–27, 1977.
31. M. G. Langseth, X. LePichon and M. Ewing, Crustal structure of midocean ridges, 5, Heat flow through the Atlantic Ocean floor and convection currents, *J. Geophys. Res.* **71**, 5321–55, 1966.
32. D. P. McKenzie, Some remarks on heat flow and gravity anomalies, *J. Geophys. Res.* **72**, 6261–73, 1967.
33. F. M. Richter, Convection and the large-scale circulation of the mantle, *J. Geophys. Res.* **78**, 8735–45, 1973.
34. D. P. McKenzie, J. M. Roberts and N. O. Weiss, Convection in the earth's mantle: towards a numerical solution, *J. Fluid Mech.* **62**, 465–538, 1974.
35. B. Parsons and D. P. McKenzie, Mantle convection and the thermal structure of the plates, *J. Geophys. Res.* **83**, 4485–96, 1978.
36. D. A. Yuen, W. R. Peltier and G. Schubert, On the existence of a second scale of convection in the upper mantle, *Geophys. J. R. Astron. Soc.* **65**, 171–90, 1981.
37. G. F. Davies, Ocean bathymetry and mantle convection, 2. Small-scale flow, *J. Geophys. Res.* **93**, 10 481–8, 1988.
38. W. F. Haxby and J. K. Weissel, Evidence for small-scale mantle convection from seasat altimetre data, *J. Geophys. Res.* **91**, 3507–20, 1986.
39. D. T. Sandwell, E. L. Winterer, J. Mammerickx, R. A. Duncan, M. A. Lynch, D. A. Levitt and C. L. Johnson, Evidence for diffuse extension of the Pacific plate from Pukapuka ridges and cross-grain gravity lineations, *J. Geophys. Res.* **100**, 15 087–99, 1995.
40. D. P. McKenzie, A. B. Watts, B. Parsons and M. Roufosse, Planform of mantle convection beneath the Pacific Ocean, *Nature* **288**, 442–6, 1980.

41. A. B. Watts, D. P. McKenzie, B. E. Parsons and M. Roufosse, The relationship between gravity and bathymetry in the Pacific Ocean, *Geophys. J. R. Astron. Soc.* **83**, 263–98, 1985.
42. D. T. Sandwell and M. L. Renkin, Compensation of swells and plateaus in the north Pacific: no direct evidence for mantle convection, *J. Geophys. Res.* **93**, 2775–83, 1988.
43. R. J. O'Connell and B. H. Hager, On the thermal state of the earth, in: *Physics of the Earth's Interior*, A. Dziewonski and E. Boschi, eds., 270–317, 1980.

PART 4

IMPLICATIONS

The potential implications of the picture of mantle convection developed in Part 3 are many and far-reaching, given that mantle convection is the fundamental tectonic driving mechanism. Some of these implications are already being explored, and presumably many other aspects will be explored in due course. Given my desire that the material of this book does not date too rapidly, there is a risk in including any such material. However, the exploration of two aspects has been under way for some time, and they provide particularly important complements to the focus, so far in this book, on the dynamical processes operating at present in the mantle. Therefore I present summaries of both the chemistry of the mantle and of the thermal evolution of the mantle and its implications for tectonic mechanisms at the earth's surface in past eras. Some aspects of these topics, particularly past tectonic mechanisms, are in a tentative stage of exploration, so you should be alert to the likelihood that the subject may move on rapidly. Nevertheless I hope it is useful to indicate some directions in this work that are apparent in 1998.

I discuss the chemistry of the mantle for two main reasons. First, through radiogenic isotopic compositions mantle chemistry gives us time information, and so constrains the evolution of the system. It is thus an important complement to the discussion of thermal and tectonic evolution. The second reason is that there have been many assertions over the past two decades that geochemical observations established one or another fact about the form of mantle convection. Many of those assertions have been overdrawn. It is certainly true that geochemistry provides very important constraints on the nature of mantle convection, but some attention to both chemical and physical processes is required for their fruitful interpretation.

The reasons for addressing the thermal and tectonic evolution are that the nature of the tectonic regime in the early part of earth's history is still quite uncertain, it is of first-order importance, and some exciting possibilities are opening up as a result of our growing understanding of the present regime. Although the exploration of these possibilities is necessarily tentative, there are some important basic constraints that limit the more unbridled kinds of speculation, and there are important concepts and possibilities that deserve to be introduced into the geological conversation.

CHAPTER 13

Chemistry

The physical process of mantle convection affects the chemistry of the mantle. The chemical changes occur mostly through melting, directly or indirectly. The resulting chemical differences are then acted upon by physical processes, such as subduction and convective stirring. As a result, mantle chemistry potentially contains a lot of information about the physical processes, and any model of mantle convection must ultimately be consistent with what is known about mantle chemistry. Also, mantle chemistry may react upon mantle convection, most directly through density and buoyancy, as discussed in Chapters 10 and 14.

A great deal of information about the mantle has been obtained from measurements of the chemical and isotopic composition of rocks derived from the mantle, and this is currently a very active field of geochemistry. The mantle, like the crust, contains minor or trace concentrations of virtually every element. Comparisons of concentrations, abetted by knowledge of crystal chemistry, have allowed geochemists to deduce some important conclusions about the mantle, such as that much of the mantle seems to be a residue, after the extraction of the atmospheric gases, the ocean and the continental crust, from a material with an initial composition like that of primitive meteorites. Further, measurements of the proportions of radioactive isotopes and their daughter products yields information on time scales of processes, and sometimes of dates of particular events. Isotopic compositions have also been used to identify distinct sources in the mantle, and such measurements have made it clear that there is a level of heterogeneity in mantle chemistry, and that much of this heterogeneity is quite ancient, of the order of two billion years old.

Observations like these are important constraints on the kind of dynamical picture developed so far in this book. Clearly the dynamical models must be capable of accommodating the source types and time scales identified from geochemistry, and ultimately they should be capable of explaining in general terms the concentrations of all the elements and isotopes. To do this, it is necessary to consider both the physical and the chemical processes that have given rise to the particular rock being measured. The highly specialised and exacting nature of much of the work makes this a daunting task. Perhaps it is not surprising, therefore, that the reconciliation of physical and chemical observations has been quite controversial. In my experience, an important factor in these controversies has been a tendency to adopt prematurely a particular interpretation of observations from an unfamiliar field, when a fuller understanding of the observations would reveal other possible interpretations.

Some general features of mantle chemistry have been established reasonably well by observations. To guide the more detailed discussion, I first outline the resulting broad picture, before presenting the key observations and arguments upon which this picture rests. I then introduce a few essential concepts and summarise some of the most important geochemical observations, followed by a discussion of the physical processes that must be considered, in conjunction with the chemical processes, in order to interpret the observations, emphasising the care with which questions must be posed. Finally I offer an assessment of the present situation. This is a very large subject, so I must necessarily be selective and concise. Background on the chemical and geochemical principles involved can be found, for example, in [1, 2]. The other references given will guide you to more complete discussions of specific topics.

13.1 Overview – a current picture of the mantle

The mantle has been depleted of those elements that are found concentrated in the crust, hydrosphere and atmosphere, relative to the original composition of the mantle, which is inferred from the compositions of meteorites. There has been some 're-enrichment' of the mantle, that is there has been re-injection of material from the shallow reservoirs. The degree and kind of depletion or re-enrichment are not uniform throughout the mantle. The geometry of these heterogeneities is not well constrained, except for one very consistent aspect: the shallowest mantle, sampled by midocean ridge magmas, tends to be the most depleted. Hotspot magmas, inferred to arise from melting of mantle plumes (Chapter 11) and thus to reflect the deeper mantle, show less depletion, and also more variability in detail. Although the mantle is being continually processed, these heterogeneities of mantle chemistry, both vertical and more general, seem to be quite old, with ages of the order of 1–2 Ga inferred from radioactive isotope systems.

All of the mantle that has been sampled has been processed or modified in one way or another from its inferred original composition, although there are some components or regions that may have been modified less than most. The heterogeneities identified from a number of radiogenic isotopic systems require at least five mantle source types to span the range of variations in the several systems. This implies that at least this number of distinct processes has operated to generate the chemical differences. The identification of particular source types with the associated processes that produced them is in a tentative stage at present, but several identifications are plausible.

There are distinctive chemical characteristics associated with continental lithospheric mantle. The continental lithosphere comprises the continental crust (to depths of 35–40 km) and a zone of mantle beneath the crust extending to various depths ranging from about 100 km to 250 km or more (Chapter 5). The mantle part of the continental lithosphere (the 'root' or 'keel' of the continent) tends to be strongly depleted, and also to be chemically quite heterogeneous. The degree of heterogeneity is uncertain because sampling is very limited and may be strongly biased, and because some heterogeneity previously attributed to the continental lithosphere may actually arise from the deeper mantle, and from mantle plumes in particular. Some of this heterogeneity has the characteristics of a relatively recent re-enrichment due to the penetration of a fluid phase rich in 'incompatible elements' (see below).

Island arc magmas have a chemical character distinct in key ways from midocean ridge or hotspot magmas. These differences are attributed partly to the influence of a hydrous fluid phase given off by hydrated minerals in the subducting oceanic crust (see below). Island arc magmas also show some important similarities to the inferred average composition of the continental crust, which has led to the hypothesis that island arcs are the sites where continental crust has been generated. While it is very likely that some of the continental crust has formed in this way, there may have been other important processes, particularly in the past (Chapter 14). There is also some remaining uncertainty in the average composition of the continental crust, since it is extremely heterogeneous, and the lower crust is poorly sampled.

13.2 Some important concepts and terms

For those not familiar with geochemistry, there are some commonly-encountered concepts and terms that may need explanation. Some of these are just useful conventions or common jargon, while others are central chemical concepts that guide the understanding of the earth's chemistry.

13.2.1 Major elements and trace elements

The so-called 'major-element composition' of the mantle has been covered in Chapter 5: the mantle is composed of magnesium-iron silicates, with lesser amounts of aluminium, calcium, and so on. These elements dictate the structure of the main minerals, and less abundant elements then have to fit in as best they can, either in minor ('accessory') minerals that occur in small amounts between the main mineral grains or in solid solution in the major mineral phases.

Elements whose concentration is much less than a per cent are generally referred to as trace elements, and the lower limit of concentrations considered has been simply the limits of detection of analytical instruments. Currently this limit is at the level of parts per trillion, and at this level most of the elements are detectable in mantle rocks or mantle-derived rocks. (Geochemists commonly use informal units like 'parts per million', by which they mean the ratio of the weight of the element to the weight of the host rock, times one million. However, the expression is ambiguous to outsiders, since it could also refer to a ratio by volume or to a molar ratio. The ambiguity can be avoided by using units like μg/g, which I do here.)

13.2.2 Incompatibility and related concepts

Loosely speaking, the 'incompatibility' of a trace element in the mantle is its tendency to move preferentially into a liquid phase, if one is present. 'Compatibility' is obviously the opposite, and refers to how well an atom fits into the crystal structure in which it finds itself. For example, an element like nickel, which forms the ion Ni^{2+}, can readily substitute for Mg^{2+} in the mineral olivine, which is a major constituent of the upper mantle and whose chemical formula is $(Mg,Fe)_2SiO_4$. (In fact olivine itself typically has iron (Fe^{2+}) substituting for about 10% of the magnesium atoms, which is why the formula is written with (Mg,Fe).) On the other hand, uranium is a much larger atom and it forms the ions U^{4+}

and U^{6+}, depending on the oxidation state. A uranium ion will fit in the olivine structure only with difficulty, and one or more substitutions will be required in adjacent sites of the crystal lattice if charge balance is to be maintained. If the mantle partially melts, the uranium will have a strong tendency to move into the liquid phase, since the structure of a liquid is irregular and can more readily accommodate exceptional ions. Nickel, on the other hand, will tend to remain in the olivine structure. As a result, uranium is called 'incompatible', while nickel is called 'compatible'. Incompatibility is a relative term, it occurs to varying degrees, and in detail it depends on the compositions of both the solid mineral phases and the melt. Although it is thus a rather loose term, the behaviour of most elements is consistent enough in many circumstances to make it a very useful concept.

The importance of incompatibility is that if a region of the mantle melts and the melt separates because of its buoyancy, the melt will preferentially remove the incompatible elements, leaving the mantle region relatively depleted in these elements. Melts accumulating near the earth's surface to form the oceanic crust or the continental crust thus tend to concentrate the incompatible elements into the crust.

The partitioning behaviour of many elements, between solid and fluid phases, depends on the kind of fluid involved. The classes of fluids important for the mantle are silicate melts, water-rich or 'hydrous' fluids, possibly also involving methane or carbon dioxide, and metallic liquids, which were presumably important during the segregation of the metallic core from the silicate mantle. The term 'incompatibility' has come to be used mainly for partitioning into silicate melts, although the distinction in behaviour between this case and a hydrous fluid does not always seem to be clearly made. Hydrous fluids are inferred to be important in subduction zones, where hydrated minerals from the oceanic crust break down under high pressure, releasing water.

A couple of other terms are commonly encountered in this subject. Elements that tend to partition into a liquid iron phase, such as the transition metals, are called 'siderophile', as in 'siderites', which are iron meteorites. (A curiosity here is that the word may derive from the Latin *sideris*, a star, evidently referring to meteorites' origin from the sky. Thus elements that would prefer to be in the core are called 'star loving'.) Elements that tend to partition into sulphide phases are called 'chalcophile' (Greek: *chalkes*, copper), as in the minerals 'chalcocite' (Cu_2S) and 'chalcopyrite' ($CuFeS_2$). Sulphides are important not only in many metallic ore bodies, but also possibly during core formation, because sulphide phases tend to dissolve into liquid iron.

13.2.3 Isotopic tracers and isotopic dating

Physical processes can affect not only the concentrations of elements but also the relative concentrations of isotopes of a given element. For example the precipitation of calcium carbonate affects the relative proportions of ^{16}O and ^{18}O, and organic and inorganic precipitation affect them by different amounts. The most widely exploited process and the one of most concern here is radioactive decay, which changes the isotopic composition of both the parent and daughter elements. Table 13.1 lists the most useful systems and some of their key properties.

The isotopic differences resulting from such processes can be used both for tracing the source of material and for radiometric dating. The main relevance of these systems here is as tracers. In other words, various source types have been recognised as having characteristic isotopic ratios that serve as fingerprints for tracing the origins of individual rock samples. Dating will arise here mainly in relation to lead isotopic data. (The U–Pb system is special because age can be inferred from the daughter isotopes alone.) The systems listed in Table 13.1 have half lives measured in billions of years, and this is pertinent for two reasons: they record changes over the long time scales of mantle processes, and if the half lives were much shorter there would be little of the parent left to generate isotopic variation in the daughter.

The fundamental relationship used in determining abundances and ages from radioactive decay expresses the exponentially declining abundance of the parent isotope (P) and the corresponding accumulation of the daughter isotope (D):

$$D = D_0 + P_0(1 - \mathrm{e}^{t/\tau}) \qquad (13.2.1)$$

where D_0 and P_0 are the initial abundances of the daughter and parent, t is time, $\tau = T_{1/2}/\ln 2$, and $T_{1/2}$ is the half life of the parent. Geochemists commonly use the 'decay constant' $\lambda = 1/\tau$, and write this equation in terms of *age* (time before the present) rather than *time* (since an initial state; and geochemists very frequently *say* time when they *mean* age.) I will not attempt to give the mathematical expressions of all of the relationships used in the following discussion, but I will have occasion to refer to some. The full expressions can be found in the references.

Table 13.1. *Radioactive decay systems used for the mantle.*

Parent nuclide	Daughter nuclide	Half life (Ga)	Tracer ratio
^{147}Sm	^{143}Nd	106	^{143}Nd/^{144}Nd
^{87}Rb	^{87}Sr	48.8	^{87}Sr/^{86}Sr
^{176}Lu	^{176}Hf	35.7	^{176}Hf/^{177}Hf
^{187}Re	^{187}Os	45.6	^{187}Os/^{188}Os
^{40}K	^{40}Ar	1.25	^{40}Ar/^{36}Ar
^{232}Th	^{208}Pb	14.01	^{208}Pb/^{204}Pb
^{238}U	^{206}Pb	4.468	^{206}Pb/^{204}Pb
^{235}U	^{207}Pb	0.738	^{207}Pb/^{204}Pb
U, Th[a]	^{4}He	—	^{3}He/^{4}He
U, Th, O, Mg[b]	^{21}Ne	—	^{21}Ne/^{22}Ne

[a] Emission of alpha particles in the three preceding reactions.
[b] Reactions induced by alpha particles and neutrons emitted by U and Th.

13.2.4 MORB and other acronyms

It has become common to use acronyms and related abbreviations to refer to various rock types and mantle source types. The ones used here are listed in Table 13.2. The first three are the principal volcanic rocks produced at the three main settings in which mantle melting occurs. The others are types of mantle composition inferred to be the sources of MORBs or OIBs of various types, and they are explained further as the observations are described below.

13.3 Observations

I present the observations initially with only limited interpretation. As I indicated in the introduction to Part 4, it is important to separate clearly the observations and their immediate implications from interpretations that depend on additional assumptions about the system. Section 13.4 will present inferences that can be made fairly directly from these data, while broader potential implications will be discussed in the last section of this chapter. Several of the illustrations used in this section are taken from the review by Hofmann [7], who gives an excellent summary of the chemistry of refractory elements and their isotopes in the mantle.

Table 13.2. *Acronyms and abbreviations.*

Rocks	
MORB	Midocean ridge basalt
OIB	Oceanic island basalt (or volcanic hotspot or plume basalt)
IAB	Island arc basalt
Mantle source types	
DMM	Depleted MORB mantle [3] or MORB source
EM-1	Enriched mantle, type 1 [3]
EM-2	Enriched mantle, type 2 [3]
HIMU	High-μ mantle, where $\mu = {}^{238}U/{}^{204}Pb$ [3]
FOZO	'Focal zone', an intermediate mantle composition [4]
PHEM	Primitive helium mantle [5]
C	'Common', an intermediate mantle composition [6]

13.3.1 Trace elements

Figure 13.1 shows element concentrations in several kinds of basaltic rocks produced by melting of the mantle, and average concentrations estimated for the continental crust. The concentrations are normalised to estimates of the concentrations in the primitive mantle, and the elements are arranged in order of increasing compatibility. This plot illustrates a number of important points.

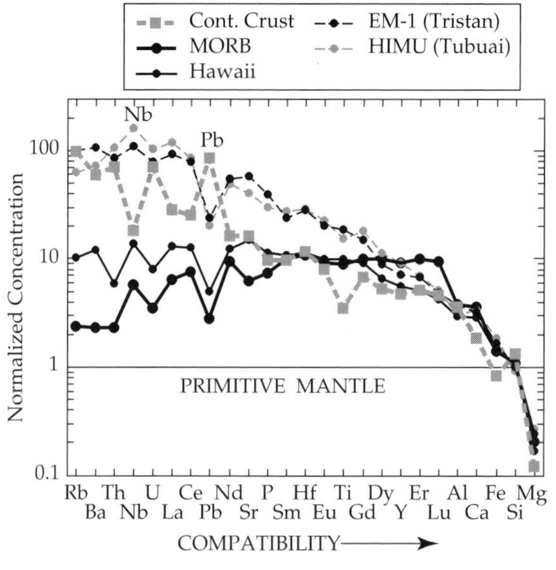

Figure 13.1. Trace element concentrations in mantle-derived rocks. Concentrations are normalised to the estimated concentrations in the primitive mantle. From Hofmann [7]. Reprinted from *Nature* with permission. Copyright Macmillan Magazines Ltd.

The fact that most elements plot above the primitive line (ratio 1) reflects the fact that incompatible elements are concentrated into the melts that form the crust. For the oceanic basalts, the melt is believed to be not substantially modified in its ascent to the surface, in which case the concentrations in the mantle source can be inferred by correcting for the degree of partial melting. For example, it is believed that about 10% of the mantle material rising under midocean ridges melts. If most of the incompatible elements are partitioned into this melt, their concentration will be about a factor of 10 higher in the melt than in the solid source. Thus the concentration of the more incompatible elements in the source of midocean ridge basalts (MORBs) is inferred to be about a factor of 10 lower than in the MORBs themselves. You can see from Figure 13.1 that with this correction, the MORB source has lower-than-primitive concentrations of the more incompatible elements. In other words, *the MORB source is depleted of incompatible elements*.

The analogous inference cannot be made so securely for the other cases plotted in Figure 13.1. The other oceanic basalts in the plot are for oceanic island basalts (OIBs), inferred to be due to melting in mantle plumes. For these, the degree of melting is believed to be usually less than under midocean ridges, perhaps 5%. Thus there is a larger, but more uncertain, correction factor required to derive the source concentrations. For the most enriched cases (EM-1 and HIMU), the source is probably enriched relative to primitive concentrations. The Hawaiian source may be slightly depleted, but the Hawaiian plume is the strongest and probably has the highest degree of partial melting of the hotspots, so its source concentrations are probably close to primitive levels (which does not mean it is pristine, as you will see below), though the isotopes considered below indicate substantial depletion in the long term. However, regardless of the concentrations relative to primitive levels, it is clear that *the plume sources are less depleted on average than the MORB source*.

The continental crust is also strongly enriched in incompatible elements. However, the continental crust is complex, very heterogeneous, and probably produced by diverse and multistage processes, so we cannot make a simple inference about its source. Nevertheless it is striking that to a first approximation the elements that are enriched in the continental crust are those that are depleted in the MORB source, and the degrees of complementary enrichment and depletion correlate with the degree of incompatibility. In fact it is remarkable that, with the exception of a few notable elements, the continental crust can be approximated quite well as the result of a single-stage partial melting of the primitive mantle by about 1%, a fact first strongly emphasised by Hofmann [8]. This does not mean that the continental crust was actually produced in this simple way, but it suggests that a complex series of processes may still yield fairly simple trace element patterns, presumably because the underlying concentrating mechanism is fairly simple. In summary, *the enrichments of trace elements in the continental crust are to a first approximation complementary to their depletions in the MORB source.*

The most notable deviations of the continental crust from a smooth pattern are for Nb, Pb and Ti. These deviations are believed to be produced by processes specific to subduction zones. Island arc volcanism is believed to result from the release under pressure of water bound in hydrated minerals in the subducted oceanic crust. The oceanic crust itself does not seem to melt, but the water migrates upwards into the hot surrounding mantle, lowering its melting temperature and producing the melt that rises under island arcs. Nb is more compatible and Pb is less compatible in the presence of the hydrous fluid released by the oceanic crust than in the presence of silicate liquid. Ti is believed to be more compatible because of the occurrence of the mineral ilmenite ($FeTiO_3$) in this environment. It is inferred from this and other evidence [9] that *island arc volcanism has contributed significantly to the growth of the continental crust.*

Complementary deviations in the concentrations of Nb, Pb and Ti are evident in the patterns of the mantle-derived basalts. This is further evidence for a complementary relationship between the continental crust and the mantle. More specifically, since this complementary pattern is thought to arise in subduction zones, it implies that the signature of the residual subducted oceanic crust, after hydrous phases have been removed, is appearing in both MORBs and OIBs. In other words, as first argued by Hofmann and White [10], *MORB, or some part of it, is being recycled from the surface through the mantle and back to the surface at spreading centres and volcanic hotspots.*

13.3.2 Refractory element isotopes

The best-established data sets are for neodymium (Nd), strontium (Sr) and lead (Pb) radiogenic isotopes (Table 13.1). Representative results are plotted in Figures 13.2–5. Melting has no effect on isotopic ratios so, unlike element concentrations, the isotopic ratios of basaltic rocks can be taken to reflect directly the ratios in the mantle sources. Again there are several important points illustrated.

There is clearly a detectable range of variation of isotopic composition in the mantle. Whether this variation is regarded as a lot or a little obviously depends on having some idea of what processes might have been involved in generating or removing such variations. The primary process creating variation is melting, which can partition radioactive parent and daughter elements differently and so change their relative proportions. Such variations in parent/daughter ratios generate isotopic variations over time. The primary processes removing isotopic variation are convective mixing and diffusion. These homogenising processes will be discussed later.

One measure of the significance of mantle isotopic variations is to compare them with variations in the crust, where the homogenising processes are likely to be less effective. The range of isotopic variation in oceanic rocks (MORB and OIB) is smaller than the total variation in continental crust, but still significant in comparison, as is illustrated for lead isotopes in Figure 13.2. The continental crust contains some highly radiogenic lead isotope ratios in some sediments with high U/Pb ratios, but the bulk of the upper crust is not so extreme (UCC, Figure 13.4). The lower continental crust contains some extremely unradiogenic lead (Figure 13.2), and the average is estimated to be moderately unradiogenic (LCC, Figure 13.4). OIBs also contain some relatively radiogenic lead, but again this is not representative, with the bulk of oceanic basalts being less radiogenic (Figures 13.2,4). The lead isotope plot in Figure 13.2 contains some other information that will be discussed later.

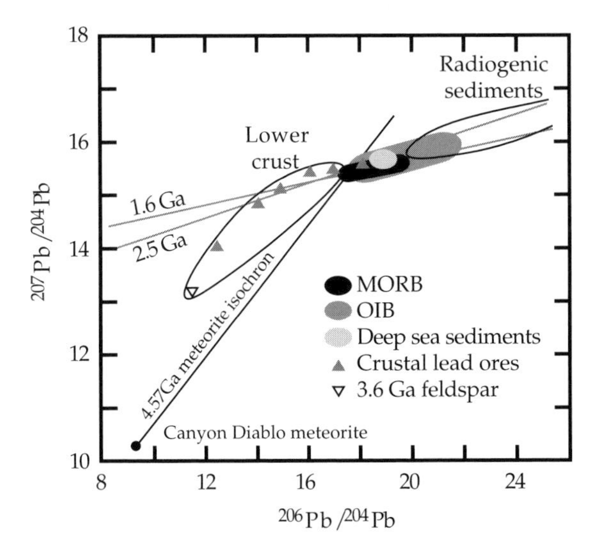

Figure 13.2. Sketch of the ranges of $^{207}Pb/^{204}Pb$ versus $^{206}Pb/^{204}Pb$ in the crust and mantle.

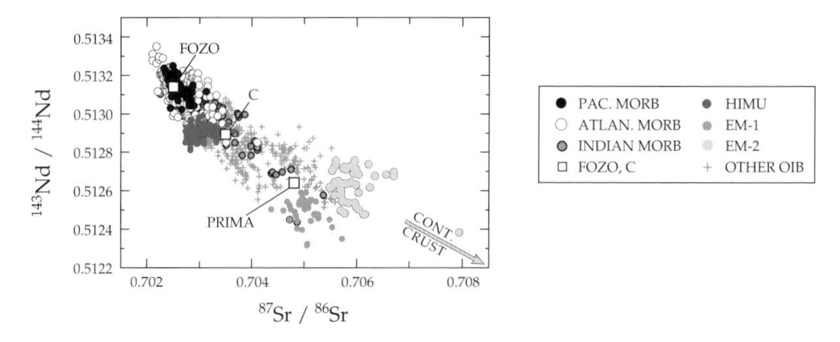

Figure 13.3. Nd and Sr isotopic ratios for oceanic mantle-derived rocks. Rock and mantle source types are identified in Table 13.2. Three groups of MORBs are shown, from the Pacific, Atlantic and Indian Ocean regions. From Hofmann [7]. Reprinted from *Nature* with permission. Copyright Macmillan Magazines Ltd.

Among the oceanic rocks, the range of isotopic variations of OIBs is greater than the range in MORBs, by a factor of about 2 (Figures 13.3,4). Otherwise *MORBs and OIBs show similar kinds of variations*. This is worth emphasising because in the history of this subject MORBs have many times been characterised as being 'remarkably uniform' and OIBs as 'diverse'. In fact the contrast is not very great, and most of the trends evident in OIBs can also be seen in MORBs, as is evident particularly in Figure 13.5a. A better characterisation is that MORBs have a slightly muted version of the diversity of OIBs.

These isotopic variations are generally consistent with the relative abundances and compatibilities illustrated in Figure 13.1. For example, the Rb/Sr ratio in the OIBs (EM-1 and HIMU) is higher than in average MORB, and the same relationship is implied in the $^{87}Sr/^{86}Sr$ ratios of Figure 13.3: OIBs have higher proportions of

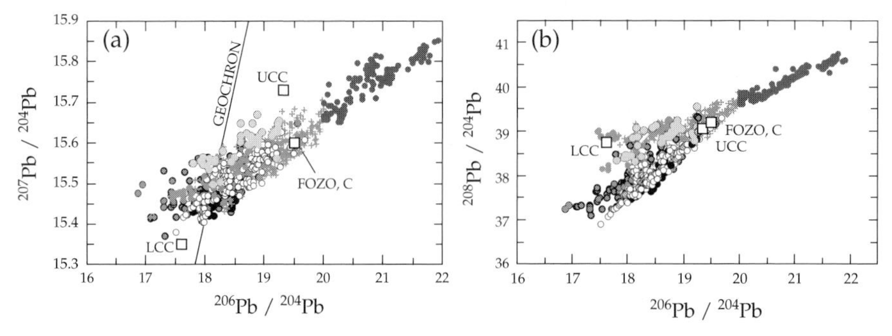

Figure 13.4. Lead isotopes in oceanic mantle-derived rocks. Symbols have the same meaning as in Figure 13.3. 'Geochron' is the meteorite isochron of Figure 13.2, which is inferred to be the locus of primitive lead in this diagram. From Hofmann [7]. Reprinted from *Nature* with permission. Copyright Macmillan Magazines Ltd.

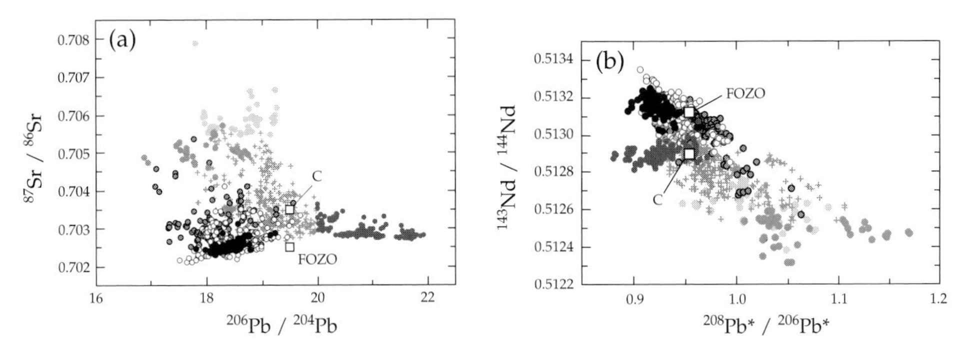

Figure 13.5. Isotopes for oceanic mantle-derived rocks. (a) Sr versus ^{206}Pb. (b) Nd versus the radiogenic components of ^{208}Pb and ^{206}Pb, denoted by a star (*). Symbols have the same meaning as in Figure 13.3. From Hofmann [7]. Reprinted from *Nature* with permission. Copyright Macmillan Magazines Ltd.

^{87}Sr, the radiogenic daughter of ^{87}Rb, than does MORB. Thus we can infer that the Rb/Sr ratio has been consistently higher in the OIB sources than in the MORB source over the time for which the isotopic differences have been accumulating (1–2 Ga, see below). Thus both the past and present relationships are consistent with OIB sources being more enriched (or less depleted) in incompatible elements than the MORB source. Thus, as with the trace element abundances, *the isotopic variations reflect the relative incompatibilities of parent and daughter elements.*

The volumetrically dominant component in Figures 13.3–5 is MORB, which is consistently at the most depleted end of the observed range. The eruption rate of MORB is about 20 km^3/a, and the combined eruption rates of OIBs are only about 1% of that. However, MORBs and OIBs are the result of different mantle processes (plate spreading and plume ascent), so we cannot translate eruption rates into the volumes of the mantle source types. Since MORBs are rather clearly derived from the shallowest mantle, as will be discussed in more detail later, the OIBs are by implication sampling deeper parts of the mantle. We must therefore look carefully at the OIB data if we are to learn about the deeper mantle.

It has been proposed that some of the OIB variation seems to radiate from intermediate isotopic compositions. This is most evident in Figures 13.4b and 13.5a. Such intermediate compositions have been called the focus zone (FOZO, [4]) or common (C, [6]) compositions. The original FOZO composition estimate has been revised to something quite close to ‘C’ [11]. The proposed interpretation is that most OIBs are mixtures of this intermediate composition and various other compositions.

Whereas Nd and Sr isotopes correlate quite strongly with each other (Figure 13.3) and with hafnium isotopes, the Pb isotopes do not correlate very well with any of these (Figure 13.5a). This could, in principle, be due to a lack of correlation of Th and U with Nd and Sr, or a lack of correlation of Pb with Nd and Sr. Figure 13.5b discriminates between these possibilities by plotting only the radiogenic components of the Pb isotopes (due to the decay of Th and U) against Nd isotopes, thus removing the influence of variations in pre-existing Pb. There is a significant correlation, indicating that Th and U do correlate with Nd, and in fact the topology of Figure 13.5b is quite similar to that of Figure 13.3. This implies in turn that it is Pb that does not correlate with Nd. In other words, it is the Pb that seems to be anomalous, and we have already seen evidence for this in Figure 13.1.

13.3.3 Noble gas isotopes

Isotopes of helium, neon, argon and xenon provide important information about the mantle. Krypton shows little variation in isotopic composition and is not discussed here. The noble gases are particularly useful because they are unreactive (hence the name 'noble') and volatile, properties which make them complementary tracers to the refractory element isotopes. (The term 'rare gases' is also often used, but it is hardly appropriate for argon, which comprises nearly 1% of the atmosphere.) Their lack of reactivity means that there will be little or no recycling of them from the atmosphere back into the mantle, whereas other components that reach the crust or atmosphere are believed to be recycled to varying degrees, as is discussed in Sections 13.4 and 13.5. It also means that they do not dissolve chemically into mantle minerals in the way that other elements do. Rather, their microscopic locations in the mantle may be determined by physical factors such as atomic size and the presence of minute fluid inclusions. Thus the chemical concept of incompatibility is not strictly applicable to them.

Although there has been a long-standing presumption that noble gases would partition strongly from solid phases into melt phases, and from melt phases into gas phases, there has been only limited direct evidence on or understanding of these processes. It is possible that standard presumptions will be found not to apply. For example, it has been suggested recently that the noble gases may exist in the deep mantle as minute inclusions of the solid phase, since their melting temperatures deeper than a few hundred kilometres are higher than the temperature of the mantle (or the core) [12]. It has also been suggested that argon solubility in silicate liquids may be quite low under the high pressures of the mantle [13]. Such possibilities could lead to some important revisions of our interpretations of their observed characteristics.

There are a total of 23 isotopes of the noble gases found in nature, and there is considerable detail in their observed variations in isotopic compositions from different sources, detail that bears on the questions of the origins of the atmosphere, the earth and meteorites as well as the structure of the mantle. Only some of the most pertinent observations can be included here. More comprehensive accounts are given by Ozima and Podosek [14], Ozima [15] and McDougall and Honda [16].

13.3.3.1 Helium

Helium occurs in two isotopes, ^{3}He that is primordial and ^{4}He that is produced principally by alpha decay of U and Th (Table 13.1). Because of its small atomic mass, helium is lost from the atmosphere quite rapidly on geological timescales, and its atmospheric abundance is low. This means on the one hand that contamination of rock samples with atmospheric helium is not a problem, but on the other hand that we don't know how much helium was incorporated into the earth at the beginning. Helium isotopic composition is usually represented by the ratio $R = {}^3\mathrm{He}/{}^4\mathrm{He}$, contrary to the more usual convention of putting the radiogenic isotope on top, and often it is referred to the atmospheric ratio ($R_A - 1.4 \times 10^{-6}$). Because of higher radioactivity in continental crust, ^{3}He/^{4}He is usually relatively low in continental crustal rocks, with $R < R_A$.

Although the mantle also contains significant U and Th, it is found that R/R_A in mantle-derived samples is usually greater than 1, indicating that there is still significant primordial helium leaking out of the earth. For MORBs, R/R_A is fairly uniform, with a value of about 8.5 (Figure 13.6). For many OIBs, R/R_A is higher, up to about 30, but for some it is lower, down to about 3 (Figures 13.6, 13.7). This indicates that some plumes tap a source that contains a higher proportion of primordial helium, but other plumes tap sources that contain less, perhaps because they contain small amounts of material derived from continental crust [5].

Some correlation between helium isotopes and refractory element isotopes has been suggested, although this is debated. Higher values of R/R_A tend to occur at intermediate values of both ^{87}Sr/^{86}Sr and ^{206}Pb/^{204}Pb (Figure 13.7). Farley *et al.* have gone so far as to suggest that the data of Figure 13.7 extrapolate to a

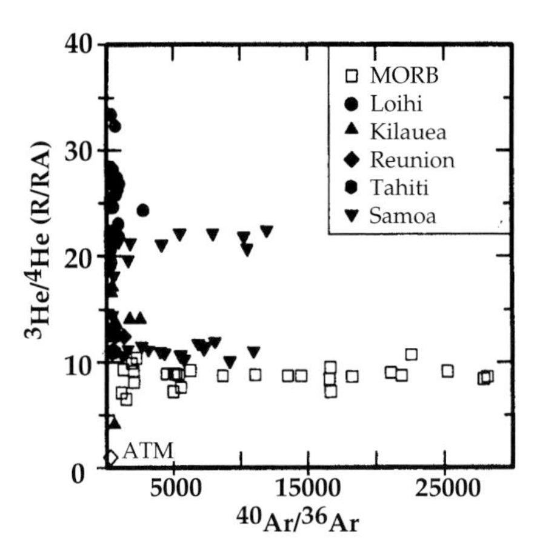

Figure 13.6. Helium and argon isotopes from MORBs and from various oceanic island groups. After Porcelli and Wasserburg [17]. Copyright by Elsevier Science. Reprinted with permission.

helium-rich source type (primitive helium mantle or 'PHEM'). However, the lead of PHEM is less radiogenic than that of FOZO/C, and the highest observed value of R, for Loihi near Hawaii, has even less radiogenic lead. As well some of the lowest values of R/R_A also tend to occur at intermediate values of $^{87}Sr/^{86}Sr$, and at medium to high values of $^{206}Pb/^{204}Pb$. It may be safer to conclude at this stage that helium does not show a clear relationship to refractory isotopic systems.

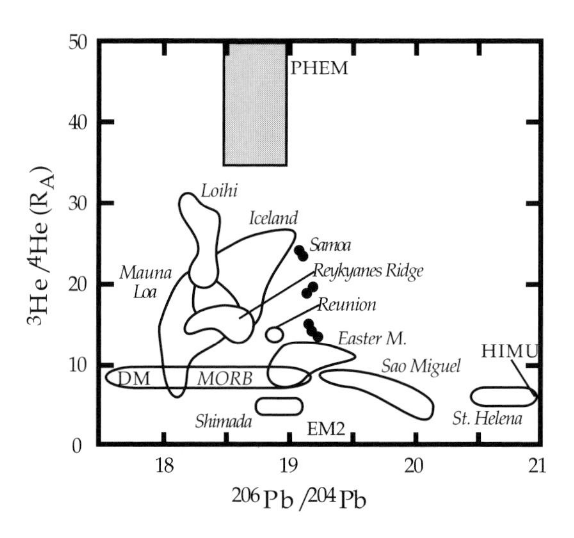

Figure 13.7. Summary of helium versus lead isotopes for mantle-derived oceanic rocks. After Farley *et al.* [5]. Copyright by Elsevier Science. Reprinted with permission.

13.3.3.2 Argon

Argon comprises nearly 1% by volume of the atmosphere, and most of that (99.6%) is ^{40}Ar, which is almost entirely radiogenic, from the decay of ^{40}K (Table 13.1) [16]. The next most abundant of the argon isotopes is ^{36}Ar, which is used as the reference isotope. The $^{40}Ar/^{36}Ar$ ratio of the atmosphere is about 300. Much higher values of $^{40}Ar/^{36}Ar$ have been found in MORBs and OIBs (Figure 13.6): up to about 44 000 in MORBs and about 13 000 in OIBs. These high ratios are taken to imply that the mantle has been rather strongly degassed, so that relatively little ^{36}Ar remains, while ^{40}Ar is continuously generated and large ratios are accumulated.

Note the contrast between helium and argon with regard to the relative compositions of the mantle and the atmosphere. Argon is presumed to have been substantially retained in the atmosphere because it is a heavier atom, so the ^{36}Ar that has emerged from the mantle has been retained and the atmospheric $^{40}Ar/^{36}Ar$ is relatively low (295.5) compared with MORB (up to about 44 000). On the other hand helium escapes continuously from the atmosphere as well as from the mantle, so much of the ^{3}He has been lost while ^{4}He is continuously added to the atmosphere, resulting in a relatively high $^{4}He/^{3}He$ ratio in the atmosphere (7×10^5) compared with MORB (8.3×10^4).

The discovery of high $^{40}Ar/^{36}Ar$ ratios in OIBs is relatively recent [18, 19], and has led to a reinterpretation of noble gas constraints. Previously the measured ratios in OIBs were much smaller, close to the atmospheric value, and it was thought that the OIBs reflected a deep mantle source with the same noble gas composition as the atmosphere. The lower ratios are now regarded as being due to the introduction of the atmospheric signature through near-surface contamination [20]. This suggests that not only the upper mantle (the MORB source) has been strongly degassed, but that the OIB source in the deeper mantle may also have been substantially degassed.

13.3.3.3 Neon

The three isotopes of neon, ^{20}Ne, ^{21}Ne and ^{22}Ne, are usually represented by a plot of $^{20}Ne/^{22}Ne$ versus $^{21}Ne/^{22}Ne$ (Figure 13.8). There are large differences in $^{20}Ne/^{22}Ne$ between the earth's atmosphere, the solar wind and the 'planetary' or meteoritic composition, with the atmosphere lying between those of some meteorites and the solar wind. Of the three neon isotopes, ^{21}Ne has the lowest abundance. Hence ^{21}Ne produced by secondary 'nucleogenic' reactions

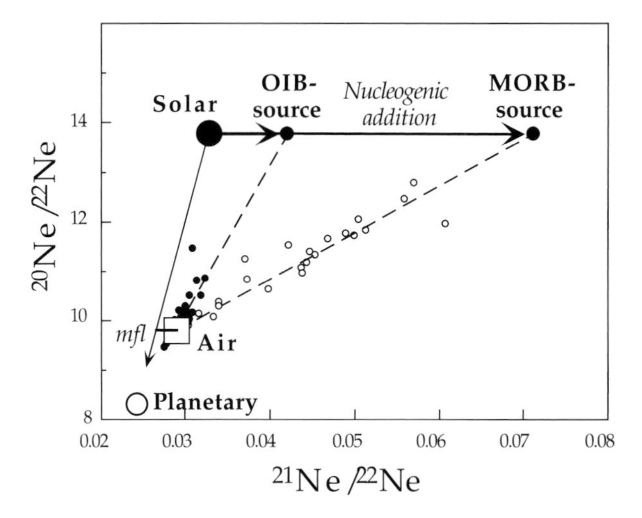

Figure 13.8. Neon isotopes from mantle-derived oceanic rocks compared with air (open square) and an inferred solar composition. 'mfl' is the trajectory due to mass fractionation. After McDougall and Honda [16, 21].

of ^{4}He and neutrons emitted by uranium and thorium with oxygen and magnesium (Table 13.1) has a non-trivial effect on the ^{21}Ne/^{22}Ne ratio. These reactions will thus tend to move compositions to the right in Figure 13.8. Observations from MORBs and OIBs are shown in Figure 13.8. Although some data cluster around the atmospheric composition, other data have distinctly higher ratios, with MORB ^{20}Ne/^{22}Ne extending closer to the solar wind value.

These observations show clearly that there are components in the mantle that are not represented in the atmosphere. This has two important implications. First, the OIB source is distinct from the atmosphere, confirming that atmosphere-like argon compositions are likely to be due to near-surface contamination, rather than to the OIB source having the same composition as the atmosphere. Second, the atmosphere cannot have been derived simply by degassing of the solid earth, and of the MORB source in particular.

The MORB and OIB data form separate linear trends in Figure 13.8, the MORB data having higher ^{21}Ne/^{22}Ne, for a given ^{20}Ne/^{22}Ne, than those of OIBs or the atmosphere. In fact the ^{21}Ne/^{22}Ne ratio has been found to correlate with the ^{4}He/^{3}He ratio (Figure 13.9). This correlation is approximately consistent with the fact that both ^{21}Ne and ^{4}He are produced from the decay of uranium and thorium [22], assuming that the initial quantities of helium and neon in the earth were in approximately the same proportions as in the solar wind (as measured by ^{3}He/^{22}Ne).

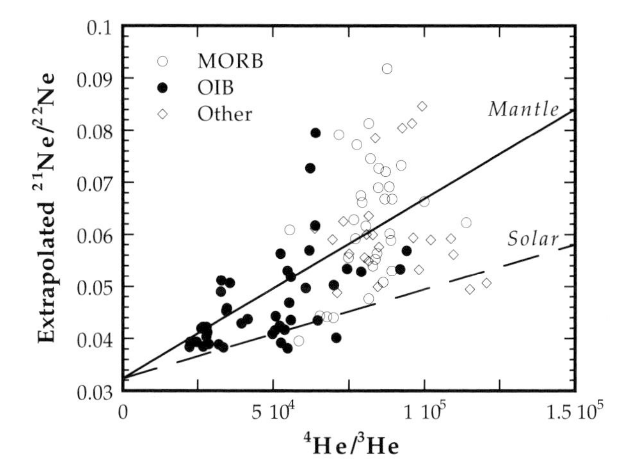

Figure 13.9. Correlation of nucleogenic neon with radiogenic helium. The $^{21}Ne/^{22}Ne$ ratio is extrapolated from the atmosphere composition to the solar $^{20}Ne/^{22}Ne$ ratio (Figure 13.8) to remove the effect of atmospheric contamination. The 'solar' line assumes a solar ratio of $^{3}He/^{22}Ne = 3.8$ and the 'mantle' line is the estimate inferred by Honda and McDougall [23] for the mantle of $^{3}He/^{22}Ne = 7.7 \pm 2.6$. Data courtesy of Honda and McDougall [23].

The latter assumption is suggested by the fact that mantle values of $^{20}Ne/^{22}Ne$ approach but do not exceed the solar wind value (Figure 13.8). It turns out that the $^{3}He/^{22}Ne$ ratio implied by this correlation is 7.7, a little higher than the measured solar wind ratio (Section 3.8, Figure 13.9) [23]. This is an important constraint on theories of the origin of the atmosphere and of the earth.

13.3.3.4 Xenon

Xenon from the mantle shows excess $^{129}Xe/^{130}Xe$ and $^{131\text{-}136}Xe/^{130}Xe$ relative to the atmosphere [16]. These observations bear on the origin of the atmosphere, but not strongly on the structure of the mantle, since no systematic differences between different mantle sources have been resolved. Therefore they are summarised only briefly here. The excess ^{129}Xe is attributed to the decay of ^{129}I, which has a half life of only 17 Ma and is 'extinct' (no longer detectable in the earth). This is usually interpreted to mean that most of the atmosphere outgassed from the mantle very early in the history of the earth, before all of the ^{129}I had decayed. However, it is also possible in principle that the excess ^{129}Xe derives from meteoritic material added to the earth over a longer period. The excess of ^{136}Xe (representative of $^{131-136}Xe$) could be due either to fission of ^{238}U (half life 4.5 Ga) or to fission of extinct ^{244}Pu (half life 72 Ma). The fact that $^{136}Xe/^{130}Xe$ correlates well with $^{129}Xe/^{130}Xe$ is usually taken to mean that the $^{136}Xe/^{130}Xe$ was established on the same time scale as the $^{129}Xe/^{130}Xe$, with the implication that it is derived from ^{244}Pu, but this is debated.

13.4 Direct inferences from observations

Some things can be inferred rather directly from the observations just summarised, with few assumptions or with assumptions that can be clearly recognised and evaluated. Examples would be that the MORB source is shallow and that mantle heterogeneities are roughly 1–2 Ga in age (see below). Other inferences depend on assumptions about or understanding of other processes, such as the way passive heterogeneities would be stirred by mantle convection, and they require the other processes to be addressed and understood before they can be fully evaluated. An example of this kind of inference would be that the difference between MORB and OIB signatures requires the mantle to convect in two layers. Some of the more direct inferences are presented in this section.

13.4.1 Depths and geometry of the MORB and OIB sources

Midocean ridges move about the earth largely in response to the dictates of lithospheric mechanics (Chapters 6, 9). The subsidence of normal sea floor in proportion to the square-root of its age implies that there is no buoyant mantle rising under normal spreading centres (Section 12.1.3). In other words, the upwelling under spreading centres is passive and represents the return flow complementary to the moving and descending lithospheric plates. The implication of these two conclusions is that midocean ridges pull up whatever mantle happens to be under them as they move around. An obvious further implication is that midocean ridges sample the upper part of the mantle. This may be contrasted with plumes, which are proposed explicitly to be buoyant upwellings that would carry material up from deeper and possibly different regions of the mantle. It is thus a straightforward inference that the MORB source is shallow and that the OIB source is deeper.

I do not want to overstate this conclusion. It is possible, for example, that there is, immediately under the lithosphere, a fairly widespread accumulation of plume material that is different from the rest of the upper mantle. We would expect that this might then be sampled transiently at new rifts, or that it would make a marginal contribution to a migrating spreading centre. Something like this has in fact been proposed [24]. However, there is a limit beyond which this picture would not be plausible. For example, it would not be plausible that spreading centres pull up material from the lower mantle without also entraining and sampling some upper mantle.

There is obviously a limit to the inferences that can be made about the three-dimensional geometry of mantle sources on the basis of chemical observations at the earth's surface. Even the inference that the MORB source is shallower than the OIB source depends on an understanding of the dynamics of plates and plumes. Beyond that, there are some recognised large-scale geographical variations of MORBs and OIBs, such as that the Indian ocean trends are distinctive (Figures 13.3–5), which can be translated into geography of mantle sources. However it is not clear, for example, what the topology of OIB sources is: similar types might be isolated from each other or they might have some sort of regional or global connection. It may be possible in the future, if models of mantle flow come to be regarded as sufficiently reliable, to trace trajectories of OIB sources more clearly, but this further illustrates the distinction I made above between direct inferences and inferences that depend on an understanding of other processes.

13.4.2 Ages of heterogeneities

Many of the radioactive systems listed in Table 13.1 are used to date rocks. The principle is embodied in Equation (13.2.1). If the concentrations of the parent and daughter isotopes are known, an age can be deduced. Usually the concentrations are taken as a ratio with a stable reference isotope of the daughter element (e.g. $^{87}Rb/^{86}Sr$ is related to $^{87}Sr/^{86}Sr$). This could be done in principle with mantle-derived rocks to determine the ages of the mantle sources, but the parent–daughter ratios cannot be accurately inferred from MORB concentrations because of the effects of melting and crystallisation on element concentrations. Nevertheless rough estimates can be made that show that the ages of mantle heterogeneities represented in Figures 13.1–4 are of the order of 1–2 Ga [25].

The only system that does not suffer from this problem is the uranium-lead system, which has the unique and long-recognised advantage of having two parents and two daughters. In this system, age can be inferred from the daughter isotopes without knowledge of the parent concentrations. This is because the relative rates of production of ^{206}Pb and ^{207}Pb depend only on time, so that at any one time all points in Figure 13.4a move along lines of the same slope. Because of the shorter half life of ^{235}U, the slope was steeper at earlier times, and steeper slopes imply greater ages. The slope of the array of MORB and OIB points in Figure 13.4a corresponds to an age of about 2 Ga ([7] ; Figure 13.2). Individual volcanic hotspots define distinct arrays whose apparent ages range from about 1 Ga to about 2.5 Ga [26].

However, the slopes in Figure 13.4a can be interpreted rigorously as ages only if there has not been mixing between mantle source types or addition of material from continental crust. Christensen and Hofmann [27] have shown that a model incorporating continuous differentiation and remixing can yield arrays of the observed slope in only 1–1.3 Ga. Thus the actual ages of mantle heterogeneities seem to be of the order of 1 Ga, with some being perhaps as old as 2 Ga. Although the ages are not very accurate, this is still a very important constraint on physical models of mantle convection, since such models must be able to reproduce the slopes of the arrays in Figure 13.4a and their implied time scale of the order of 1 Ga for mantle recirculation. It is also notable that this time scale is considerably shorter than the mean age of the continental crust, which is about 2.5 Ga [9].

13.4.3 Primitive mantle?

The possibility that some of the mantle may have remained unaltered from very early in earth history has been an important issue in understanding the earth's heat budget (e.g. [28]) and in interpreting isotopic observations, particularly since measurements of the samarium-neodymium system led to the inference of a primitive lower mantle reservoir (e.g. [29]). The term 'primitive' here needs careful qualification. Since the formation of the earth is likely to have been violent, protracted over tens of millions of years, and to have left the earth in a hot and dynamic state, it is unlikely that any portion of the earth emerged unaltered. For some refractory elements, it is plausible that their relative abundances did not change much, and there is evidence for this, as we will see. For more volatile or siderophile components there might have been substantial changes.

There would, during and soon after formation, have been a vigorous competition between processes tending to segregate and differentiate components and other processes tending to homogenise everything [30, 31], and the outcome cannot be reliably predicted. An important observation is that the only stratifications of the earth that are reliably thought to date from that time are the separations of the mantle, core and atmosphere. The mantle seems to have survived with remarkably little internal stratification, at least no more than could be substantially eradicated by subsequent mantle dynamics to the point that we have had difficulty deciding whether any compositional stratification remains.

The differentiation of some elements would depend on whether an early crust formed and what kind of crust it was, and this is poorly constrained. The processes by which the atmosphere attained its present composition are not well understood but probably involved degassing of the mantle that persisted at least for tens of millions of years [32]. Given these possibilities and uncertainties, it is not obvious what one might define as 'primitive', and so it is important that some meaning be defined for the term whenever it is used. Two common and useful meanings are 'the same as the estimated average composition of the crust plus mantle' [33, 34] and 'the same as in the chondritic meteorites', which is equivalent to the former for many refractory elements. Another meaning sometimes implied is 'containing more ^{3}He than the MORB source'.

For two of the refractory isotopic systems discussed earlier (U–Pb and Sm–Nd) there are good constraints on what the present isotopic composition of primitive mantle (estimated average crust plus mantle) would be. Since both Sm and Nd are highly refractory and are not siderophile, it is believed that there would have been little loss of either element from the silicate part of the earth, either by being vaporised during the formation of the earth or by being partitioned into the core. Their mean concentrations in the average or 'bulk' silicate part of the earth are therefore assumed to be in the same proportions as in meteorites, which are well constrained by measurements (their mean silicate concentrations are actually higher than in meteorites by a factor of 2.5–2.7 due to the presumed removal of volatiles and core material [34]). Consequently the ^{143}Nd/^{144}Nd ratio of the bulk silicate earth (BSE) is believed to be well constrained, at 0.512 638, and this composition is marked in Figure 13.3 as 'PRIMA' (primitive mantle). (The Sr isotopic composition of this point is not determined independently, but is assumed to lie within the array of mantle points shown.) There is also a constraint on lead isotopes, based on the isotopic composition of lead in iron meteorites and on the unique relationships in the U–Pb double system (Table 13.1): primitive mantle should lie close to the 'geochron' line of Figure 13.4(a), which corresponds to an age of 4.5 Ga. This is the estimated time at which the mantle ceased to interact with the core.

It appears from Figures 13.3 and 13.4(a) that some compositions overlap this inferred primitive composition, in particular the

EM-1 component. Also the highest $^3He/^4He$ ratios correlate with $^{206}Pb/^{204}Pb$ ratios of 18.5–19 (Figure 13.7), not much above the geochron value of about 18. This has led to the suggestion that some mantle component or components are less processed and more primitive than the rest, or even that some primitive mantle survives and is mixed with other source types.

A test of the existence of primitive mantle was proposed by Hofmann *et al.* [35] on the basis of the anomalous abundances of Nb and Pb evident in Figure 13.1. The ratio of Nb to its neighbour U is systematically high in oceanic rocks and low in continental rocks, as noted in Section 13.3.1. This is illustrated in Figure 13.10, which shows Nb/U concentration ratios plotted against $^{143}Nd/^{144}Nd$ (expressed as ε_{Nd}, see caption). All of the oceanic data, MORBs and OIBs, have similar and high Nb/U values averaging 47, compared with a primitive ratio of 30 and an average for continental crust of about 10. The OIB points, which are averages for particular islands, do not show a significant trend towards the primitive point, even for EM-1. Instead they are similar to MORB values. The few oceanic EM-2-type points with lower values have high $^{87}Sr/^{86}Sr$ and are better explained as having a component of recycled continental sediment in them. Analogous results have been demonstrated for Ce/Pb and Nd/Pb. Hofmann *et al.* [7, 35] conclude that OIB sources do not include a substantial primitive component.

This result was presented in the context of a debate about whether the lower mantle is entirely or substantially primitive [36–40]. It argues rather strongly against that proposition, as

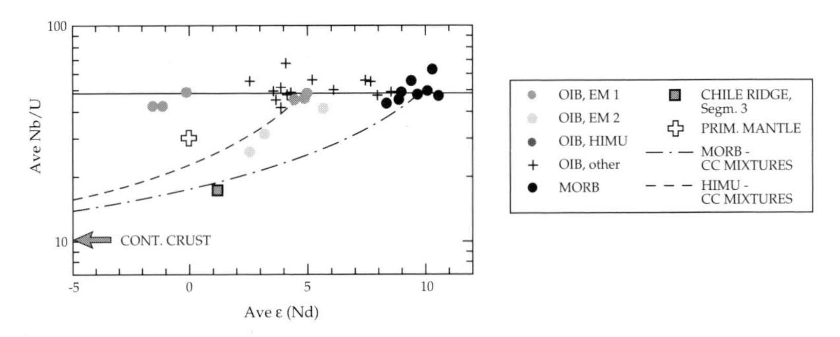

Figure 13.10. Average Nb/U versus ε_{Nd} for suites of mantle-derived oceanic rocks. $\varepsilon_{Nd} = 10\,000[(^{143}Nd/^{144}Nd)/(^{143}Nd/^{144}Nd)_p - 1]$, where subscript 'p' denotes 'primitive'; ε_{Nd} is thus the deviation of the measured isotopic ratio, in parts per 10 000, from the present-day mean silicate composition. From Hofmann [7]. Reprinted from *Nature* with permission. Copyright MacMillan Magazines Ltd.

does the fact that most of the refractory element isotopes from the mantle are different from the primitive compositions (Figures 13.3–4). Presumably these constraints do not totally exclude the possibility of some primitive mantle still surviving, nor the possibility that some parts of the mantle are more primitive in some characteristics. However it is difficult to quantify the amount of primitive mantle that is permitted. This point will be taken up in Section 13.4.5, where mass balances are discussed.

13.4.4 The mantle–oceanic lithosphere system

Hofmann *et al.* [7, 35] also emphasise that Figure 13.10 implies that the oceanic basalts (MORBs and OIBs) resemble each other much more closely than they resemble either primitive mantle or continental crust. Since MORBs and OIBs are derived from the mantle (shallow and deeper, respectively), this implies that the mantle system (including the recycling oceanic lithosphere, oceanic crust and oceanic islands) operates as a relatively closed system, with only minor input either from continental crust or from any putative primitive mantle reservoir.

The implication of this is that the oceanic crust may be best thought of as part of the convecting mantle system, rather than as a distinct reservoir like the continental crust. This is certainly consistent with the physical picture developed earlier in this book. It implies that the oceanic lithosphere is a part of the convecting mantle, but from the chemical point of view it is the part where the mantle is differentiated into compositionally distinct components (the oceanic crust and its refractory residue). These distinct components are then re-injected into the mantle. Thus the oceanic lithosphere continuously introduces chemical heterogeneity into the mantle. The similarity of Nb/U for OIBs and MORBs shows that not all chemical characteristics are modified in this process. Thus Nb/U reflects the unity of oceanic crust with 'normal' mantle, in contrast with other chemical characteristics that reflect the distinctions.

13.4.5 Mass balances

It has long been presumed that the continental crust was extracted from the mantle. The complementary trace element patterns of continental crust and MORB (Section 13.3.1), and their derived isotopic patterns (Section 13.3.2, Figure 13.3) support this idea [8, 41]. If it is true, then it should be possible to test it by comparing the present inventories of any chemical species in the crust and mantle with estimates of the original quantity incorporated into the silicate part of the earth. The latter estimates have been made for many of the so-called refractory lithophile elements, such as uranium, samarium or neodymium [33, 34].

A complication is that the differences between MORB and OIB data indicate that the mantle is not uniform. The MORB source is strongly depleted of incompatible elements, whereas the OIB source is less depleted (Section 13.3.1), suggesting that the shallow mantle is strongly depleted while the deeper mantle is less strongly depleted. In order to estimate the present quantity of a species in the mantle, we need to know for each part of the mantle both its concentration and the mass of the relevant part of the mantle:

$$^{s}m_i = {}^{s}n_i M_i \tag{13.4.1}$$

where $^{s}m_i$ is the number of moles of species s, $^{s}n_i$ is its molar concentration in reservoir i, and M_i is the mass of reservoir i. Suppose we adopt the simplified picture that the silicate part of the earth comprises the crust (c), the MORB-source mantle (m) and the OIB-source mantle (o). Conservation of the species then requires

$$^{s}m_{\mathrm{p}} = {}^{s}m_{\mathrm{c}} + {}^{s}m_{\mathrm{m}} + {}^{s}m_{\mathrm{o}} \tag{13.4.2}$$

where p denotes the primitive value, while conservation of mass requires

$$M = M_{\mathrm{c}} + M_{\mathrm{m}} + M_{\mathrm{o}} \tag{13.4.3}$$

where M is the mass of the silicate earth. It is useful to define the fraction $^{s}f_i$ of the total inventory $^{s}m_{\mathrm{p}}$ of species s in reservoir i:

$$^{s}f_i = {}^{s}m_i/{}^{s}m_{\mathrm{p}} = {}^{s}n_i M_i/{}^{s}n_{\mathrm{p}} M = {}^{s}n_i X_i/{}^{s}n_{\mathrm{p}} \tag{13.4.4}$$

where $X_i = M_i/M$ is the mass fraction of reservoir i. Then

$$1 = {}^{s}f_{\mathrm{c}} + {}^{s}f_{\mathrm{m}} + {}^{s}f_{\mathrm{o}} \tag{13.4.5}$$

Analogous relationships apply to concentrations by weight ($^{s}c_i$) and the weight of a species in a reservoir ($^{s}w_i$). Masses and mass fractions of parts of the silicate earth are shown in Table 13.3.

The basic mass balance idea has been applied in a number of variations: directly using element abundances or slightly more elaborately using observed ratios of elements or isotopes. Mass balances have been done for refractory elements and isotopes and for

Table 13.3. *Masses and mass fractions of silicate reservoirs.*

Reservoir	Total	Crust	Upper mantle	Lower mantle
Mass, M (10^{22} kg)	400	2.4	130	270
Mass fraction, X	1	0.006	0.33	0.67

argon. The results are not all mutually consistent. I summarise the more important cases here.

13.4.5.1 Refractory incompatible elements

A straightforward application of Equations (13.4.2–5) is to calculate the fractions of highly incompatible elements that currently reside in the crust, using estimates of average crustal concentrations. For example Rudnick and Fountain [42] estimate that between about 58% (Rb) and 35% (Ba) of the earth's complements of Rb, Pb, Th, U, K and Ba reside in the continental crust. Mass balances for Rb and Ba are summarised in Table 13.4. From Figure 13.1, Rb and Ba, after being corrected downward from the MORB points to allow for 10% melting, are depleted in the MORB source by a factor of about 4 relative to primitive mantle. If the entire mantle were depleted to this degree, then the mantle would contain only about 25% of these elements, and the crust should then contain the remaining 75%. The implication of this mismatch is that not all of the mantle is depleted to the same degree as the MORB source. Presumably some or all of the balance is contained in the OIB source. For example Hofmann and White [10] proposed that the OIB source is predominantly old subducted oceanic crust, in which case the missing 20–40% could be accounted for if it comprised 7–16% of the mass of the mantle (Table 13.4, assuming OIB source concentrations about 2.5 times primitive: Figure 13.1).

On the other hand, with the hypothesis that the OIB source has primitive concentrations, Equation (13.4.5) yields

$$X_{\rm m} = \frac{f_{\rm c}}{(1 - n_{\rm m}/n_{\rm p})} \qquad (13.4.6)$$

Results from this relationship are shown in the second line for each element in Table 13.4. These imply that the OIB source comprises between about 20% and 50% of the mantle mass.

Table 13.4. *Refractory incompatible element mass balances.*

Species (s)	${}^s c_p$ (μg/g)	${}^s c_c$ (μg/g)	f_c (%)	X_m	${}^s c_m$ (μg/g)	f_m (%)	f_o (%)	${}^s c_o$ (μg/g)	X_o
Rb	0.60	58	58	<1	0.15	<25	>17	$2.5{}^s c_p$	>0.07
				0.78		19	22	${}^s c_p$	0.22
Ba	6.60	390	35	<1	0.83	<25	>40	$2.5{}^s c_p$	>0.16
				0.47		12	53	${}^s c_p$	0.53

13.4.5.2 Size of the OIB reservoir

A very simple and basic point here is that the mass of the OIB source can only be estimated if the concentration of a species in it is known or assumed. This is emphasised here because it has very commonly been assumed that the fraction of a species remaining in the mantle is the same as the mass fraction of the OIB source: $f_i = X_i$. This is only true if the OIB source has primitive concentrations (Equation (13.4.4)), and we have seen in Section 13.4.3 that there is substantial evidence against this. The point applies equally to estimates based on the ratio method described below. Early estimates yielded a OIB-source mass of about $\frac{2}{3}$ of the mantle [36], which is similar to the mass of the lower mantle. This was taken as support for the hypothesis that the lower mantle is primitive. However, later estimates have tended to cluster around the mantle containing only about 50% of the inventory of various species [35, 40, 43, 44], which requires more than just the upper mantle to have been depleted. As we have just seen, as much as 90% of the mantle might have been depleted to the same degree as the MORB source if the OIB source contains concentrations higher than primitive.

13.4.5.3 Refractory ratios

An alternative form of the mass balance equation (13.4.2) is

$$ {}^s c_p = {}^s c_c X_c + {}^s c_m X_m + {}^s c_o X_o \qquad (13.4.7) $$

using here concentrations by weight (${}^s c_i$) rather than molar concentrations (${}^s n_i$). If a species b is known through its ratio R with species a ($R_i = {}^b c_i / {}^a c_i$), then combining the relationships (13.4.7) for each, substituting ${}^b c_i = {}^a c_i R_i$, and eliminating ${}^a c_m X_m$ yields

Table 13.5. *Refractory ratio mass balances.*

Ratio (b/a)	${}^{a}c_{\mathrm{c}}$ (μg/g)	R_{c}	R_{m}	${}^{a}c_{\mathrm{p}}$ (ng/g)	R_{p}	${}^{a}c_{\mathrm{o}}$ (ng/g)	R_{o}	X_{o}
^{143}Nd/^{144}Nd	20	0.5108	0.5132	1250	0.51263	$10{}^{a}c_{\mathrm{p}}$	0.5126	0.06
						${}^{a}c_{\mathrm{p}}$	R_{p}	0.60
Nb/U[a]	1.53[a]	10	47	20	30	—	—	—
Nb/U[b]	1.4	10	47	20	30	${}^{a}c_{\mathrm{p}}$	R_{p}	0.085

[a] From Equation (13.4.9).
[b] Reinterpreting reservoir ‘m’ to include the OIB source and reservoir ‘o’ as a primitive residue.

$$X_{\mathrm{o}} = \frac{{}^{a}c_{\mathrm{p}}(R_{\mathrm{p}} - R_{\mathrm{m}}) - {}^{a}c_{\mathrm{c}}X_{\mathrm{c}}(R_{\mathrm{c}} - R_{\mathrm{m}})}{{}^{a}c_{\mathrm{o}}(R_{\mathrm{o}} - R_{\mathrm{m}})} \tag{13.4.8a}$$

For the special case where the OIB reservoir is primitive (${}^{a}c_{\mathrm{o}} = {}^{a}c_{\mathrm{p}}$, $R_{\mathrm{o}} = R_{\mathrm{p}}$), this becomes

$$X_{\mathrm{o}} = 1 - X_{\mathrm{c}}\frac{{}^{a}c_{\mathrm{c}}(R_{\mathrm{c}} - R_{\mathrm{m}})}{{}^{a}c_{\mathrm{p}}(R_{\mathrm{p}} - R_{\mathrm{m}})} \tag{13.4.8b}$$

These relationships are applied to the Nd isotopes in Table 13.5. If the OIB source were recycled MORB, then its Nd concentration would be about 10 times primitive (Figure 13.1) and it would comprise only about 6% of the mantle mass. On the other hand if it had primitive concentrations, as was assumed when this approach was first used [36], it would comprise about 60% of the mantle mass.

Hofmann *et al.* [35] used ratios of Nb/U to calculate a mass balance. Equation (13.4.8a) is indeterminate if $R_{\mathrm{o}} = R_{\mathrm{m}}$, as they observed, so it cannot be used directly. However if Equation (13.4.8a) is multiplied by ${}^{a}c_{\mathrm{o}}(R_{\mathrm{o}} - R_{\mathrm{m}})(= 0)$, then an expression for ${}^{a}c_{\mathrm{c}}$ can be obtained:

$${}^{a}c_{\mathrm{c}} = \frac{{}^{a}c_{\mathrm{p}}(R_{\mathrm{m}} - R_{\mathrm{p}})}{X_{\mathrm{c}}(R_{\mathrm{m}} - R_{\mathrm{c}})} \tag{13.4.9}$$

The result, shown in Table 13.5, is that the calculated U concentration in the crust is higher than estimates from other observations. For example, Rudnick and Fountain [42] estimate the U concentration in the crust to be about 1.4 μg/g, which is higher than several previous estimates but still less than the value of 1.53 μg/g resulting from Equation (13.4.9). Hofmann *et al.* concluded that not all of the mantle has been differentiated, implying that there is a

Table 13.6. ^{40}*Ar mass balance.*

	Quantity	Percentage
Total generated in 4.5 Ga	375×10^{16} mol	100
Undegassed concentration in mantle	940 pmol/g	
Atmosphere	165×10^{16} mol	44
Continental crust	13×10^{16} mol	3.5
Upper mantle concentration	25 pmol/g	
Upper mantle total	3.3×10^{16} mol	0.9
Total atmosphere + crust + upper mantle	181×10^{16} mol	48
Balance	194×10^{16} mol	52
Concentration if in lower mantle	720 pmol/g	

third, primitive mantle reservoir whose signature is not observed in either MORBs or OIBs. With the crustal U values given in Table 13.5, this would imply that the MORB and OIB sources comprise about 92% of the mantle, leaving about 8% primitive. Earlier estimates of crustal U concentrations were lower, implying a larger primitive reservoir, up to about 40% [35].

Another way to treat this case is to use Equation (13.4.8a) with the assumption that the MORB and OIB reservoirs have similar concentrations, as well as ratios ($R_m = R_o$), and to reinterpret the 'o' reservoir as the primitive residue. This approach is included in Table 13.5. The remaining primitive reservoir would then comprise only about 8% of the mantle, which is the same answer as obtained above.

13.4.5.4 Argon

The mass balance of radiogenic ^{40}Ar has also been used to constrain mantle structure. The argument is simple, and has been presented several times [17, 40, 45]. Qualitatively, the argument is that the atmosphere and crust contain only about half of the ^{40}Ar that should have been generated in the earth, while the upper mantle contains only about 1%. If the balance is assumed to be in the lower mantle, it must have a concentration that is about 75% of what it would be if the lower mantle were primitive. Alternatively, the mass balance could be accommodated if about half of the mantle is primitive and the rest is like the MORB source. The quantitative balance is given in Table 13.6.

This argument depends on estimates of the total amount of potassium in the earth, since the ^{40}Ar is generated from the decay of ^{40}K, and the conventional estimate is about 240 μg/g of K in the silicate earth [33]. This is based on the relative constancy of the K/U ratio in the crust and mantle [46] and on the estimate of the total U abundance in the earth [33]. The mass balance also depends on an estimate of the ^{40}Ar concentration in the upper mantle, which can be done in two ways [40]. One is to combine the flux of ^{3}He detected in sea water near midocean ridges [47, 48] with the observed ratios of ^{36}Ar/^{3}He and ^{40}Ar/^{36}Ar to get a flux of ^{40}Ar from the midocean ridge system. If this is assumed to come from a depth range of about 100 km under ridges, then the ridge spreading rate (totalling about 3 km^3/a) gives a volume of mantle per year from which the ^{40}Ar is being extracted. The other approach is to take the K content of the upper mantle inferred from MORBs and a likely residence time of ^{40}Ar in the upper mantle to estimate how much ^{40}Ar will have accumulated. The latter approach gives larger values, but the total ^{40}Ar in the MORB source is still small.

An important implication of this result is that the lower mantle would have a ^{40}Ar concentration 30–40 times greater than the upper mantle. This would strongly limit the permissible flux of material from the lower mantle into the upper mantle [38]. On the other hand, the ^{40}Ar/^{36}Ar isotopic ratio of the OIB source is not so very different from that of the MORB source (Figure 13.6), and this suggests the contrary conclusion that the degrees of degassing of the lower mantle and the upper mantle are not so very different. These issues will be discussed in Section 13.7.

13.4.5.5 Summary

The mass balances presented here yield a wide range of results. The MORB source is estimated to comprise anything from 40% to 94% of the mantle. The OIB source may comprise the rest, or some of the rest may be primitive, according to the mass balance arguments. If the trace element and isotope evidence is interpreted to mean that the OIB source is not primitive, and contains little primitive component (Section 13.4.3), then it would comprise only about 10% of the mantle. If it is assumed to have primitive concentrations, then it would comprise up to 60% of the mantle. All mass balances except that for argon seem to permit either possibility. However the argon constraint, as presented above, seems to require at least 50% of the mantle to be relatively undegassed, and presumably this implies that it would be primitive in other respects also. The argon mass balance thus seems to be seriously inconsistent with the other evidence (discussed in Section 13.4.3) that both the MORB source and the OIB source(s) are non-primitive. This point will be taken up again when the interpretation of the geochemistry is discussed.

13.5 Generation of mantle heterogeneity

The depletion of incompatible elements from the MORB source is attributed to their extraction during the formation of continental crust (Section 13.3.1; [8]). How this occurs is not completely settled, but it is commonly thought that plate tectonics plays a major role, at least in the younger half of earth history. A two-stage process is usually envisaged. First, about 10% partial melting at midocean ridges concentrates the incompatible elements into the oceanic crust by a factor of about 10. Second, the oceanic crust dehydrates during subduction, and the resulting fluid carries incompatible elements into the adjacent overlying mantle which, being hotter, melts in the presence of the water-rich fluids to produce magmas that erupt in island arc volcanoes. These subsequently become incorporated into continents. More recently a more direct extraction process has been proposed involving melting of plumes and incorporation of the melt product into continental crust [49, 50]. This may have been a secondary contributor during the last 500 Ma, but may have been more important in the Archean [49].

We saw in Section 13.3 that the continental crust carries a distinctive signature of depletions of Nb and Ti and enrichment of Pb (Figure 13.1), and noted that this is thought to be generated during subduction by the dehydration-melting process. This is why the plate-tectonic mechanism is thought to have been a major contributor to the formation of continental crust [7, 8, 9].

We saw also that a complementary pattern, such as would be expected in the residue from crustal extraction, shows up in both OIBs and MORB, and that the U–Th–Pb isotopic system provides evidence that the Pb depletion in MORB and OIBs has existed for a long time (Figure 13.5b). This suggests that the residual subducted MORB, after the continental crustal components have been extracted, has been recycled through the mantle, to reappear in OIBs and in a new generation of MORB. Such a proposal, based more generally on the trace element and isotopic evidence, was made by Hofmann and White in 1982 [10].

The formation of oceanic crust at midocean ridges yields a compositional layering in oceanic lithosphere, with the crust on top and a residual region below it that must be more depleted than the average MORB source. The oceanic crust is about 7 km thick, and if it is due to about 10% partial melting of underlying mantle [8], then the region from which melt was extracted would be about 70 km thick. The subduction of oceanic lithosphere thus continuously introduces compositional heterogeneity into the mantle, including components that are both enriched and depleted in incompatible trace elements. This provides a straightforward physical mechanism that could account for the chemical inference that MORB has been recycled through the mantle.

The proportions of each trace element that are extracted during subduction and that remain in the residue are not well known, and a lot of work is being done to better constrain them [51, 52]. If, as proposed by Hofmann and White [7, 10], a significant proportion of these incompatible elements is returned to the mantle, then the resulting heterogeneities in trace elements would generate isotopic heterogeneities after sufficient time has elapsed.

This general picture has become well-accepted, and it has been pursued to try to determine in more detail the origins of the different kinds of isotopic anomaly that have been identified. The isotopic data of Figures 13.3–5 are subdivided into groups labelled MORB, HIMU, EM-1 and EM-2. These groupings and names were suggested by Zindler and Hart [3] on the basis that there are characteristic combinations of signatures: if the data were plotted in a five-dimensional space, they would form distinct groupings. HIMU stands for 'high-μ', where $\mu = {}^{238}\mathrm{U}/{}^{204}\mathrm{Pb}$: their characteristically high ${}^{206}\mathrm{Pb}/{}^{204}\mathrm{Pb}$ is inferred to have arisen from high U/Pb ratios in their sources. HIMU occurs in St Helena in the Indian Ocean, the Austral Islands in the South Pacific and the Azores in the Atlantic Ocean. EM stands for 'enriched mantle': their higher values of ${}^{87}\mathrm{Sr}/{}^{86}\mathrm{Sr}$ relative to primitive mantle imply enrichment in incompatible elements. EM-1 and EM-2 occupy discernibly different regions, most notably in Figure 13.3. Examples of EM-1 are Pitcairn in the Pacific, Tristan in the South Atlantic and Kerguelen in the Indian Ocean. Examples of EM-2 are Samoa and the Society Islands in the Pacific. Zindler and Hart also identified a component DMM, 'depleted MORB mantle', this being the depleted extreme of the MORB data.

The fact that the same group or kind occurs in completely different parts of the world most likely implies that the same process has operated in different places to produce similar isotopic signatures. This then raises the question of what each process is. Current thinking is that the overall depletion of the mantle, observed particularly in MORBs, is due to the formation of continental crust, and that the more diverse isotopic signatures of OIBs are due to re-enrichment of the mantle by the subduction of various crustal components. I summarise these ideas here.

It seems that the Pb isotopic characteristics of HIMU can be accounted for by the preferential extraction of lead from oceanic crust during subduction [7], as attested by its excess abundance in continental crust and its relatively low abundance in the mantle (Figure 13.1). This would leave high U/Pb and Th/Pb ratios in the residual subducted oceanic crust that can account for the highly radiogenic lead isotopes that are the defining characteristic of HIMU (Figure 13.4).

The characteristics of EM-1 and EM-2 can be accounted for if a small amount of sediment is carried down with subducting lithosphere. Sediments contain much higher concentrations of the relevant incompatible elements (by two orders of magnitude or so), so the incorporation of only a small percentage of sediment can alter the isotopic signature of a source. Two kinds of sediment occur on the sea floor. Pelagic sediments, occurring characteristically on the deep sea floor far from continents, contain a large proportion of biogenic material, such as siliceous and phosphatic skeletons of plankton. Terrigenous sediments comprise material washed off continents, and hence tend to occur close to continents. Pelagic sediments have relatively low U/Pb and Sm/Nd ratios, whereas terrigenous sediments have relatively high Rb/Sr ratios, and these characteristics can account for the isotopic differences evident in Figures 13.3–5 [53, 54, 55].

There is an alternative proposal that EM-1 may be due to detached portions of continental lithosphere that have been recycled through the deep mantle. There are indeed similarities between EM-1 and some xenoliths derived from subcontinental mantle [56, 57], but it is not clear in which direction the influence might have operated. Bearing in mind the diversity of plume (OIB) signatures, the occurrence of metasomatism in the subcontinental mantle (that is, the migration of enriched fluids into the lithosphere) and the possible role of plumes in continental flood volcanism and kimberlitic eruptions, it is possible that the continental lithosphere has acquired the EM-1 signature from plumes, rather than vice versa. Osmium isotopes may discriminate between these alternatives [7].

13.6 Homogenising processes

The chemical heterogeneities introduced into the mantle at subduction zones will tend to be homogenised by mantle convection. The interpretation of the heterogeneities observed in MORBs and OIBs requires an understanding of the homogenising processes as well as the generating processes just discussed. There are several distinct homogenising processes. There are also important subtleties in the way homogenisation occurs in a very viscous flow like mantle convection and in the way such heterogeneity is characterised and measured. There have been some very divergent claims made concerning the durability of chemical heterogeneities in the mantle, and their reconciliation requires an understanding of these things. I will thus digress back into some physics, in preparation for a discussion of the implications of mantle chemistry for mantle dynamics.

13.6.1 Stirring and mixing

Stirring can be distinguished from mixing as follows. Stirring is the intermingling of different fluids. Mixing is the homogenising of two different fluids to form a single intermediate kind of fluid. In the mantle context, mantle convection stirs chemical heterogeneities, while homogenisation requires the transport of chemical species from one component to another. The transport of chemical species must involve solid-state diffusion, and may also involve transport in fluid phases, by diffusion or by fluid percolation. The mantle has key physical properties that make the homogenising process distinctive.

The mantle's high viscosity limits the flow to large scales, and this strongly affects the rates at which stirring can occur. The mantle's flow regime is *laminar* flow. This can be distinguished from *turbulent* flow, in which a large-scale flow generates eddies at smaller scales. The transition between these regimes occurs when the Reynolds number is near unity. The Reynolds number is $Re = \rho v L/\mu$, where ρ is the fluid density, v is a typical flow velocity, L is the size of the fluid body, and μ is the fluid viscosity (not, for the moment, the U/Pb ratio). Using $v = 30$ mm/a $= 10^{-9}$ m/s, $L = 3 \times 10^6$ m, $\mu = 10^{21}$ Pa s and $\rho = 3300$ kg/m^3, we get $Re = 10^{-20}$. Thus the mantle is emphatically not in the turbulent flow regime. This means that any analogy from experience in the atmosphere, ocean or a coffee cup is irrelevant. The reason that cream and coffee efficiently homogenise is that the flow induced by a moving spoon generates a turbulent cascade of smaller and smaller eddies, so that cream and coffee quickly become intermingled at a fine scale, and diffusion operates much more efficiently at small scales, as we have seen in Chapter 7 with the diffusion of heat. In the mantle, in contrast, the flow remains at the scale of the generating agent – a plate or a plume – and no smaller-scale eddies are generated. Try stirring cream into cool honey to get some appreciation of the difference.

Chemical species in mantle rocks have very small solid-state diffusivities, of the order of 10^{-17} m^2/s, while diffusivities in basaltic liquids might be of the order of 10^{-12} m^2/s, [58]. Such low diffusivities, D, mean that after a billion years ($t = 3 \times 10^{16}$ s), diffusion length scales $l = \surd(Dt)$ are of the order of 1 m in the solid state and 100 m in liquids. This means that, in the absence of migrating fluid phases, heterogeneities have to be stirred down to a scale of 100 m or less before diffusion can homogenise the components.

Thus in the mantle stirring depends on flow that remains very large in scale while diffusion operates only at small scales. Together, these properties mean that stirring and mixing in the mantle are quite inefficient compared with more familiar liquids, even when the very different flow rates are accounted for.

13.6.2 Sampling – magma flow and preferential melting

If there is fluid phase migration, then more efficient mixing becomes possible. For example, under a midocean ridge melting occurs through a volume tens of kilometres in dimension and the melt phase migrates to the surface, emerging in a narrow rift zone only a few kilometres across. The magma migration probably involves percolation between mineral grains, channelling into progressively larger conduits, possibly ponding in a magma chamber at a few kilometres depth, and rapid ascent through a fissure or dike before eruption onto the sea floor. There is clearly much more opportunity for *mixing* (mingling plus diffusive homogenisation) during this process.

The way in which the mantle is sampled may thus have a large effect on the amount of heterogeneity observed. Mantle xenoliths (fragments of the mantle carried to the surface in erupting magmas) can be quite heterogeneous at centimetre and smaller scales. Mid-ocean ridge basalts are much more homogeneous, even at much larger scales.

Although melting and magma flow can be efficient homogenising processes, they still may not yield a sample that is identical to the average composition of the source if different mantle components melt at different temperatures. Consider, for example, that if oceanic crust has been injected into the mantle for billions of years it is not implausible that any given volume of mantle will have streaks of former oceanic crust through it. Oceanic crust does have a lower melting temperature than the more refractory residue (peridotite or lherzolite) left when it forms. It may seem straightforward then that the first melt product to emerge at the surface from such mantle would come from the old crust component. However there are two complications, each involving interaction with surrounding material.

The first interaction to consider is chemical. Magma produced in the old oceanic crust would be in chemical equilibrium with that material, and would not be in chemical equilibrium with adjacent peridotite. Thus if the magma migrated through the peridotite it would react with it and change its composition. It might all freeze again as a result of the changed chemical equilibria. This is potentially a complex process. The composition of any magma that reached the surface would depend on how much it had interacted with peridotite, and this would depend on the size of the channels it had flowed through, the time it was in transit and whether magma had previously flowed through each channel and formed a hybrid, less reactive aureole.

The second interaction is thermal, and was pointed out by Sleep [59]. Melting uses latent heat and causes the temperature of the melting material to drop. If melting occurs preferentially in a lens of old crust, then its temperature will drop below that of surrounding peridotite. Heat will then conduct from the peridotite into the old crust. This extra heat will permit more melting to occur. The net effect can be to approximately double the amount of melting, until the old crust is used up. The magnitude of the effect will depend on the relative proportions of old crust and peridotite and the size of the old crust bodies: thermal diffusion will be less effective for large bodies.

The thermal interaction tends to increase the rate of melting, while the chemical interaction tends to modify the magma composition and probably to reduce the amount. The net effects are difficult to estimate because of the potential complexity. Bounding estimates are possible, and the maximum effect, when the thermal interaction is greatest and the chemical interaction is least, can be a substantial increase in the amount of magma produced [60].

In any case the magma would give a biased sample of the source region, the degree of bias depending on the net degree of preferential melting.

13.6.3 Stirring in viscous flows

An example of how passive tracers might be stirred in the mantle is shown in Figure 13.11. The flow has a number of features that might be found in mantle flow, though it is still idealised, particularly in being two-dimensional. The model is a convection

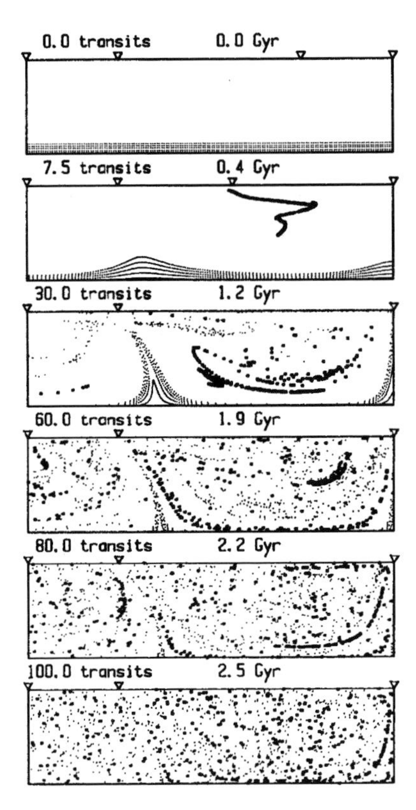

Figure 13.11. Stirring of passive tracers in a mantle convection model incorporating thermal convection and a plate evolution sequence (see text). Heavy tracers are introduced progressively, marking 'subducted' material. Light tracers mark material initially near the base. Triangles mark plate margin locations. After Gurnis and Davies [67, 69]. Copyright by Elsevier Science. Reprinted with permission.

model, though the flow structure is modified by the imposition of piece-wise uniform horizontal velocities on the top, simulating the presence of three plates. The plates are separated by a spreading centre (left) and a subduction zone (right) that migrate through a sequence that repeats every 10 transit times (plate model 1 of Gurnis and Davies [61]; a transit time is the time it takes to traverse the depth of the fluid layer at a characteristic plate velocity). The model also features an increase of viscosity with depth by a factor of 1000. Tracers were inserted in two ways: the small dots mark fluid that was originally in a layer at the bottom of the model, and the large dots simulate material subducted during the first 10 transits. The model times are scaled to real mantle times, in gigayears ('Gyr'), taking into account faster convection in the past.

The distribution of tracers remains quite heterogeneous even after 1.9 Ga. The survival of heterogeneities is enhanced by the higher viscosities at depth, which cause the flow to be slower. At intermediate times (1.2 Ga) the heterogeneities are both small-scale, in the form of tight clusters of tracers, and large-scale: the subduction tracers are concentrated in the right half of the model. The heterogeneities observed in MORBs similarly range from small-scale to differences between ocean basins [3].

Several important aspects of slow viscous stirring are evident in this simple model. I note them briefly here, and more extensive discussions can be found in [27, 61–68].

Some of the subduction tracers are in the left side of the model at 1.2 Ga, and although the coarse tracer distribution of this model does not reveal it, there would be a very thin sheet or tendril of subducted material connecting the material on the two sides, as was demonstrated explicitly by Gurnis [65]. This illustrates the extreme variation in the way heterogeneities are deformed. Some parts remain in clusters and are only moderately deformed. Other parts become extremely stretched and convoluted. This means that heterogeneities will exist in two distinct forms simultaneously: as fine streaks that permeate much of the fluid and as relatively concentrated blobs or locally much thicker sheets.

This is better illustrated in Figure 13.12, which shows how initial blocks of tracers are stirred by a simple kinematic flow. (A *kinematic* flow is one in which the velocities are prescribed directly by a formula. Convection, on the other hand, is a *dynamic* flow in which the sources of buoyancy are prescribed and the flow velocities are calculated from these.) The important implication for the mantle is that we might expect heterogeneities to be expressed both as relatively large deviations from the norm, as in OIBs, and as a pervasive but subdued component in virtually any sample of the mantle, which is how MORB isotopic data can be interpreted (Section 13.3.2).

A crucial feature of Figures 13.11 and 13.12 is that the flow is unsteady. In steady flow, heterogeneities just go round and round the same streamline. They become progressively sheared, but the amount of shearing (or equally, their lengths or perimeters) increases only linearly with time. On the other hand, if the flow is unsteady the streamlines move through the fluid, and heterogeneities can come to straddle the boundary between two cells. If this happens, then part of the heterogeneity is transferred to an adjacent, counter-rotating cell and the two parts move rapidly apart (as has happened in Figure 13.11 at 1.2 Ga). However they do retain a connection through a thin sheet. Once this has happened, the sheet

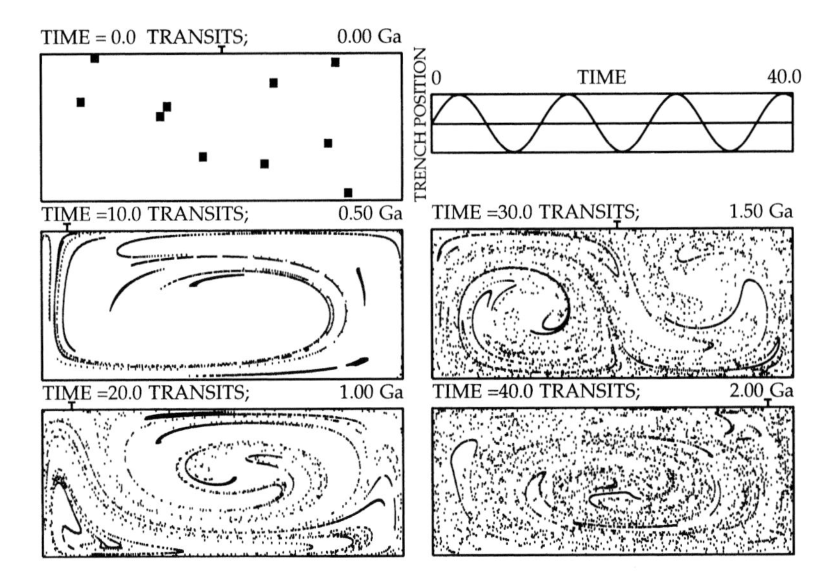

Figure 13.12. Stirring of passive tracers in a kinematic flow (Flow type 1, below). There are two flow cells with upflows at the sides of the box. The location of the downflow ('trench') oscillates left and right, as plotted in the top right-hand panel. Trench position is marked on each flow panel by a 'T'. The initial 10 squares are each blocks of 20 × 20 tracers, so there are 4000 tracers total.

is frequently carried across cell boundaries and it rapidly becomes convoluted [65]. In this phase of behaviour, the perimeter of the heterogeneity doubles almost every overturn of the fluid, and consequently it increases exponentially. Conversely, its *average* thickness decreases exponentially. By this means a heterogeneity can be stirred down to a thickness at which diffusion or fluid flow can homogenise it with its surroundings. Note, however, that part of the heterogeneity may still remain relatively undeformed, as illustrated in Figure 13.12.

13.6.4 Sensitivity of stirring to flow details

The rate of stirring can depend very sensitively on details of the flow [62], although exactly what characteristics of the flow determine the rate of stirring is not well understood. An illustration of this is given in Figure 13.13, which compares the distribution of tracers after 30 transit times in three different flows, specified below. It is clear that the degree of stirring is substantially different in the three cases, even though the average velocities are similar.

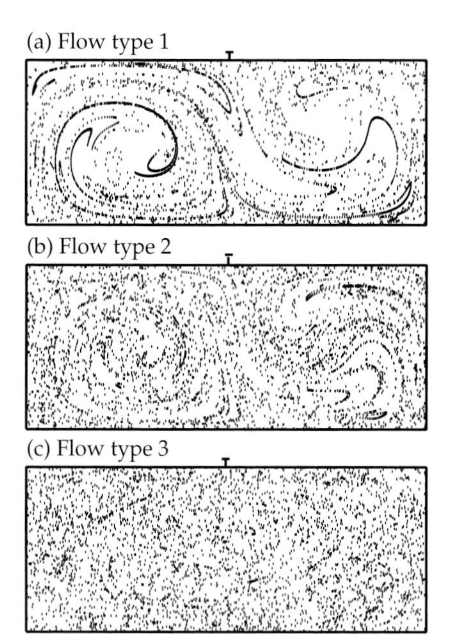

Figure 13.13. Comparisons of stirring in three kinematic flows that differ in relatively minor ways, as described in the text. The passive tracers are shown after 30 transit times of each flow. The rate of stirring of the tracers is quite sensitive to the type of flow.

In flow type 1 there are two cells, and the boundary between them oscillates horizontally and sinusoidally with time. The stream function for this flow is specified by

$$\Psi = -\frac{b}{\pi} x_t \sin\left(\pi \frac{x}{x_t}\right) \sin(\pi y), \qquad x \leq x_t \tag{13.5.1a}$$

$$\Psi = \frac{b}{\pi}(L - x_t) \sin\left(\pi \frac{(L-x)}{(L-x_t)}\right) \sin(\pi y), \qquad x \geq x_t \tag{13.5.1b}$$

where

$$x_t = (1 + a)L/2 \tag{13.5.2}$$

$$a(t) = c \sin(2\pi t/T) \tag{13.5.3}$$

$$b(t) = \left[\frac{2L}{x_t^3 + (L - x_t)^3}\right]^{1/2} \tag{13.5.4}$$

where x and y are horizontal and vertical coordinates, t is time, L is the length of the box of fluid, which has unit depth, c is the amplitude and T is the period of oscillation. x_t is the position of the boundary between the two cells. The factor b is time-dependent because it is adjusted to keep the root-mean-square surface velocity equal to one.

Flow type 2 is very similar except that the horizontal variation of the velocity components is prescribed by a superposition of two sinusoids of wavelength L and $2L$, with the longer-wavelength component varying sinusoidally with time. This also results in two cells with an oscillating boundary between them. Flow 2 is specified by

$$\Psi = -b\left[2a(t)\sin\left(\frac{\pi x}{L}\right) + \sin\left(\frac{2\pi x}{L}\right)\right]\sin(\pi y) \tag{13.5.5}$$

where a also is given by Equation (13.5.3) and

$$b(t) = \frac{\sqrt{2}}{\pi\sqrt{1+4a^2}} \tag{13.5.6}$$

In flow type 3, two sinusoids are superposed, this time of wavelengths $2L$ and $2L/3$. Both are time-dependent in such a way that the flow changes from one cell to three and back. It is specified by

$$\Psi = \frac{\sqrt{2}}{\pi}\left[C\sin\left(\frac{\pi x}{L}\right) + D\sin\left(\frac{3\pi x}{L}\right)\right]\sin(\pi y) \tag{13.5.7}$$

$$C = \cos\left(\frac{2\pi t}{T}\right), \qquad D = \sin\left(\frac{2\pi t}{T}\right) \tag{13.5.8}$$

Further discussion of how the character of the flow influences the rate of stirring can be found in the references given earlier.

13.6.5 Separation of denser components

Through much of the depth of the mantle, the oceanic crust component of subducted lithosphere is likely to be denser than average mantle, while the complementary depleted residue is likely to be less dense ([70] and Section 5.3). This has led to conjectures that these components would separate into distinct bodies or layers in the mantle [10, 24, 71, 72]. In some of the conjectured scenarios the separation is assumed to occur as the lithosphere descends through the upper mantle, which takes only 10–20 Ma. It is unlikely that there would be significant separation in such a short time, especially as the lithosphere is still cold and stiff so soon after subduction [73]. However separation on longer time scales is more plausible.

Some accumulation of dense material at the base of the mantle may occur on a billion-year time scale due to slow separation and settling in the body of the mantle, after it has thermally equilibrated with normal mantle [64]. However, separation would be enhanced if the old lithosphere were further heated, so that its viscosity would be further reduced. Christensen and Hofmann [27] have shown that this can occur in the lower, hot thermal boundary layer of the mantle. The more buoyant depleted residue then rises, leaving the old oceanic crust component behind. In this way a more efficient separation and accumulation of the denser material is accomplished.

13.6.6 Summary of influences on stirring and heterogeneity

Stirring in slow viscous (low Reynolds number) laminar flow is inefficient because the flow has only the largest-scale component determined by the driving buoyancy, and does not generate the smaller-scale components that occur in turbulent flows. Stirring is least efficient in steady flow. Unsteady laminar flow generates heterogeneities of two kinds, one modestly sheared and stretched, like those in steady flow, and the other exponentially stretched and convoluted, that comes to permeate the fluid. As long as the less deformed concentrations persist, heterogeneities exist on all scales from very small to very large.

The rate of stirring in laminar flow depends significantly on subtleties of the flow that are not well understood. There has been some consequent debate as to whether flow in the mantle would yield the faster or slower stirring rates [62, 63, 68]. The differences in rates involved are perhaps a factor of three, much smaller than the orders-of-magnitude differences between laminar and turbulent flows.

The increase in viscosity with depth in the mantle (Chapter 6) will extend the life of heterogeneities, particularly in the deeper parts of the mantle. An indication of this can be seen by comparing the models of Figures 13.11 and 13.12 each at 30 transit times. A direct implication is that the upper mantle, where viscosities are lower, is likely to be better stirred than the lower mantle.

The density differences between the oceanic crust and its complementary residue are likely to produce some separation and accumulation of the crustal component at or near the bottom of the mantle.

An important factor, considered in some models of chemical evolution but not yet evaluated in the kind of stirring models considered here, is that heterogeneities are removed or reset at the top of the mantle (by melting at spreading centres and subduction zones) but nowhere else, so far as we know. This will tend to reduce the concentration of incompatible species in the upper mantle, while concentrations in the lower mantle will remain higher, particularly as the residence time is longer due to the higher viscosity.

Some other factors may be significant, but have not been extensively investigated. If an isolated region of higher viscosity existed in the mantle, it would persist longer (in proportion to its viscosity ratio to normal mantle) for the same reason that lumps persist in porridge: the stresses from surrounding lower-viscosity material deform it only slowly [74]. If the buoyancy effects of the phase transformations near 660 km depth inhibit vertical flow to any degree, differences between the upper mantle and lower mantle would accumulate in response.

13.7 Implications of chemistry for mantle dynamics

This topic remains, at the time of writing, probably the most debated aspect of mantle dynamics, although some consensus seems to have begun to emerge. This means that the following discussion necessarily takes account of current uncertainties and debates. Presumably these things will be resolved in due course, and it may be that this happens relatively quickly, in which case some of this discussion will as quickly become dated. This means that the reader must be alert to the possibility that the story has moved on. I start with the things that seem more certain.

It is clear that the top of the mantle has different trace element and isotopic characteristics than the deeper mantle. This is implicit in the differences between MORBs and OIBs: MORBs are generally more depleted of incompatible elements and somewhat less heterogeneous than OIBs. The straightforward geometric interpretation is that the MORB source is at the top of the mantle and the OIB source is deeper. Thus some kind of vertical stratification of trace elements and isotopes is clearly implied. This might take the form of two or more layers with relatively sharp boundaries, or it might imply a more-or-less continuous variation from top to bottom. I will return to this question below.

There are distinct isotopic heterogeneities that are 1–2 Ga old, and the several kinds identified can be plausibly explained as due to the subduction of oceanic lithosphere that has been variably affected by hydrothermal alteration and dehydration and that has dragged down variable, small amounts of pelagic or terrigenous sediments. These origins are not fully settled, and some involvement of detached continental lithosphere is possible, for example. There is a component inferred indirectly from converging isotopic trends that has an intermediate refractory isotopic composition. This component may have a noble gas component with a lower $^{4}He/^{3}He$ ratio, but this is not established. In any case these properties are consistent with this component being less processed than the others.

The MORB source occupies at least 40% of the mantle, and possibly as much as 95%, depending on the nature of the OIB sources and the concentrations of trace elements in them, according to mass balances simplified by assuming uniform reservoirs. Only the balance of ^{40}Ar, as usually interpreted, does not permit the MORB source to be larger than about 50%.

With the exception of the ^{40}Ar mass balance, these constraints on time scales and volumes of mantle sources are much less stringent than was commonly thought a decade ago. Whereas the early Nd isotopic data were interpreted in terms of a primitive mantle component, it is now clear that the refractory trace element and isotope signatures of OIBs are not primitive, though some less processed contribution to some OIBs is suggested. It used to be thought that the noble gas isotopic signature of the OIB source was the same as the atmosphere, and that a large and nearly primitive noble gas reservoir was therefore required, with the implication that the atmospheric noble gas composition is primitive. However, the recognition that both MORB and OIB sources contain a solar-like component of neon that is not present in the atmosphere implies that the primitive mantle could not have been like the atmosphere. It has also been recognised that atmosphere-like noble gas isotopic signatures may all be due to near-surface contamination just prior to eruption, and not to any mantle component with an atmospheric signature. It follows then that the persistence of ^{129}Xe and ^{136}Xe anomalies in MORBs from very early in earth history says only that the mantle is different from the atmosphere, rather than that the MORB source has remained different from a putatively primitive, atmosphere-like OIB source since very early in earth history. This would only be true if OIBs were found to have *different* xenon anomalies than MORBs, but so far no xenon anomalies have been detected in OIBs.

There is thus no longer any direct requirement that primitive mantle has survived to the present. The ^{40}Ar mass balance, as usually interpreted, does not strictly require any primitive mantle, since the ^{40}Ar was all generated after the earth formed, though it does require about half of the mantle to have avoided substantial degassing for much of earth history.

The nature of the stratification of trace elements and isotopes in the mantle has been a contentious issue for two decades [7, 40, 41]. The major question has been whether there are two distinct layers separated at the 660-km seismic discontinuity. Recently there has been a major crystallisation of opinion, with many mantle geochemists concluding that the images from seismic tomography show clearly that subducted lithosphere is currently descending into the lower mantle (Figure 5.13) and that there must therefore be a substantial mass flux across the 660-km discontinuity. There is less agreement on how far back in earth history this situation might have existed.

Most of the geochemical evidence is broadly consistent with the picture of mantle flow that was developed in preceding chapters and summarised in Chapter 12. Thus all of the geochemical data except the ^{40}Ar mass balance require or permit most or all of the mantle to have been processed. (By processed I mean degassed or depleted of refractory incompatible elements.) Even the observed ^{40}Ar/^{36}Ar ratios suggest this, since the values observed in OIBs are now within a factor of two or three of the values observed in MORBs, though they do not require it. On the other hand the ^{40}Ar mass balance seems to require that there is a large reservoir that has retained most of its ^{40}Ar, and it is not clear how this can be reconciled with either the geophysical or the other geochemical evidence. I will return to this question shortly.

The MORB source has been strongly depleted and degassed, and the OIB sources can be interpreted plausibly as due to subducted oceanic crust, variously modified by hydration–dehydration or minor sediment addition. There are hints of an OIB component (C or FOZO) that is less depleted, and it might also be less degassed than MORB, by a factor of 2–5 according to the observed ratios of ^{4}He/^{3}He, ^{21}Ne/^{22}Ne and ^{40}Ar/^{36}Ar. The OIB source(s) might comprise between 5% and 20% of the mantle.

All of this seems comfortably compatible with the preferred picture of the mantle presented in Chapter 12. We know that the plate-scale flow has been injecting oceanic crust into the mantle for a long time, at a present rate of about 1.5% of the mantle mass per billion years. The MORBs would be derived from the top of the mantle and OIBs, via plumes, from the bottom of the mantle. The differences between MORBs and OIBs might be attributed to the increase of viscosity with depth, probably augmented by some gravitational settling of denser oceanic crust [27]. Thus the upper mantle would convect faster (at about the same speed as the plates) and be stirred faster and have incompatible elements extracted more often by melting under spreading centres. The deeper mantle would convect more slowly and be stirred more slowly, so it would be more heterogeneous and the heterogeneities would be older. Convection models of this type have yielded residence times of 1–2 Ga (Figure 13.11, [27, 66]), comparable to the apparent ages of mantle heterogeneities. While it is unlikely that a substantial fraction of the mantle would have escaped some processing, it is not unlikely that some portions of the mantle would have undergone less depletion and degassing, sufficient to account for the C/FOZO component and the distinctive noble gas signatures in some OIBs. A sketch that I made some time ago [75] remains a useful impression of this kind of mantle (Figure 13.14). I should be clear here that the plausibility of this kind of picture has been argued (e.g. [75]), but it remains to be demonstrated that the geochemical observations can be quantitatively accounted for. This question remains as an important subject of numerical convection modelling.

The only geochemical inference that is in obvious conflict with this general interpretation is the ^{40}Ar mass balance. There are three ways in principle in which the standard ^{40}Ar mass balance could be modified. First, there might be less potassium in the earth than has

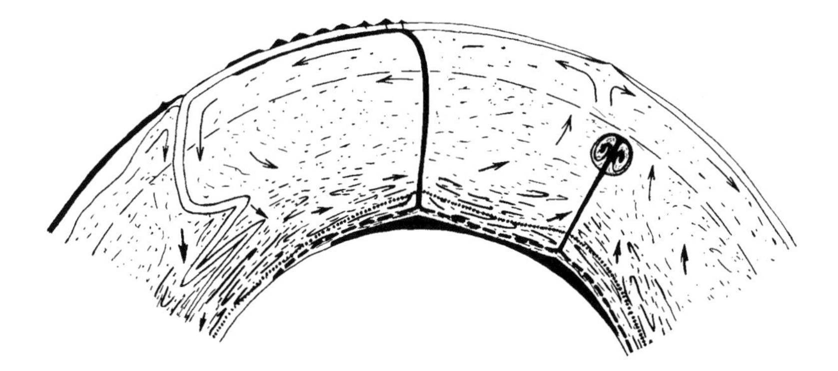

Figure 13.14. Sketched impression of important features of the mantle. A plate subducts on the left, passing through the 660-km seismic discontinuity (long-dashed) and buckling as it encounters higher viscosity in the deep mantle (cf. Figure 10.12). Plate-scale flow rises passively under a spreading centre on the right. A new plume rises on the right and an older plume continues to rise in the centre, having created a flood basalt province and a hotspot track on the plate passing over it. Chemical heterogeneity (dots and streaks) is greater near the bottom of the mantle. An irregular and possibly discontinuous layer of 'dregs' (black) lies at the base of the mantle, overlain by a velocity boundary layer (dashed) that feeds the plumes, which is in turn embedded in a thermal boundary layer (dotted) due to heat conducting from the core. From Davies [75]. Copyright by Elsevier Science. Reprinted with permission.

been thought, so that less ^{40}Ar would have been generated. Second, ^{40}Ar might have been lost from the earth entirely, presumably due to some mechanism of atmosphere loss. Third, ^{40}Ar might be sequestered in the core. Each of these possibilities deserves to be explored, and such exploration has begun (e.g. [12, 76, 77]).

I leave this story at this point, since it is likely to develop rapidly and is therefore best followed in the specialist journals. I only make two general points here. One is that a major discrepancy or apparent incompatibility between different kinds of evidence indicates that there is something important about the earth that we don't understand. Thus important insights may await those who are willing seriously to address both sides of such a controversy. The second point is that the physics of the noble gases in the earth is poorly understood: how they were incorporated, where and in what state they reside, how they escape from the interior and what processes may have affected them in the atmosphere [12, 13, 32]. We may note here another opportunity to learn important things, and we might also conclude for the moment that a discrepancy involving one of these poorly understood species is perhaps neither too surprising nor too disturbing.

13.8 References

1. A. E. Ringwood, *Composition and Petrology of the Earth's Mantle*, 618 pp., McGraw-Hill, 1975.
2. F. Albarede, *Introduction to Geochemical Modeling*, 543 pp., Cambridge University Press, Cambridge, 1995.
3. A. Zindler and S. Hart, Chemical geodynamics, *Annu. Rev. Earth Planet. Sci.* **14**, 493–570, 1986.
4. S. R. Hart, E. H. Hauri, L. A. Oschmann and J. A. Whitehead, Mantle plumes and entrainment: isotopic evidence, *Science* **256**, 517–20, 1992.
5. K. A. Farley, J. H. Natland and H. Craig, Binary mixing of enriched and undegassed (primitive?) mantle components (He, Sr, Nd, Pb) in Samoan lavas, *Earth Planet. Sci. Lett.* **111**, 183–99, 1992.
6. B. B. Hanan and D. W. Graham, Lead and helium isotope evidence from oceanic basalts for a common deep source of mantle plumes, *Science* **272**, 991–5, 1996.
7. A. W. Hofmann, Mantle chemistry: the message from oceanic volcanism, *Nature* **385**, 219–29, 1997.
8. A. W. Hofmann, Chemical differentiation of the Earth: the relationship between mantle, continental crust, and oceanic crust, *Earth Planet. Sci. Lett.* **90**, 297–314, 1988.
9. S. R. Taylor and S. M. McLennan, *The Continental Crust: Its Composition and Evolution*, 312 pp., Blackwell, Oxford, 1985.

10. A. W. Hofmann and W. M. White, Mantle plumes from ancient oceanic crust, *Earth Planet. Sci. Lett.* **57**, 421–36, 1982.
11. E. H. Hauri, J. A. Whitehead and S. R. Hart, Fluid dynamic and geochemical aspects of entrainment in mantle plumes, *J. Geophys. Res.* **99**, 24 275–300, 1994.
12. A. P. Jephcoat, Rare-gas solids in the Earth's deep interior, *Nature* **393**, 355–8, 1998.
13. E. Chamorro-Perez, P. Gillet, A. Jambon, J. Badro and P. McMillan, Low argon solubility in silicate melts at high pressure, *Nature* **393**, 352–5, 1998.
14. M. Ozima and F. A. Podosek, *Noble Gas Geochemistry*, 367 pp., Cambridge University Press, Cambridge, 1983.
15. M. Ozima, Noble gas state in the mantle, *Rev. Geophys.* **32**, 405–26, 1994.
16. I. McDougall and M. Honda, Primordial solar noble-gas component in the earth: Consequences for the origin and evolution of the earth and its atmosphere, in: *The Earth's Mantle: Composition, Structure and Evolution*, I. N. S. Jackson, ed., Cambridge University Press, Cambridge, 159–87, 1998.
17. D. Porcelli and G. J. Wasserburg, Mass transfer of helium, neon, argon and xenon through a steady-state upper mantle, *Geochim. Cosmochim. Acta* **59**, 4921–37, 1995.
18. H. Hiyagon, M. Ozima, B. Marty, S. Zashu and H. Sakai, Noble gases in submarine glasses from midoceanic ridges and Loihi seamount: constraints on the early history of the earth, *Geochim. Cosmochim. Acta* **56**, 1301–16, 1992.
19. R. J. Poreda and K. A. Farley, Rare gases in Samoan xenoliths, *Earth Planet. Sci. Lett.* **113**, 129–44, 1992.
20. D. B. Patterson, M. Honda and I. McDougall, Atmospheric contamination: a possible source for heavy noble gases in basalts from Loihi Seamount, Hawaii, *Geophys. Res. Lett.* **17**, 705–8, 1990.
21. M. Honda, I. McDougall, D. B. Patterson, A. Doulgeris and D. A. Clague, Noble gases in submarine pillow basalt glasses from Loihi and Kilauea, Hawaii: A solar component in the earth, *Geochim. Cosmochim. Acta* **57**, 859–74, 1993.
22. M. Honda, I. McDougall and D. Patterson, Solar noble gases in the Earth: The systematics of helium–neon isotopes in mantle derived samples, *Lithos* **30**, 257–65, 1993.
23. M. Honda and I. McDougall, Primordial helium and neon in the Earth – a speculation on early degassing, *Geophys. Res. Lett.* **25**, 1951–4, 1998.
24. D. L. Anderson, Lithosphere, asthenosphere, and perisphere, *Rev. Geophys.* **33**, 125–49, 1995.
25. C. Brooks, S. R. Hart, A. Hofmann and D. E. James, Rb-Sr mantle isochrons from oceanic regions, *Earth Planet. Sci. Lett.* **32**, 51–61, 1976.

26. C. G. Chase, Oceanic island Pb: two-stage histories and mantle evolution, *Earth Planet. Sci. Lett.* **52**, 277–84, 1981.
27. U. R. Christensen and A. W. Hofmann, Segregation of subducted oceanic crust in the convecting mantle, *J. Geophys. Res.* **99**, 19 867–84, 1994.
28. H. C. Urey, The cosmic abundance of potassium, uranium and thorium and the heat balances of the earth, moon and Mars, *Proc. Natl. Acad. Sci. U.S.A.* **42**, 889–91, 1956.
29. D. J. DePaolo and G. J. Wasserburg, Inferences about mantle sources and mantle structure from variations of $^{143}Nd/^{144}Nd$, *Geophys. Res. Lett.* **3**, 743–6, 1976.
30. G. F. Davies, Heat and mass transport in the early earth, in: *Origin of the Earth*, H. E. Newsome and J. H. Jones, eds., Oxford University Press, New York, 175–94, 1990.
31. W. B. Tonks and H. J. Melosh, The physics of crystal settling and suspension in a turbulent magma ocean, in: *Origin of the Earth*, H. E. Newsom and J. H. Jones, eds., Oxford University Press, New York, 151–74, 1990.
32. M. Ozima and K. Zahnle, Mantle degassing and atmospheric evolution: Noble gas view, *Geochem. J.* **27**, 185–200, 1993.
33. W. F. McDonough and S.-S. Sun, The composition of the Earth, *Chem. Geol.* **120**, 223–53, 1995.
34. H. S. C. O'Neill and H. Palme, Composition of the silicate Earth: implications for accretion and core formation, in: *The Earth's Mantle: Composition, Structure and Evolution*, I. N. S. Jackson, ed., Cambridge University Press, Cambridge, 3–126, 1998.
35. A. W. Hofmann, K. P. Jochum, M. Seufert and W. M. White, Nb and Pb in oceanic basalts: new constraints on mantle evolution, *Earth Planet. Sci. Lett.* **79**, 33–45, 1986.
36. S. B. Jacobsen and G. J. Wasserburg, The mean age of mantle and crustal reservoirs, *J. Geophys. Res.* **84**, 7411–27, 1979.
37. R. K. O'Nions, N. M. Evenson and P. J. Hamilton, Geochemical modelling of mantle differentiation and crustal growth, *J. Geophys. Res.* **84**, 6091–101, 1979.
38. R. K. O'Nions and I. N. Tolstikhin, Limits on the mass flux between the lower and upper mantle and stability of layering, *Earth Planet. Sci. Lett.* **139**, 213–22, 1996.
39. C. J. Allegre and D. L. Turcotte, Geodynamic mixing in the mesosphere boundary layer and the origin of oceanic islands, *Geophys. Res. Lett.* **12**, 207–10, 1985.
40. C. J. Allegre, A. Hofmann and K. O'Nions, The argon constraints on mantle structure, *Geophys. Res. Lett.* **23**, 3555–7, 1996.
41. G. J. Wasserburg and D. J. DePaolo, Models of earth structure inferred from neodymium and strontium isotopic abundances, *Proc. Natl. Acad. Sci. U.S.A.* **76**, 3594–8, 1979.

42. R. L. Rudnick and D. M. Fountain, Nature and composition of the continental crust: a lower crustal perspective, *Rev. Geophys.* **33**, 267–309, 1995.
43. C. J. Allegre, T. Staudacher and P. Sarda, Rare gas systematics: formation of the atmosphere, evolution and structure of the earth's mantle, *Earth. Planet. Sci. Lett.* **81**, 127–50, 1987.
44. S. J. G. Galer, S. L. Goldstein and R. K. O'Nions, Limits on chemical and convective isolation in the earth's interior, *Chem. Geol.* **75**, 257–90, 1989.
45. G. Turner, The outgassing history of the earth's atmosphere, *J. Geol. Soc. London* **146**, 147–54, 1989.
46. K. P. Jochum, A. W. Hofmann, E. Ito, H. M. Seufert and W. M. White, K, U and Th in midocean ridge basalt glasses and heat production, *Nature* **306**, 431–6, 1986.
47. H. Craig, W. B. Clarke and M. A. Beg, Excess ^{3}He in deep water on the East Pacific Rise, *Earth Planet. Sci. Lett.* **26**, 125, 1975.
48. K. A. Farley, E. Maier-Reimer, P. Schlosser and W. S. Broecker, Constraints on mantle ^{3}He fluxes and deep-sea circulation from an ocean general circulation model, *J. Geophys. Res.* **100**, 3829–39, 1995.
49. R. I. Hill, I. H. Campbell, G. F. Davies and R. W. Griffiths, Mantle plumes and continental tectonics, *Science* **256**, 186–93, 1992.
50. M. Stein and A. W. Hofmann, Mantle plumes and episodic crustal growth, *Nature* **372**, 63–8, 1994.
51. J. D. Morris, W. P. Leeman and F. Tera, The subducted component of island arc lavas: Constraint from Be isotopes and B–Be systematics, *Nature* **344**, 31–6, 1990.
52. T. Ishikawa and E. Nakamura, Origin of the slab component in arc lavas from across-arc variation of B and Pb isotopes, *Nature* **370**, 205–8, 1994.
53. D. Ben Othman, W. M. White and J. Patchett, The geochemistry of marine sediments, island arc magma genesis, and crust-mantle recycling, *Earth Planet. Sci. Lett.* **94**, 1–21, 1989.
54. C. Chauvel, A. W. Hofmann and P. Vidal, HIMU-EM: the French Polynesian connection, *Earth Planet. Sci. Lett.* **110**, 99–119, 1992.
55. M. Rehkamper and A. W. Hofmann, Recycled ocean crust and sediment in Indian Ocean MORB, *Earth Planet. Sci. Lett.* **147**, 93–106, 1997.
56. D. P. McKenzie and R. K. O'Nions, Mantle reservoirs and ocean island basalts, *Nature* **301**, 229–31, 1983.
57. D. McKenzie and R. K. O'Nions, The source regions of ocean island basalts, *J. Petrol.* **36**, 133–59, 1995.
58. A. W. Hofmann and S. R. Hart, An assessment of local and regional isotopic equilibrium in the mantle, *Earth Planet. Sci. Lett.* **38**, 4–62, 1978.
59. N. H. Sleep, Tapping of magmas from ubiquitous mantle heterogeneities: an alternative to mantle plumes?, *J. Geophys. Res.* **89**, 10 029–41, 1984.

60. M. J. Cordery, G. F. Davies and I. H. Campbell, Genesis of flood basalts from eclogite-bearing mantle plumes, *J. Geophys. Res.* **102**, 20179–97, 1997.
61. M. Gurnis and G. F. Davies, Mixing in numerical models of mantle convection incorporating plate kinematics, *J. Geophys. Res.* **91**, 6375–95, 1986.
62. U. R. Christensen, Mixing by time-dependent convection, *Earth Planet. Sci. Lett.* **95**, 382–94, 1989.
63. U. R. Christensen, Reply to comment on 'Mixing by time-dependent convection', *Earth Planet Sci. Lett.* **98**, 408–10, 1990.
64. M. Gurnis, The effects of chemical density differences on convective mixing in the earth's mantle, *J. Geophys. Res.* **91**, 11407–19, 1986.
65. M. Gurnis, Stirring and mixing in the mantle by plate-scale flow: large persistent blobs and long tendrils coexist, *Geophys. Res. Lett.* **13**, 1474–7, 1986.
66. M. Gurnis and G. F. Davies, The effect of depth-dependent viscosity on convective mixing in the mantle and the possible survival of primitive mantle, *Geophys. Res. Lett.* **13**, 541–4, 1986.
67. G. F. Davies and M. A. Richards, Mantle convection, *J. Geol.* **100**, 151-206, 1992.
68. G. F. Davies, Comment on 'Mixing by time-dependent convection' by U. Christensen, *Earth Planet. Sci. Lett.* **98**, 405–7, 1990.
69. M. Gurnis, Convective mixing in the earth's mantle, Ph.D. Thesis, Australian National University, 1986.
70. A. E. Ringwood, Phase transformations and their bearing on the constitution and dynamics of the mantle, *Geochim. Cosmochim. Acta* **55**, 2083–110, 1991.
71. A. E. Ringwood and T. Irifune, Nature of the 650-km discontinuity: implications for mantle dynamics and differentiation, *Nature* **331**, 131–6, 1988.
72. A. E. Ringwood, Phase transformations and differentiation in subducted lithosphere: implications for mantle dynamics, basalt petrogenesis, and crustal evolution, *J. Geol.* **90**, 611–43, 1982.
73. M. A. Richards and G. F. Davies, On the separation of relatively buoyant components from subducted lithosphere, *Geophys. Res. Lett.* **16**, 831–4, 1989.
74. R. J. O'Connell, Mantle structure, material transport and geochemical reservoirs, *Eos, Trans. Amer. Geophys. Union* **77**, F780, 1996.
75. G. F. Davies, Mantle plumes, mantle stirring and hotspot chemistry, *Earth Planet. Sci. Lett.* **99**, 94–109, 1990.
76. F. Albarede, Time-dependent models of U–Th–He and K–Ar evolution and the layering of mantle convection, *Chem. Geol.* **145**, 413–29, 1998.
77. G. F. Davies, Geophysically constrained mantle mass flows and the ^{40}Ar budget: a degassed lower mantle?, *Earth Planet Sci. Lett.* **166**, 149–62, 1999.

CHAPTER 14

Evolution

14.1 Tectonics and heat

Tectonics and the transport of heat through the mantle are intimately related. In the picture developed in Part 3, plate tectonics and plumes are forms of mantle convection, each is therefore a form of heat transport, and each is also a tectonic mechanism. This is the most direct connection. Plumes are the mechanism by which heat from the core is transported into the mantle and plates are the mechanism by which heat is removed from the mantle. Plate tectonics is the dominant tectonic mechanism of the earth and plumes are an important secondary mechanism.

If we understand the mechanisms by which the mantle transports heat, then it is possible to calculate the rate at which the mantle's heat will change, under given assumptions. That is, we can calculate the temperature as a function of time, or the thermal history of the mantle. A different aspect of the relationship between heat and tectonics then comes into play, because there are reasons to suspect that the present tectonic mechanisms might not have been able to operate in the past when the mantle was probably hotter. We must then ask what tectonic mechanisms might have operated instead, or in other words, how might the mantle have transported its heat. Any proposed tectonic mechanism must then satisfy two fundamental requirements: it must be dynamically viable, that is there must be appropriate forces available to drive it, and it must be capable of transporting heat at a sufficient rate to cool the mantle. The latter requirement arises from the geological evidence that the mantle was hotter in the past, and it is sufficiently stringent to throw some dynamically attractive possibilities into doubt.

This question of past tectonic mechanisms is a first-order question about the earth. Since the advent of the theory of plate tectonics, there has been a lively debate amongst geologists as to how far back in earth history it can be traced. Initially the dominant approach was minimalist: plate-tectonic signatures could be seen in the Phanerozoic, but there was scepticism that it could be seen in earlier eras. More recently opinion seems to have swung in the other direction, supposing that plate tectonics can be identified right back through the Archean. I regard the question of the nature of Archean tectonics to be quite open, observationally and dynamically. If there was plate tectonics of a sort, then it must have been significantly different from the present. The tectonic regime may have been quite different. It may have been episodic. Indeed, it may have been catastrophically episodic, although that is quite conjectural at present.

So, we have a good theory of the present tectonic regime, and it seems to have operated for at least the past billion years, perhaps the last two billion years [1]. Beyond that we do not have a reliable theory, yet much of the present continental crust dates from those early eras, and that crust contains important suites of mineral deposits. We are lacking a theory of the fundamental context in which they formed. Our understanding of the present regime is by now sufficiently good, I claim, to allow us sensibly to propose and test quantitative theories of past regimes.

In this chapter we discuss the thermal evolution of the mantle and its implications for tectonic evolution. Thermal evolution is controlled by the way heat is transported into and out of the mantle by convection, so we start with that topic. We then look at how the mantle's temperature might have changed with time under various alternative assumptions about how convection might have operated in the past. Finally we look at potential tectonic implications of some of these possibilities.

14.2 Review of heat budget, radioactivity and the age of earth

We saw in Chapter 10 that nearly 90% (36 TW) of the total heat lost from the earth (41 TW) emerges from the mantle, the balance being generated in the upper continental crust by radioactivity. Most of the mantle heat is lost through the plate-tectonic cycle, the rest conducting through the continental lithosphere.

As I noted in Sections 7.5 and 12.1.1, the sources of this heat are not entirely clear. A summary of the earth's heat budget is given in Table 14.1. Some of the heat is from the slow cooling of the

Table 14.1. *The earth's heat budget.*

		Heat Flow (TW)	%
Earth heat loss		41	100
Upper crust radioactivity		5	12
Mantle heat loss		36	88
Heat sources:			
Mantle cooling (70 °C/Ga)	9.3		
Core cooling (carried by plumes)	3.5		
Upper mantle radioactivity	1.3		
Total		14	34
Balance unaccounted for		22	54
Lower mantle radioactivity hypotheses:			
Chondritic heat sources ($K/U = 2 \times 10^4$)		11	
80% chondritic, 20% MORB		14	
30% chondritic, 70% MORB		22	
100% MORB		27	

mantle. From the kind of thermal evolution models to be presented in Sections 14.5 and 14.7, the mantle is estimated to be cooling at a rate of about 70 °C/Ga. If plumes come from the base of the mantle, then they carry heat from the core, presumably due to cooling of the core, as discussed in Chapter 11. The other identified source of heat is radioactivity. The radioactivity of the upper mantle is inferred from mantle rocks carried to the surface and from the radioactivity of midocean ridge basalts (MORBs). It is only a small source of heat. The various contributions to the mantle heat budget are summarised in Table 14.1.

The radioactivity of the lower mantle is not very well constrained. Several possibilities have been or might be considered. Hotspot basalts, inferred to be derived from mantle plumes, carry significantly higher concentrations of 'incompatible' trace elements (Chapter 13), and a plausible reason for this is that the deep mantle has a higher proportion of subducted oceanic crust, perhaps because it is denser and has settled towards the bottom of the mantle [2]. It is thus possible that a proportion of the lower mantle has a radioactive heat generation rate of about 10 pW/kg, like MORB (oceanic crust, Table 7.2). On the other hand it was widely assumed in the 1980s that the lower mantle had abundances of refractory trace elements like those of chondritic meteorites. The justification for this is now not considered to be strong, but there are arguments that the lower mantle, or much of it, is less depleted in these elements than the upper mantle [3]. An upper limit on this possibility would then be the chondritic values.

Table 14.1 includes some estimates of lower mantle radioactivity that illustrate these hypotheses. If the lower mantle radioactivity were close to chondritic, it would generate 11 TW. It is plausible, from present subduction rates, that the lower mantle could contain as much as 20% of subducted oceanic crust. Including 20% MORB component would increase the lower mantle heat generation to 14 TW. If the lower mantle were 100% subducted MORB, it would generate 27 TW. To make up the balance of 22 TW required to match the surface heat loss would require the lower mantle heat production to be like that of 30% chondritic and 70% MORB. Such a composition of the lower mantle is considered to be unlikely by many geochemists, but it has its defenders [4].

This apparent mismatch could mean any of three things. Either the composition of the lower mantle is indeed different from what many geochemists currently infer, or there is another source of heat, or the earth at present is cooling more rapidly than is assumed for Table 14.1. Although it is physically conceivable that there is still iron settling out of the mantle into the core, with an accompanying release of gravitational energy, this is not usually considered capable of meeting strong geochemical constraints. The possibility that the earth is far from thermal steady state will be taken up in Section 14.7.

Table 14.1 gives estimates of present rates of heat generation, but we can also infer past rates from the known rates of radioactive decay of the main heat producing elements. These are given in Table 7.1, in terms of radioactive half lives of the relevant isotopes. The total heat production rate as a function of time is then

$$H = \sum_{i=1}^{4} h_i \exp[\lambda_i(t_E - t)] \tag{14.2.1}$$

where h_i is the current rate of heat production by the ith isotope, λ_i is its half life, t_E is the age of the earth and t is time.

The age of the earth is about 4.57 Ga. Although this is a well-known result, it is not known very directly. It is inferred from the fact that the estimated lead isotope composition of the earth is similar to that of many meteorites, whose lead isotopes define an isochron of the above age. This would then be the age of condensation of meteoritic material. Estimates of the lead isotopic composition of the silicate parts of the earth have some uncertainty, but are discernibly different from the meteorite isochron, and correspond to an age range of 4.47–4.52 Ga [5, 6]. This would plausibly correspond to the age of separation of the mantle from the core, at which time the uranium–lead ratio of the silicates would have been established. Current theories of the formation of the earth make it seem likely that core separation coincided with the later stages of the accretion of the earth from meteoritic material. Thus a reasonable estimate of the age at which the earth attained close to its final mass is about 4.5 Ga [6].

14.3 Convective heat transport

The simple theory of convection developed in Section 8.3 was used to estimate the velocity of convection, the boundary layer thickness and the surface heat flux. We can use this theory to calculate the way these quantities change as the temperature of the mantle changes. It is thus possible to calculate the rate at which heat is removed from the mantle, by the action of plate tectonics, as a function of mantle temperature. This is straightforward and the formula is given below. However, we must also consider the other identified mode of mantle convection, the plume mode, and its role in heat transport. It is possible to develop a theory analogous to that of Section 8.3 for the plate mode, and this is also done below. It turns out that there are significant differences between the results for the plate and plume modes. The strong temperature dependence of the viscosity of mantle rocks (Section 6.10) is the main contributor to the temperature dependence of mantle heat transport, so this is reviewed here.

14.3.1 Plate mode

The total heat loss through the earth's surface is $Q_\mathrm{m} = 4\pi R^2 q$, where R is the earth's radius and q is the average surface heat flux. Equation (8.3.9) for the heat flux transported by the plate mode of convection can then be used to write

$$Q_\mathrm{m} = 4\pi R^2 K \left(\frac{g\rho\alpha}{\kappa}\right)^{1/3} \frac{\Delta T^{4/3}}{\mu^{1/3}} \qquad (14.3.1)$$

The expression is put in this form to highlight the dependence on the temperature difference, ΔT, across the top thermal boundary layer and on the mantle viscosity μ.

14.3.2 Effect of temperature dependence of viscosity

The temperature dependence of the ductile rheology of rocks was expressed in the form of Equation (6.10.3), and this implies that in the case of linear rheology the viscosity has the form

$$\mu = \mu_0 \exp(T_A/T) \tag{14.3.2}$$

where $T_A = H^*/R = (E^* + PV^*)/R$, E^* is the activation energy, P is pressure, V^* is the activation volume, H^* is the activation enthalpy, R is the gas constant and μ_0 is a reference viscosity. The variation of viscosity with temperature was illustrated in Figure 6.18 for activation energies of 200 kJ/mol and 400 kJ/mol. However, in the deep mantle the variation is likely to be even stronger because of the pressure term in the activation enthalpy. H^* is possibly in the range 500 to 800 kJ/mol near the base of the mantle, yielding T_A in the range 6×10^4 °C to 10^5 °C. The variation of viscosity resulting from an activation enthalpy of 600 kJ/mol is illustrated in Figure 14.1, with the low-pressure curves of Figure 6.18 included for comparison. The deep mantle curve is normalised to a viscosity of 10^{23} Pa s at the reference temperature of 1300 °C, compared with a viscosity of 10^{21} Pa s for the shallow mantle curves.

We will see in Section 14.5 that the effect of the strong temperature dependence of the viscosity is to limit the range of variation of mantle temperature. In effect, large changes in heat transport can be accomplished by moderate changes in mantle temperature. We have already seen in Chapter 11 that this strong temperature dependence is the reason that plume tails are believed to be narrow.

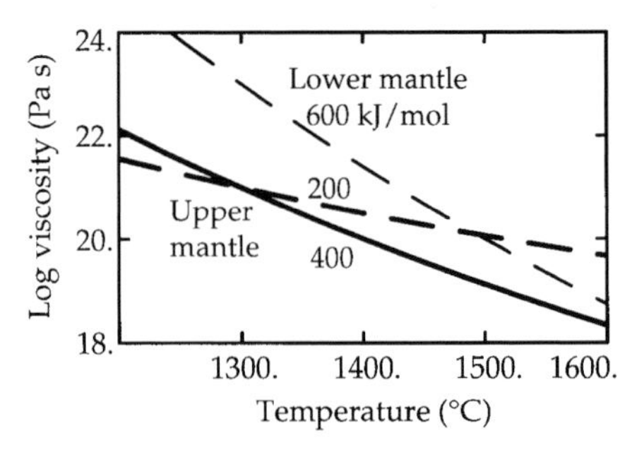

Figure 14.1. Estimates of viscosity dependence on temperature in the deep mantle (light dashed) and shallow mantle (heavy). Assumed activation energies are shown. Reference viscosities at 1300 °C are 10^{23} Pa s (lower mantle) and 10^{21} Pa s (upper mantle).

14.3.3 Plume mode [*Intermediate*]

An expression analogous to Equation (14.3.1) can also be derived for the heat transported by a plume tail, but the geometry is different and it matters that the viscosity within the plume is likely to be lower than the surroundings by up to a factor of 100 or so. The following approach was developed by Stacey and Loper [7], though it is slightly simplified here. A key point is that the viscosity is assumed to be strongly temperature-dependent, as is appropriate for rocks (Section 6.10.2). This means not only that the hot plume material and the hot bottom boundary layer will have lower viscosity than the surrounding mantle, but that the flow in the boundary layer will be concentrated strongly towards the base of the layer, where the temperature is highest and the viscosity lowest. We should therefore distinguish the thermal boundary layer, with thickness δ, from a velocity boundary layer, of thickness h, in which the horizontal flow is concentrated. This is sketched in Figure 14.2.

The volumetric flow rate, Φ_{p}, up the plume can be taken from the expression (6.7.5) for viscous flow through a pipe of radius a (Figure 14.2):

$$\Phi_{\mathrm{p}} = \frac{\pi g \rho \alpha \Delta T a^4}{8\mu} \qquad (14.3.3)$$

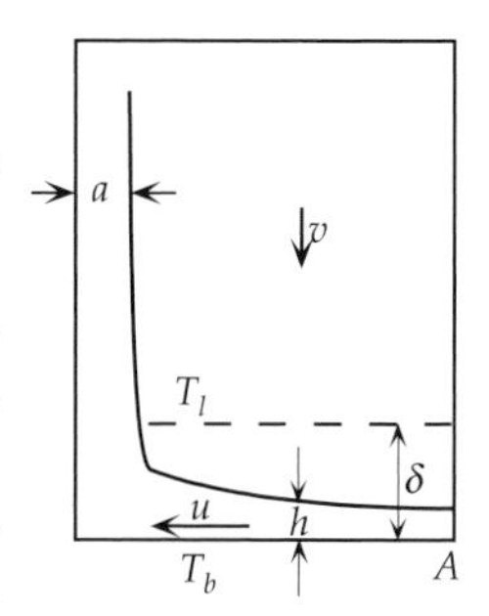

Figure 14.2. Sketch of a plume (radius a) fed by a low-viscosity boundary layer of thickness h. The plume axis is on the left. The thickness of the thermal boundary layer is δ. The hot, low-viscosity fluid in the velocity boundary layer flows into the plume with velocity u, and the fluid returns outside the plume with velocity v.

where μ is the plume viscosity, $\Delta T = (T_{\mathrm{b}} - T_{\mathrm{l}})$ is the temperature difference between the bottom boundary and the interior of the mantle, and $\rho\alpha\Delta T$ is the resulting density difference.

The plume is fed by a flow through the bottom boundary layer (Figure 14.2) that is driven by a pressure gradient set up by the plume. We saw in Section 6.7.1 that the volumetric flow rate driven by a pressure gradient, P', in a thin layer is proportional to the cube of the thickness, h, of the layer. For the case with a no-slip top boundary and a free-slip bottom boundary, the result has a different numerical factor: $P'h^3/3\mu$. Applying this result around a perimeter at distance r from the base of the plume, the flow rate in the boundary layer through that perimeter is

$$\Phi_{\mathrm{b}} = \frac{2\pi r h^3 P'}{3\mu}$$

One way to think of the pressure gradient is that it is due to the sloping interface between the lower-density boundary layer and the overlying mantle. This slope is in turn due to the pull from the base of the buoyant plume. We only need the way the pressure gradient scales, rather than an accurate expression for it. If h is a representative thickness, then at a distance r a representative gradient of the interface is h/r. The pressure gradient is related to this though

$$P' = g\Delta\rho h/r$$

The boundary layer flow rate is then

$$\Phi_b = \frac{2\pi g\Delta\rho h^4}{3\mu} \tag{14.3.4}$$

Conservation of mass requires that Φ_p and Φ_b are equal, from which a simple relationship follows:

$$a^4 = 16h^4/3 \tag{14.3.5}$$

Conservation of mass also requires that there be a downward return flow outside the plume. If it has an average velocity v out to a radius A (Figure 14.2), then

$$\Phi_p = \pi A^2 v \tag{14.3.6}$$

The thickness of the bottom thermal boundary layer is controlled in this situation by a competition between the diffusion of heat upward from the bottom boundary and advection of heat downwards by the slow return flow. We looked at this situation in Section 7.8.2. The relevant length scale is

$$\delta = \kappa/v \tag{14.3.7}$$

The thickness of the velocity boundary layer is proportional to δ, and for the moment I leave it just in terms of their ratio, ε:

$$h = \varepsilon\delta \tag{14.3.8}$$

The heat transported by a plume is the volume flow rate, Φ_p, times the heat content, $\rho C_P \Delta T$ (Equation (7.7.1)). There are approximately 40 identified plumes. I express their number as the ratio of the surface area of the earth's core, $4\pi R_c^2$, to the average area of a plume feeding zone, πA^2. In other words, $4\pi R_c^2/\pi A^2 = 40$, which yields $A = 1100$ km. Then the rate at which plumes transport heat is

$$Q_p = 4R_c^2 \rho C_P \Delta T \Phi_p / A^2$$

Using Equations (14.3.3–8), this can be written

$$Q_p = 4\pi R_c^2 K \left(\frac{2g\rho\alpha}{3\kappa A^2}\right)^{1/5} \varepsilon^{4/5} \frac{\Delta T^{6/5}}{\mu^{1/5}} \tag{14.3.9}$$

This has the same general form as Equation (14.3.1), although the exponents are smaller. This means that heat transport by plumes is somewhat less sensitive to temperature than heat transport by plates, but the dependence is still fairly strong.

I have derived elsewhere [8] an analogous formula for the rate of heat transport by plume heads. It has the same form except for a small factor, and there should as well be a factor of the order of 1 that accounts for details of the derivation not included in the approximations of the kind made here. Hill *et al.* [9] estimated from the frequency of known flood basalts in the geological record, assumed to be caused by the arrival of plume heads (Section 11.5), that plume heads transport roughly 50% of the heat that is carried by plume tails. Thus the total heat transported by plumes (heads and tails) at present is about 3.5 TW, or 10% of the mantle heat loss (Table 14.1). We can therefore lump the heat transport by plume heads and plume tails together and assume they are both governed by a formula of the form of Equation (14.3.9).

It remains to specify the ratio $\varepsilon = h/\delta$ of the velocity to thermal boundary layer thicknesses. This depends on the viscosity contrast between the boundary and the interior, which depends in turn on the temperature contrast. Stacey and Loper showed that it is inversely proportional to the change in the logarithm of the viscosity, $\Delta[\ln(\mu)]$, which is approximately $\Delta T[\partial \ln(\mu)/\partial T]$. Then

$$\varepsilon = -1/\Delta T[\partial \ln(\mu)/\partial T]$$

With the dependence of μ on T given by Equation (14.3.2), ε becomes

$$\varepsilon = T_1^2/T_A(T_b - T_1) \tag{14.3.10}$$

14.4 Thermal evolution equation

The rate of change of temperature of a layer of the earth (the mantle, part of the mantle, or the core) can be calculated if the net rate of gain or loss of heat can be calculated. The rate of change of heat content due to a change of temperature is $MC\partial T/\partial t$, where M is the mass of the layer, C is the specific heat (at constant pressure) of the material comprising the layer, T is temperature and t is time. Thus if the rate of heat loss through the top of the layer is Q_{out}, the rate of heat input through the base of the layer is Q_{in}, and the rate of radioactive heating per unit mass is H, then

$$\frac{\partial T}{\partial t} = \frac{H}{C} + \frac{Q_{in} - Q_{out}}{MC} \tag{14.4.1}$$

For example, for the mantle Q_{in} might be identified with the plume flux, Q_p (Equation (14.3.9)), and Q_{out} with the heat loss due to plate-scale convection, Q_m (Equation (14.3.1)). H can be estimated from Equation (14.2.1). For the core, H is usually taken to be zero and no other heat input is known, so Q_{in} would also be zero, while Q_{out} might be identified with Q_p, the plume flux. Since Q_m and Q_p are given as functions of temperature, Equation (14.4.1) can be integrated to give temperature as a function of time (given also a starting temperature).

14.5 Smooth thermal evolution models

The results of such a calculation are shown in Figure 14.3. This example quantifies the sequence depicted in Figure 11.15, in which the core and mantle start off with equal high temperatures at the core–mantle boundary and the mantle subsequently cools more quickly. Three temperatures are shown in Figure 14.3a: the temperature near the top of the mantle (T_u), near the bottom of the

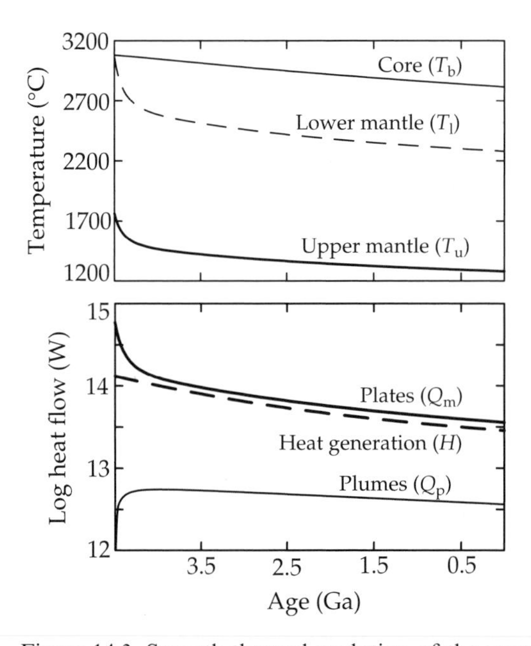

Figure 14.3. Smooth thermal evolution of the mantle and core.

mantle (T_l) and at the core–mantle boundary (T_b). T_l is greater than T_u because of the adiabatic temperature gradient that exists if the mantle convects (Section 7.9.3), and the behaviour of T_l is similar to the behaviour of T_u. In this model, parameters have been adjusted so that the present heat flows match the observed values (Table 14.1) and the present T_u matches the inferred upper mantle temperature of about 1300 °C. The core temperature, T_b, is not very well constrained at present, and so the particular value used in this model does not have great significance; the significant feature of the model is how T_b changes with time (see below).

Initially $T_l = T_b$ but the mantle cools rapidly within the first few hundred million years. This phase of rapid cooling then gives way to slower cooling. The reason can be seen in Figure 14.3b, which shows heat flow versus time. The heat loss from the mantle is given by Q_m, and the radioactive heat generation by H. Initially Q_m is much higher than H, and the mantle cools rapidly, until Q_m approaches H. If H were constant, the mantle would approach a steady state, with Q_m equal to H. However H declines slowly and Q_m follows it, the difference between them reflecting the rate of cooling. During the slow cooling phase, the mantle temperature declines by only about 200 °C over 4 Ga, while the heat loss rate drops by a factor of about 3. The small change in temperature is the result of the strong temperature dependence of the mantle viscosity, which allows the change in Q_m to be accommodated by only a 15% change in temperature: as the temperature drops, the buoyancy force decreases marginally, but the viscosity increases by a factor of about 30 and convection slows accordingly.

Also shown in Figure 14.3b is the heat transported by plumes coming from the thermal boundary layer at the base of the mantle. Initially there is no thermal boundary layer, because T_l equals T_b. The rapid cooling of the mantle then creates a temperature difference that will give rise to a thermal boundary layer and to plumes (Section 11.7). Correspondingly, the plume heat flow (Figure 14.3b) rises rapidly from zero. It then stays remarkably constant for the remainder of the evolution of the model, decreasing by only about 40% from its early maximum. This is because the rates of cooling of the core and the mantle happen to be very similar, so that ($T_b - T_l$) remains fairly constant. This in turn is because plumes carry a relatively modest amount of heat, so the core cools only slowly, approximately matching the slow cooling of the mantle, which is paced by the decline of radioactive heating. There is no fundamental requirement for this result, and models in which the plume flow varies substantially are possible [8].

This model demonstrates several important points. There is a transient adjustment period lasting a few hundred million years during which the heat loss rate declines towards the heat generation rate. Thereafter the thermal regime of the mantle is controlled by the slow decline of radioactive heat sources. During this phase, the mantle temperature declines by only about 200 °C, reflecting the strong temperature dependence of mantle viscosity. If the plume heat transport is small at present, then it may have been fairly constant through earth history.

14.6 Age distribution of the continental crust

In contrast to the thermal evolution model of Figure 14.3, the age distribution of the continental crust is not at all smooth, as Figure 14.4 shows. The age distribution is quite uneven, and it has some pronounced peaks, notably at about 1.9 Ga and 2.7 Ga. This unevenness has been remarked upon particularly since about 1960, when sufficiently reliable radiometric dates became available in sufficient quantity [10]. It is easy to jump to the conclusion that the earth's tectonic activity has been episodic, with bursts of activity corresponding to the peaks in the age distribution. However, we must be cautious in interpreting this observation. There are in principle three possible explanations, which I will discuss shortly.

The uneven age distribution is perhaps more surprising since it has been a common idea that the growth of continental crust is closely related to plate tectonics. The idea is that new crust is created mainly at island arcs, where subducting lithosphere generates magmas from the mantle that ultimately become processed and incorporated into the continental crust [12]. If plate tectonics is the main expression of mantle convection, and mantle convection

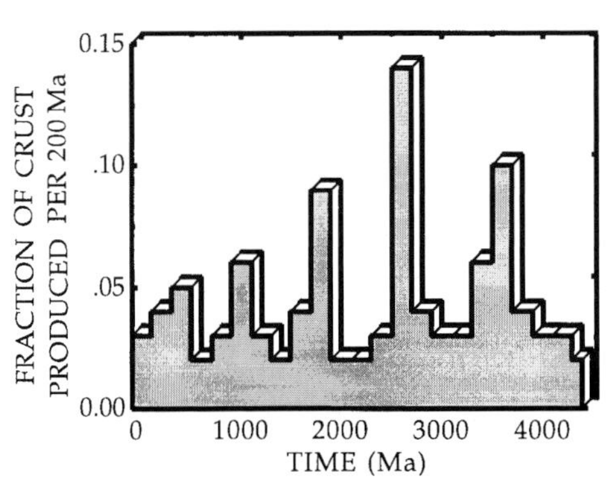

Figure 14.4. Age distribution of the continental crust. From McCulloch and Bennett [11]. Copyright by Elsevier Science. Reprinted with permission.

proceeds as in Figure 14.3, then we might expect the crust to have accumulated rather smoothly.

The first possible explanation for the uneven age distribution is that the sampling of the earth's continental crust might be incomplete. The age distribution might be more uniform, but the parts with the ages not represented in Figure 14.4 may not yet have been sampled. Some decades ago, this was a serious possibility, but by now most continents have been sampled by sufficiently reliable dating that it seems unlikely that there are large tracts of crust with ages between the peaks of Figure 14.4. The main features of the age distribution have not changed substantially over the past two decades or so as new data have accumulated. It is commonly found, as well, that the continental crust is divided into quite large blocks of fairly uniform age, with dimensions of the order of a few hundred kilometres, so the chances of missing large parts of the age distribution seem even smaller.

A second possibility is that rate of formation of the continental crust has been uniform in time, but its preservation has been uneven. Thus a large and intense tectonic event, such as the uplift of Tibet cause by the collision of India with Asia, may cause the erosion, dispersal and/or age-resetting of one or more pre-existing age provinces, leaving a gap in the age distribution. It is less obvious that this is unimportant in creating the uneven observed age distribution. However, the number of age provinces is still moderately large, and it seems hard to account for the larger features of Figure 14.4 in this way.

The third possibility is that the continental crust has not been generated uniformly in time. According to the above discussion, we must take this possibility seriously. In that case, it is not obvious how a very smooth thermal evolution of the mantle like that in Figure 14.3 could be consistent with such episodic tectonic activity. Some possibilities will be presented in the following sections.

14.7 Episodic thermal evolution models

It seemed once that the thermal evolution of the earth's interior would inevitably be very smooth and monotonic, like that depicted in Figure 14.3, and that the explanation for the non-uniform age distribution of the continental crust must lie elsewhere. However, it has been demonstrated that a phase transformation in the transition zone with a negative Clapeyron slope can induce episodic layering in some models of mantle convection [13]. An example of this is given in Figure 10.6.

That example is an extreme case, in which for a time there is nearly complete separation of the flow above and below the phase barrier, followed by a complete overturn, in which the cooler upper layer completely drains into the lower layer. Such complete overturns tend to occur in two-dimensional models with constant viscosity. In three-dimensional, constant-viscosity models, the separation is less complete, and 'breakthroughs' occur more frequently and more locally [14]. At the time of writing, three-dimensional models that include reasonable simulations of plates and plumes have not been presented, so it is not clear where in this range of behaviour the mantle might fall. Nor is it clear, as discussed in Chapters 5 and 10, that the spinel–perovskite transformation actually has a sufficiently negative Clapeyron slope to induce layering. There is, however, another possible mechanism that will be mentioned in the next section.

With these cautions, it is instructive to look at the potential implications of such episodic layering, and to enquire whether it can lead to behaviour that has any resemblance to the geological record. One way to do this is to incorporate criteria for whether the mantle would be layered or not into the kind of thermal evolution calculation presented above. How this should be done is not completely established. Rather than give a detailed discussion of this, I only want here to illustrate some possibilities. Accordingly I only summarise the main ideas. The details of these particular calculations are given elsewhere [15].

A phase transformation with a negative Clapeyron slope tends to resist the penetration of cool fluid (from above) or hot fluid (from below). If penetration is sufficiently resisted to induce layering (Figure 10.6), then there will be, in general four thermal boundary layers in the system: one each at the top and bottom of each layer. Each of these might be the agent that breaks through the phase barrier, as illustrated in Figure 14.5. Fluid from the top or bottom boundary layer might impinge with sufficient force to break through, or fluid from either of the internal boundary layers might detach with sufficient force to pull material through. In the models to be presented here, only two of these mechanisms were included. The effect of plumes from the base of the mantle was neglected because plumes carry only a secondary heat flow (though this might not be a sufficient reason). The effect of the boundary layer at the base of the upper layer was neglected because it would have a relatively low viscosity, whereas the other internal boundary layer, immediately beneath the interface, would have a relatively high viscosity. The latter boundary layer would tend to be more sluggish, thicker and it would couple more strongly to the interface.

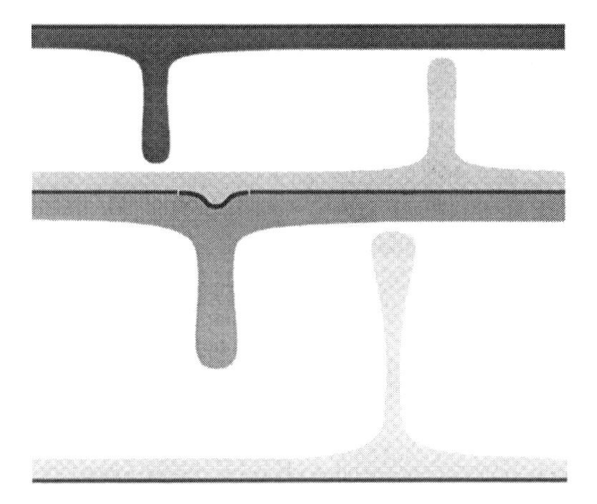

Figure 14.5. Mechanisms for breaching a phase barrier in the layered mode of convection. Rising or descending columns from any of the four thermal boundary layers might breach the internal boundary between the two fluid layers.

For these reasons, the boundary layers at the top of each fluid layer were assumed to be the most important in controlling the timing of overturns.

Three important parameters in this system are (i) the age of lithosphere that can penetrate the phase barrier, (ii) the temperature difference between the layers at which an overturn is triggered by the internal boundary layer, and (iii) the efficiency of heat transport through the interface between the layers. Lithosphere that is older and thicker at the time of subduction is less likely to buckle when it impinges on the phase barrier, and it is then more likely to penetrate. Now the maximum age of lithosphere at subduction, τ_s, depends on the rate of convective overturn, which depends ultimately on the temperature of the mantle, so it turns out that a particular value of τ_s corresponds to a particular value of upper mantle temperature. Nevertheless, I express this criterion for breakthrough in terms of τ_s.

The vigour of the internal thermal boundary layers depends on the temperature difference between the fluid layers. We would thus expect that as the temperature difference increases, a breakthrough and overturn becomes more likely. For the calculations, it was assumed in effect that breakthroughs occur when the temperature difference reaches a critical value, $\Delta T_c = (T_l - T_u)$, of about 250 °C.

The third important parameter, the efficiency of heat transport through the interface, depends on the thickness of the internal thermal boundary layers, and in particular on the thickness of the top boundary layer of the lower fluid layer. Because this boundary layer is cooler than the interior of the lower fluid layer, it will be stiffer. If it is sufficiently cool, its cooler parts may become effectively static. Convection driven by this kind of thermal boundary layer has been studied by Davaille and Jaupart [16]. They found that the heat transport is given by a formula like Equation (14.3.1), but with a temperature difference, ΔT, corresponding only to the hotter, mobile part of the thermal boundary layer. This value depends on the temperature dependence of the viscosity, and they found that a useful measure is $\Delta T = -\mu/(\partial\mu/\partial T)$. (This is equivalent to the same fraction, ε, of the total thermal boundary layer as was used in Section 14.3.3; see precursor to Equation (14.3.10).) There is also an empirical multiplying factor, which they found to have a value of $b = 0.47$. Because the precise conditions of the mantle may not match those of the experiments from which the value of b was derived, we can regard the value of b as a moderately adjustable parameter, and its effect is to control the efficiency of heat transfer between the fluid layers.

In order to clarify the behaviour of the system, it is useful initially to simplify it even further by including only one of the mechanisms for breaching the phase barrier. Thus the lithosphere penetration of the phase barrier is effectively turned off for the moment by taking τ_s to be very large. Then Figure 14.6 shows three thermal evolution models with different values of b, measuring the efficiency of heat transport through the interface. In Figure 14.6a, the experimentally determined value $b = 0.47$ is taken, and the result is a series of episodes of transient layering terminated by an overturn. Initially, the temperature of the upper mantle drops, because it is cooled efficiently by heat loss at the earth's surface. This behaviour is like the early transient cooling in the smooth model of Figure 14.3. However, the lower mantle does not lose heat very efficiently through the interface, and it warms by radioactive heating until the temperature difference between the layers is great enough to trigger an overturn. At that point in the calculation the temperatures are reset: the upper mantle is set to the former temperature of the lower mantle, while the lower mantle is set to the weighted mean of the former upper mantle and the remainder of the lower mantle. The sequence of layering and overturn then repeats in this model. The interval between overturns increases with time due to the slow decline in the level of radioactive heating.

If the heat transfer between the layers is greater ($b = 0.6$), there comes a time when the temperature difference between the layers

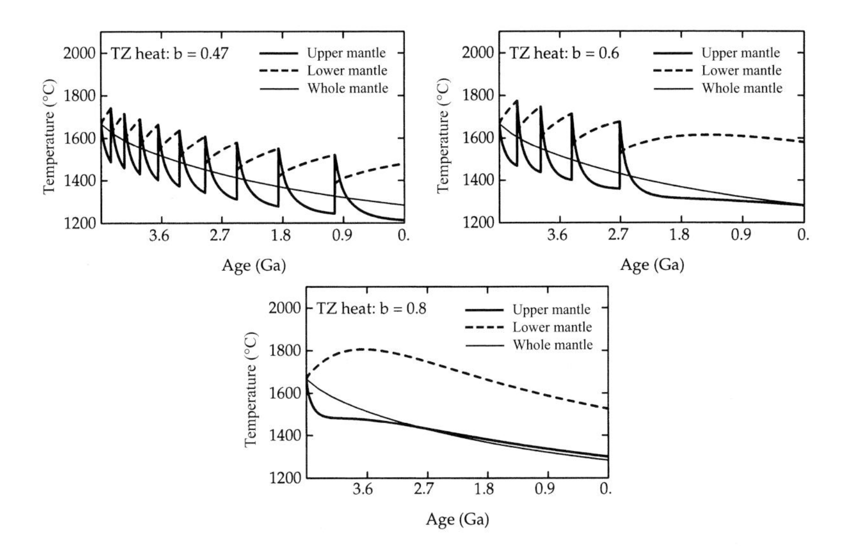

Figure 14.6. Three episodic thermal evolution models in which overturns are triggered only by the temperature difference between the fluid layers. The three cases correspond to different efficiencies of heat transport through the transition zone (TZ) that intermittently forms the internal interface between two fluid layers. Curves for smooth, whole-mantle evolution are included for comparison.

does not rise above the critical value and the layering becomes permanent (Figure 14.6b). In other words the lower mantle can lose heat efficiently enough through the interface that it does not become overheated. If b is even larger (0.8, Figure 14.6c) there are no overturns at all in this approximation: the mantle is permanently layered.

The behaviour in Figure 14.6 is controlled only by the internal boundary layer. If now we add back the possibility that subducted lithosphere can also break through and trigger an overturn, a new style of behaviour occurs. Thus in Figure 14.7 it is assumed that lithosphere can break through if it is older than 70 Ma at the time of subduction. The value of b is 0.6, as in Figure 14.6b, and the early part of the evolution in Figure 14.7 is the same. At about 1.9 Ga, the upper mantle becomes cool enough that τ_S exceeds the critical value of 70 Ma and there is an overturn. Now because the whole system is cooling, the upper mantle reaches the temperature at which τ_S is critical more quickly, and there is a second overturn of this kind before the lower mantle has heated as much as before. Consequently, this and subsequent overturns occur more and more frequently, until the lower mantle temperature is reduced essentially to the upper mantle temperature. At this point it is assumed

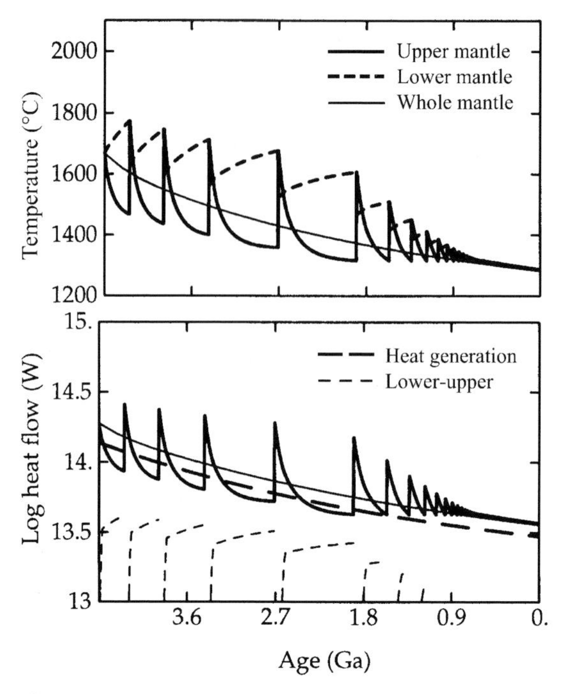

Figure 14.7. A particular episodic thermal evolution model with resemblances to the geological record. From Davies [15]. Copyright Elsevier Science. Reprinted with permission.

that the mantle will convect as a single layer, and the calculation proceeds like that of Figure 14.3. You can see that the series of late overturns triggered by lithosphere penetration has a different pattern than the early overturns triggered by the internal boundary layer.

I posed the question earlier as to whether a model with episodic layering can yield behaviour resembling the geological record. The model shown in Figure 14.7 is notable for showing three phases of behaviour: the early series of overturns triggered by the internal boundary layer, the later series triggered by lithosphere penetration, and a final phase of whole-mantle convection. These phases roughly correspond with the three main geological eras: Archean (before 2.5 Ga), Proterozoic (2.5–0.6 Ga) and Phanerozoic (after about 0.6 Ga). As well, the last of the early series occurs at 2.7 Ga, and the first of the later series at 1.9 Ga, close to the times of the largest peaks in the crustal age distribution (Figure 14.4). Finally, the model moves progressively into whole-mantle convection over a period that corresponds to the emergence of the plate tectonic signal in the geological record.

Although these results are encouraging, the models are still exploratory and the interpretation is conjectural. The particular values of the model parameters were chosen to yield the particular timing of this model, though the parameter values are within the range of plausibility. I should also reiterate that the assumption of complete dichotomy between layering and overturning is the end-member form of the behaviour of numerical models. Thus the significance of these models is not that they make detailed predictions of timing of tectonic events (they do not), but that they yield styles of behaviour that have some important resemblances to the geological record. In particular, I am not aware of any previous models that yield long phases of behaviour in such a straightforward way with such a resemblance to the geological eras. While the potential for episodic overturns was clearly built into the model formulation, these phases were something that emerged unanticipated. I will return to the possible tectonic implications of this kind of model in the later discussion of tectonic evolution.

14.8 Compositional effects on buoyancy and convection

So far in this book I have considered only thermal buoyancy. In other words, I have only considered convective flow driven by density differences due to thermal expansion. From the point of view of understanding how a system like the earth's mantle works, it is better to take things one step at a time rather than to launch directly into the potentially complicated interactions between thermal and compositional buoyancy. Even with thermal buoyancy, the interaction with phase transformations introduces considerable complication into the model behaviour, as the previous section illustrates. As well, it seems that the most fruitful approach to understanding the mantle is to consider thermal buoyancy first.

It is nevertheless clear that density differences due to differences in composition modify mantle dynamics in important ways. It is possible that mantle dynamics is modified in more radical ways that are unverified or unrecognised, particularly in the context of the earlier phases of earth history.

An obvious example, recognised soon after plate tectonics was formulated [17] (though the idea goes at least as far back as Holmes [18]), is that continental crust does not subduct *en masse*. Continental crust has a mean density of about 2700 kg/m^3, compared with the mantle density of about 3300 kg/m^3, and the reason is their different compositions. This means that when a continent is carried into a subduction zone, the plate system has to change. Either the relative motion of the two plates must stop, or the other plate must begin to subduct, or a new subduction zone must form elsewhere. Thus continental buoyancy modifies the details of the flow pattern of the plate-scale convection. In the short term, this may make little difference to the average behaviour of the system. In the longer term, there may be more subtle effects that are important for the history of the continenal crust. This will be taken up below.

There are other possible effects of compositional buoyancy. The oceanic crust is also buoyant relative to the mantle, and this may have determined the viability of plate tectonics in the past. The interaction of subducted oceanic crust with the transition zone is potentially complex, particularly in the past, and it provides another mechanism for episodic layering, distinct from the thermal interactions considered in Section 14.7. There is good evidence for a compositionally distinct layer at the base of the mantle (the D″ layer), and this may also have significant dynamical effects. There may be other possibilities.

What has been emerging is a realisation that the interaction of thermal and compositional buoyancy may have major consequences for mantle dynamics. I briefly summarise some of the possibilities here, but this is a relatively new topic and I want only to indicate some stimulating directions, as this topic is likely to evolve rapidly.

14.8.1 Buoyancy of continental crust

Since the earth's surface is finite, if a continent is carried in any direction sufficiently far, it will eventually encounter another continent. It is thus a likely outcome of plate tectonics that all of the continental crust will accumulate into one 'supercontinent'. This accords with interpretations of the geological record of continents that there have been at least two phases of supercontinent accumulation, and subsequent breakup [19]. It has been conjectured for some time that the presence of a large continent would affect the thermal state of the mantle and hence alter the convective flow pattern.

The mechanism usually appealed to is thermal blanketing due to the higher radioactive heat production in continents, but there are two other mechanisms that are likely to be more important, though the effect will be qualitatively similar. The second mechanism is that the presence of a continent excludes subduction, and thus prevents the local cooling due to the introduction of cold lithosphere into the mantle. Subduction causes the largest temperature changes in the mantle. The third mechanism is that the mobility of the mantle under a large continent is relatively restricted by the non-subducting surface boundary condition that the continent imposes. In contrast, mantle under an oceanic plate can be returned *en masse* to the deep mantle.

The effect of a non-subducting boundary condition on mantle convection has been investigated by Gurnis and Zhong [20, 21]. Their models do not include plates as such, so their results will not strictly be quantitatively accurate, but they are likely to give a good indication of the effect. They find that heat does accumulate under a large simulated continent, that the mantle flow eventually changes to upwelling under the supercontinent in response, and that the continent can be rifted and fragmented by the combined effect of the diverging mantle flow and gravitational sliding off the topographic high that the warmer mantle generates. An example of their results is given in Figure 14.8.

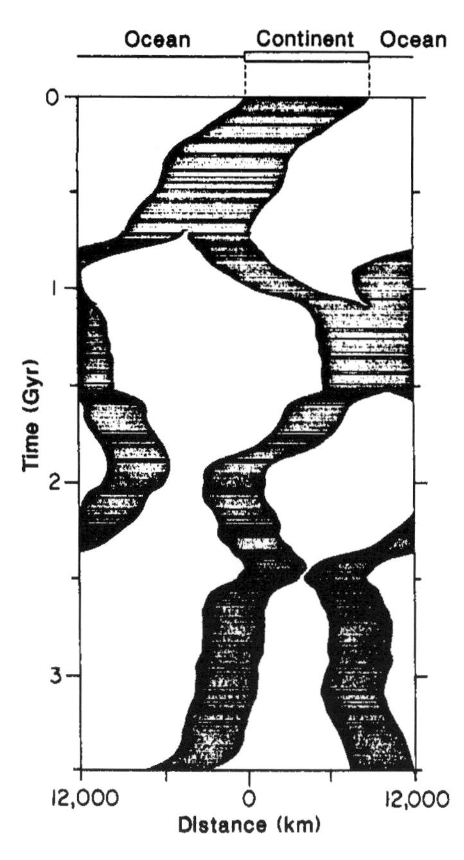

Figure 14.8. Continental aggregation and dispersal. Results from a numerical convection model in which a single large continent (top) fragments and recombines by the action of underlying convection. From Gurnis [20]. Reprinted from *Nature* with permission. Copyright Macmillan Magazines Ltd.

14.8.2 Interaction of oceanic crust with the transition zone

The difference in composition of oceanic crust from the mantle causes the sequence of pressure-induced phase transformations through the transition zone to be different, as was discussed in Chapter 5 (see Figures 5.8, 5.10). As a result, the density of the oceanic crust component of subducted lithosphere is greater than the mantle density at some depths and less at other depths (Figure 5.11; [22–25]). For example, at depths of 60–80 km, the crust transforms to eclogite (density 3500 kg/m^3), which is substantially denser than the mantle (density 3300 kg/m^3). However, at depths immediately below the 660 km discontinuity the oceanic crust component is less dense, and the lithosphere is estimated to have a small net positive compositional buoyancy. Ringwood has conjectured that this would be sufficient to prevent subducted lithosphere from penetrating into the lower mantle [22, 23].

Whether this compositional buoyancy is greater in magnitude than the negative thermal buoyancy in the present mantle is doubtful [26], but in the past it might have been, for two reasons. First, plates would have been younger at the time of subduction and so had lower negative thermal buoyancy, and second the oceanic crust component would have been thicker, so that its contribution to compositional buoyancy would have been greater. We will discuss this in Section 14.8.4. There are substantial uncertainties, but this mechanism might have served to induce layered mantle convection in the past [26].

14.8.3 The D″ layer

The nature and origin of the D″ layer at the base of the mantle is not completely clear. Its properties are unlikely to be accounted for just as a thermal boundary layer [27, 28], so it is likely to be a layer of different composition, and by immediate implication of greater density than the overlying mantle. It has been proposed that it is an accumulation of subducted oceanic crust that has settled to the base of the mantle [2, 29], or that it may be a product of chemical reactions between the mantle and the core [30]. It is conceivably a relic of the earth's formation.

Several possible dynamical roles have been proposed for the D″ layer. It may be entrained into mantle plumes in small amounts, with the potential to modify the ascent of plumes [31, 32]. It would be expected to be swept around by the large-scale mantle flow and affect the topography of the core–mantle boundary and the pattern of heat loss from the core [33].

An important effect of the D″ layer would be less direct: it would partially insulate the core from the mantle. The temperature difference between the core and the mantle would then be accommodated in two steps, the first from the core into the D″ layer, and the second from the D″ layer into the mantle. As a result, the thermal boundary layer from which plumes arise would involve a temperature drop much less than the total temperature difference between the mantle and the core [32]. This may explain a discrepancy between plume models, which require a temperature drop of only 400–500 °C to explain their inferred excess temperature at the top of the mantle, and estimates of the temperature excess of the core of over 1000 °C [34].

Campbell and Griffiths [35] have proposed that the D″ layer may not have begun to accumulate until the late Archean, and that this may explain why inferred plume temperatures were considerably higher during the Archean than in later times. The evidence for this change in temperatures is that the occurrence of komatiites, which are the product of very high degrees of partial melting, is confined almost completely to the Archean and early Proterozoic.

14.8.4 Buoyancy of oceanic crust

The oceanic crust has a thickness of about 7 km and a density of about 2950 kg/m^3, and the formation of the oceanic crust by decompression melting of the mantle under spreading centres leaves a melt residue that is depleted in iron and slightly less dense than normal mantle [22]. As a result, the oceanic lithosphere is initially positively buoyant. Only as it ages and thickens does its negative thermal buoyancy grow to outweigh the compositional buoyancy of the crust and depleted zone. At present, oceanic lithosphere passes through a state of neutral buoyancy at an age of about 20 Ma. Thereafter its net buoyancy is negative, and it is able to subduct if the mechanical conditions permit it; in other words, it could subduct if it arrived at a subuction zone.

In the past, two factors would have conspired to magnify the effect of the compositional buoyancy of oceanic lithosphere. Each factor is a consequence of the mantle being hotter. First, the oceanic crust and the depleted zone would both have been thicker, because there would be a greater degree of decompression melting under spreading centres. Second, the mantle viscosity would have been lower, convection would have been faster, and plates would have arrived at a subduction zone sooner. (The latter conclusion is independent of the size of plates. It can be deduced from the areal rate of seafloor spreading.)

The age at which a plate becomes neutrally buoyant can be estimated as follows (ignoring the buoyancy of the depleted residue; a more detailed estimate is given in [36]). Suppose the oceanic crust has density ρ_c and thickness h, while the lithospheric mantle has density ρ_L and the total plate thickness is d. The mean density of the plate will equal the mantle density, ρ_m, when

$$d\rho_m = h\rho_c + (d - h)\rho_L$$

so that the plate thickness at neutral buoyancy, d_n, is

$$d_n = \left(\frac{\rho_L - \rho_c}{\rho_L - \rho_m}\right) h \tag{14.8.1}$$

From Equation (7.3.3), the age, τ_n, at which the plate reaches this thickness is

$$\tau_n = \frac{d_n^2}{4\kappa} \tag{14.8.2}$$

This can be cast as a function of mantle temperature by using an estimate of thickness of the oceanic crust as a function of mantle temperature [36, 37] :

$$h = h_0 + b(T - T_0) \tag{14.8.3}$$

where $h_0 = 7$ km is the present crustal thickness, $b = 0.085$ km/ °C, T is mantle temperature and T_0 is the present mantle temperature (about 1300 °C).

A corresponding estimate of the mean age of subduction as a function of mantle temperature can be obtained from Equation (14.3.1), which gives the mantle heat flow, Q_m, as a function of mantle temperature, and Equation (10.6.1), which gives the heat flow as a function of the age at subduction, τ_s. The latter can be rewritten, using Equation 7.3.4 for heat flux versus age, as

$$\tau_s = \left(\frac{2Aa}{Q_m}\right)^2 \tag{14.8.4}$$

where $a = KDT/\sqrt{(\pi\kappa)}$ and A is the area of the plates. These estimates are plotted in Figure 14.9.

Surprisingly, the two trends cross at a mantle temperature only about 60 °C above the present mantle temperature. At this temperature, on average, a plate arriving at a subduction zone would

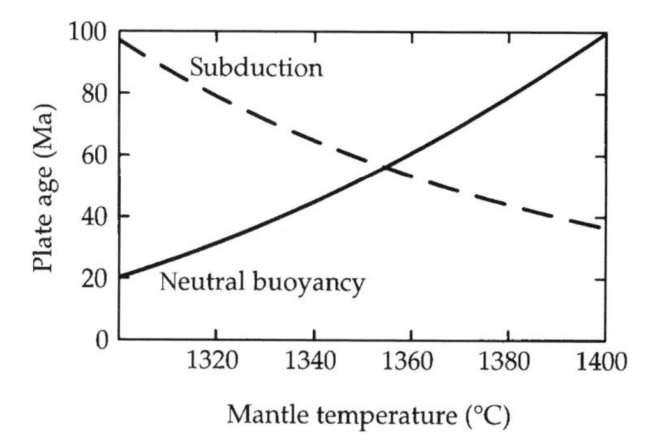

Figure 14.9. Average age of plates at subduction and age at which plates reach neutral buoyancy, as a function of mantle temperature. After [36].

be neutrally buoyant. At higher temperatures the paradoxical implication is that a plate arriving at a subduction zone would still be positively buoyant and would not subduct. The way out of the paradox is that plates would not be moving as fast as assumed in generating the 'subduction' curve of Figure 14.9 (which is based on Equation (8.3.4) for the convection velocity, from boundary layer theory), since the required driving force (their negative buoyancy) would not be present. Plates could move more slowly, such that they were negatively buoyant upon arrival at a subduction zone. Thus plate tectonics would still be possible, but at a slower rate than predicted by the boundary layer theory of Chapter 8, which takes account only of thermal buoyancy.

This may seem to be an effective resolution of the problem of the greater compositional buoyancy of oceanic plates in the past, but it is not. This is because a further implication is that the plate-scale mode of mantle convection would not be able to remove heat at a sufficient rate to cool the mantle. Instead of the mantle heat flux being given by Equation (14.3.1), above the critical mantle temperature it would be given by

$$Q_\mathrm{m} = \frac{2Aa}{\sqrt{\tau_\mathrm{n}}} \tag{14.8.5}$$

The resulting variation of Q_m with mantle temperature is compared with that resulting from free plate convection in Figure 14.10.

The problem is more severe further back in time, when radioactive heat generation is greater but the plate-scale mode is progressively less efficient. Unless there were another way to cool the mantle, it would get hotter instead of cooler. This will be illustrated in Section 14.10.

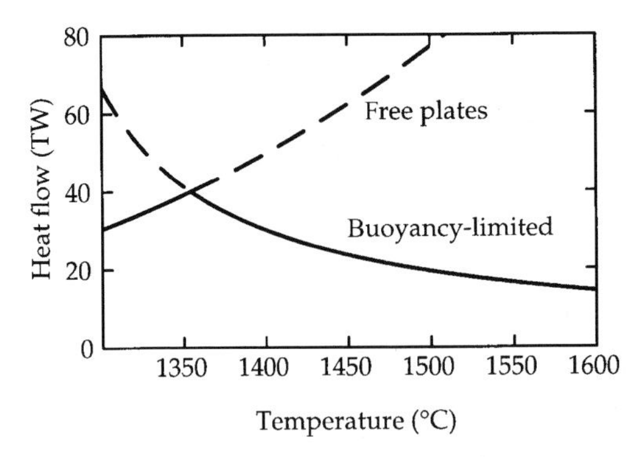

Figure 14.10. Comparison of mantle heat loss due to free plate convection with the heat loss when subducting plates are limited to be older than the age of neutral buoyancy. After [36].

A clear implication of the compositional buoyancy of oceanic lithosphere is that plate tectonics may not have been the dominant tectonic mechanism earlier in earth history. The mantle may have lost its heat by a different mechanism, which would have determined the tectonic style. Such a mechanism would involve different dynamical behaviour of the top thermal boundary layer of the mantle. Some possibilities will be discussed in Section 14.10.

14.8.5 Alternatives to plates

If subduction of oceanic lithosphere was not viable, then some other way for the mantle to lose heat would be required if the mantle were not to overheat. One possibility would be for the mantle part of the lithosphere to founder, leaving the lighter crustal component at the surface. Two variations on this are sketched in Figure 14.11 [36]. The distinction between them is that in case (a) the mantle part of the lithosphere is assumed still to behave like a plate, while in case (b) it is assumed to be more deformable. In the latter case, the asymmetric subduction characteristic of strong, brittle plates might be replaced by symmetric foundering or 'dripping'. There would presumably be a corresponding difference in the tectonic imprint left by these mechanisms.

Case (b) would be more likely at higher mantle temperatures, since then the plates would be thinner, the oceanic crust thicker and the mantle part of the lithosphere thinner on both accounts, so that it might not have the strength to form large plates. Case (a) would presumably be an intermediate stage between this and plates proper.

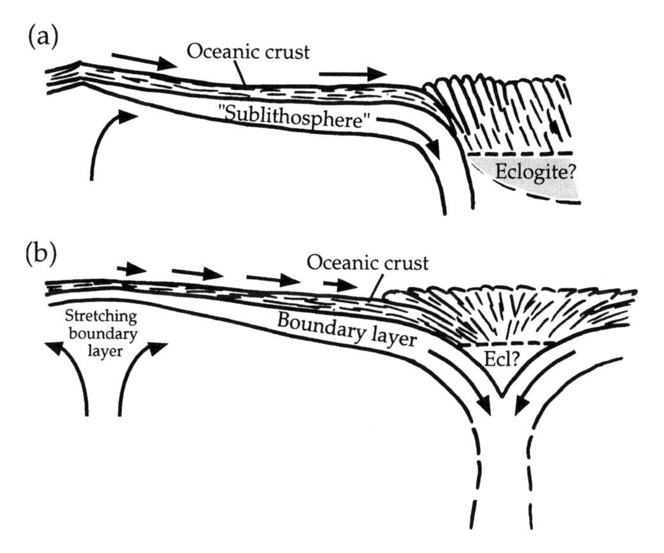

Figure 14.11. Conjectured alternatives to plate subduction. (a) 'Subplate tectonics'. (b) 'Drip tectonics'. From [36].

Unfortunately, these modes suffer from the same limitation as plate tectonics in a hot mantle: they cannot remove heat at a sufficient rate. The reason is that it is only the thin mantle part of the lithosphere that is available both to drive mantle convection and to cool the mantle. Thus convection would be slower and the efficiency of cooling would be less. The rate of heat loss as a function of temperature can be estimated by modifying the boundary layer theory presented in Section 8.2 [38].

Suppose the oceanic crust in Figure 14.11 has a thickness h and the lithosphere has a thickness d. If the mantle temperature, at the base of the lithosphere, is T, then the temperature at the base of the crust will be approximately $T_h = hT/d$. The average temperature deficit, relative to the mantle, within the subcrustal part of the thermal boundary layer will be $\Delta T = (T - T_h)/2 = T(1 - h/d)/2$. It is this layer, of thickness $(d - h)$, whose negative buoyancy is available to drive convection. Thus, following Section 8.2, the buoyancy force is

$$\begin{aligned} B &= -gD(d-h)\rho\alpha\Delta T \\ &= -g\rho\alpha D\frac{T}{2}d\left(1-\frac{h}{d}\right)^2 \end{aligned}$$

The viscous resistance force is again $R = 2\mu v$. Using $v = \kappa D/d^2$ from Equation (8.2.2) and requiring B and R to sum to zero yields

$$d(d-h)^2 = \frac{4\mu\kappa}{g\rho\alpha T} = \frac{4D^3}{Ra} \equiv c$$

This has a rather messy algebraic solution, and the surface heat flux can then be obtained from $q = KT/d$. Suitably scaled to the mantle, the total surface heat flow, $Q_s = 4\pi R^2 q$, is included in Figure 14.12 as a function of T. The thickness h of the oceanic crust was estimated as a function of T using Equation (14.8.3). You can see that Q_s reaches a maximum and then declines with increasing mantle temperature, in the same way as the buoyancy-hindered plate flow of Figure 14.10.

14.8.6 Foundering melt residue

When the mantle melts, there are two opposing effects on the density of the solid residue. First, iron is partitioned preferentially into the melt, leaving the residue slightly depleted in iron and slightly less dense (e.g. [22]). Second, the latent heat of melting reduces the temperature, increasing the density of the residue. Bickle [39] has estimated that the iron depletion substantially outweighs the latent heat effect, but Niu and Batiza [40] estimate the effects to be much closer in magnitude, though still leaving the residue slightly buoyant. However, there are significant uncertainties in the experimental constraints at the large melt fractions and higher pressures relevant to the early earth, and it remains possible that there is a regime in which the residue is denser than normal mantle.

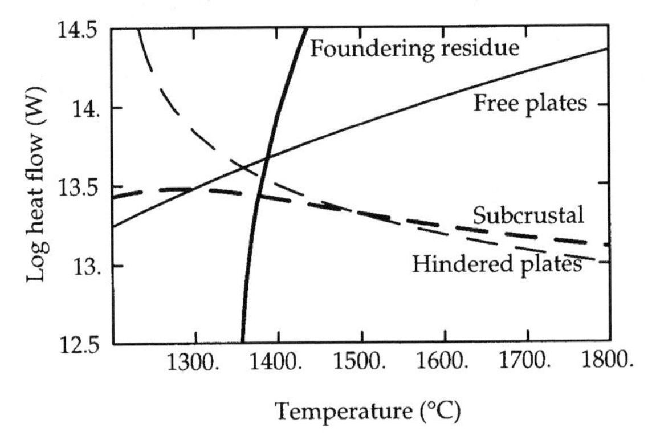

Figure 14.12. Dependence of heat loss on temperature for several of the modes discussed here. Free plates: Equation (14.3.1). Hindered plates: plates limited by the buoyancy of the oceanic crust (Equation (14.8.5)). Subcrustal: the modes of Figure 14.11. Foundering residue: the mode of Section 14.8.6 (Equation (14.8.8)).

Although quite conjectural, I include this possibility here because of the dramatic effect it could have on cooling the mantle. I summarise a development given elsewhere [41], but include here the additional possibility that the melt residue has some compositional buoyancy. Suppose that as mantle rises towards the surface it starts to melt at a depth d, and by the time it reaches the surface its temperature has been reduced by δT from the effect of latent heat, so that it is δT cooler than its 'potential temperature', T (the temperature it would have been if it had not melted). Suppose, for simplicity, that d varies in simple proportion to $(T - T_{sol})$, where T_{sol} is the temperature of the mantle solidus at the surface: $d \propto (T - T_{sol})$. We expect that δT will be some fraction of $(T - T_{sol})$ so again, for simplicity, assume

$$d = \beta \delta T \tag{14.8.6}$$

where β is a constant. Suppose also that the composition of the residue is changed in proportion to the degree of melting, such that there is a reduction in density at the surface of $\Delta\rho_c$. Then the net *increase* in density due to both the thermal and compositional effects, averaged down to depth d, will be

$$\Delta\rho = (\rho\alpha\delta T - \Delta\rho_c)/2 \tag{14.8.7}$$

We can think of the layer in which melting occurs as a melt boundary layer, in the sense that its density is altered, but it has a key property that is different from a thermal boundary layer produced by thermal diffusion. A diffusion thermal boundary layer is thinner if the mantle is hotter, because the mantle viscosity is lower, the convection velocity is faster and material spends less time cooling at the surface. On the other hand, the thickness of the melt boundary layer is independent of convection velocity, and it actually increases as the mantle gets hotter.

Again following the development in Section 8.2, if the melt boundary layer is denser and sinks, it will generate a buoyancy force

$$B = -gDd\Delta\rho$$

that is balanced by a resistance force $R = 2\mu v$. These should sum to zero. The rate of heat removal, H, from the melt layer is given by the volume flux through the melt layer times the heat deficit per unit volume:

$$H = dv\rho C_P \delta T/2$$

Averaging this over the top surface, which we can assume to be of width D, gives the average heat flux, $q = H/D$. The total heat flow loss rate out of the earth is then $Q = 4\pi R^2 q$. Combining these relationships yields

$$Q = \frac{\pi R^2 g\, \rho^2 \alpha C_P \beta^2}{2\mu}\left(1 - \frac{\Delta\rho_c}{\rho\alpha\delta T}\right)\delta T^4 \qquad (14.8.8)$$

If $\Delta\rho_c$ is small, this expression is proportional to δT^4 and, more importantly, to μ^{-1}. This means that the full effect of the strong temperature dependence of the viscosity, μ, is present in Equation (14.8.8), whereas in the forms of convection considered so far (Sections 8.2, 14.3.2, 14.8.5) it occurs through a fractional power ($-1/3$ or $-1/5$). If $\Delta\rho_c$ is larger than $\rho\alpha\delta T$, then the melt boundary layer will be buoyant and Equation (14.8.8) will not apply. As δT becomes larger than $\Delta\rho_c/\rho\alpha$, Q will rise rapidly from zero. An example of this behaviour is included in Figure 14.12, where you can see that Q is very strongly dependent on temperature. The implication of this for thermal evolution will be taken up in Section 14.10.

14.9 Heat transport by melt

If melt is generated in the mantle, it can remove heat very efficiently. This raises the question of whether melting might have enhanced the rate of cooling of the mantle, particularly if the mantle was hotter in the past. While this might seem an obvious and attractive possibility, it turns out to be not so simple. This is because the limitation on heat removal from the mantle becomes the rate at which melt can be generated, which returns us to the question of what buoyancies are available to drive the mantle and how fast it goes in response. Except for one possibility, it seems that melting may change the details of the temperature variation near the earth's surface without changing the underlying rate of heat loss. The remaining possibility is that the melt residue layer might be denser than normal mantle, as was discussed in the last section.

The efficiency of heat loss from mafic and ultramafic magma in contact with a cold surface of the earth can be appreciated from a few rough estimates. Such magmas have quite low viscosities (as low as 10 Pa s [42]) and commonly spread in flows of the order of 1 m thick. Such flows will cool within a few days. With a density of about 3000 kg/m^3, a specific heat of about 1000 J/kg °C, a temperature of about 1000 °C, and a latent heat of melting of about 500 kJ/kg [43], the heat loss per unit area of a flow is about 5×10^9 J/m^2, and the average heat flux is about 5×10^3 W/m^2. This is far greater than the present average heat flux out of the sea floor (0.1 W/m^2).

Another estimate comes from observations of basaltic lava lakes at Hawaii [44], upon which a solid crust was maintained at only a few centimetres thick because it continuously foundered into the magma. The heat flux through this crust would then be about $q = K\Delta T/d = 10^5$ W/m^2. Slightly more sophisticated treatments of a convecting magma lake yield heat fluxes of the order of 10^4 W/m^2 [41] [45].

These rates of heat loss are so high that in terms of the rates of solid mantle processes, the rate of cooling of melt that is erupted onto the surface of the earth can often be considered to be instantaneous. This has the interesting implication that it is very difficult to form or maintain a magma ocean unless the earth's surface is maintained at a high temperature. For example, a surface heat flux of 10^4 W/m^2 would freeze a magma layer at a rate of about 200 m/a, fast enough to freeze a 100 km-deep magma ocean in about 500 years [41].

I want to note, in passing, that the efficiency of heat removal by melt is not due to the latent heat of freezing of the magma, as I have sometimes heard supposed. From the values used above, the heat lost in cooling a magma from 1000 °C to 0 °C is made up of 5×10^5 J/kg from latent heat of freezing and 10^6 J/kg from cooling the resultant solid. Thus the latent heat contribution is significant, but not dominant. Melt removes heat efficiently because it is so mobile, which is because it has a low viscosity.

14.10 Tectonic evolution

We have, in the present earth, two rather clearly identified tectonic agents: plates and plumes. Based on the ideas outlined so far in this chapter, we should consider the possibilities that the plate and plume modes each may have changed, or that one or both may not have operated in the past, and that in any case other mechanisms may also have operated.

There is an important principle that limits the range of possibilities. It is that a mechanism deriving from one thermal boundary layer will not substitute for a mechanism deriving from another thermal boundary layer. This is because the dynamics of the lower thermal boundary layer is the means by which heat enters a fluid layer, while the dynamics of the upper thermal boundary layer is the means by which heat is lost from the fluid layer. This means, in particular, that plumes are not a possible substitute for plate tectonics. If plate tectonics were not viable, then the top thermal boundary layer would have had to operate in some other dynamical mode in order to remove heat from the mantle, and that mode would determine the style of the associated tectonics.

Among the possibilities raised in this chapter are that plate tectonics operated more vigorously in the recent past, but that it may have been minor or absent early in earth history. Plumes may have been more vigorous or less, depending on various factors, but they have probably operated throughout earth history. Major or catastrophic mantle overturns may have occurred, with corresponding tectonic effects. If plates did not operate, then the top thermal boundary layer would have operated in some other dynamical mode or modes whose nature is still conjectural. At times of high mantle temperature, notably during and soon after the formation of the earth, heat transport by melt might have been important, with a distinctive associated tectonic style. I will discuss some of the tectonic consequences that can be envisaged at present, and then briefly discuss how we might be able to discriminate between them.

14.10.1 Plumes

Whether plumes have been more vigorous or less in the past depends very much on the thermal history of the mantle. Thus in the smooth thermal evolution model of Figure 14.3, the plume flux is fairly constant, because the core cools at about the same rate as the mantle, keeping the temperature difference across the lower thermal boundary layer nearly constant. On the other hand, in the episodic model of Figure 14.7 the plume flux would generally be lower before 1 Ga ago because the lower mantle temperature is 100–200 °C greater than in a whole-mantle convection model. As well, the plume flux would fluctuate, being low when the lower mantle temperature is high, and vice versa. Yet another possibility follows from the proposal of Campbell and Griffiths [35] that the insulating D$''$ layer was not present in the Archean (Section 14.8.4). In that case, I have shown that the plume flux may have been an order of magnitude greater during the Archean because of the larger temperature difference available to drive plumes [38].

14.10.2 Mantle overturns

The potential consequences of a full-scale mantle overturn, like the models in Figures 10.6 and 14.7, are dramatic, and I have discussed them in more detail elsewhere [15]. In the model of Figure 14.7, the relatively cool upper mantle is replaced by lower mantle material that is 200–300 °C hotter. This temperature excess is characteristic of that of plumes, and the volume of material involved is about 200 times the volume of a large plume head. If flood basalt eruptions are caused by the arrival of a plume head, then such a mantle overturn would be like 200 flood basalt eruptions occurring within a few million years.

In regions of continental crust, the consequence might be much like a flood basalt eruption (Figure 14.13a), which suggests a possible mechanism for the formation of greenstone belts. In oceanic regions, if plate tectonics was operating, as is plausible, then initially the thickness of new oceanic crust would increase because of the greater degree of melting from the hotter material under spreading centres (Figure 14.13b), and the plate system might then choke and stop because of the buoyancy of this thicker

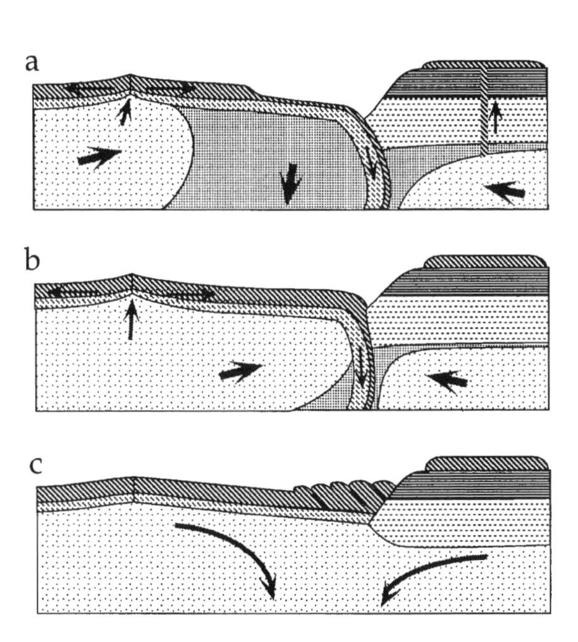

Figure 14.13. Possible tectonic events accompanying a mantle overturn. Hot material from the lower mantle (stippled) replaces cooler material in the upper mantle (grey). As a result, thicker oceanic crust is formed, and flood-basalt-like eruptions occur on continents. The thickened oceanic crust blocks subduction in (c). From Davies [15]. Copyright by Elsevier Science. Reprinted with permission.

oceanic crust (Figure 14.13c). The magmatic result might thus be the eruption of the equivalent of several kilometres' thickness of basalt over the earth's surface.

Other tectonic effects would be likely. As the hotter material flooded the upper mantle, it would raise the earth's surface by 3–5 km. The resulting gravity sliding forces would temporarily change plate motions. Relative sea level would change dramatically. After emplacement of the hot material, plate tectonics might no longer be viable (Section 14.8.4) and some other tectonic mode might operate (14.8.5) until the upper mantle cooled again. The large volume of oceanic-type crust might be tectonically thickened (Figure 14.13c), remelted in the same or a later overturn, and possibly founder if its roots transformed to eclogite.

The atmosphere and life might well be affected. Degassing of the mantle during such voluminous eruptions might significantly affect the composition of the atmosphere. This may provide an observational discriminant, since there are constraints from isotope geochemistry on the timing of mantle degassing (Chapter 13). The combination of sea level changes, atmospheric changes and the physical effects of the tectonics might have affected the viability of life, perhaps severely.

Even if full mantle layering did not occur, inhibition of flow through the transition zone could have led to more localised break-throughs of temporarily blocked hot or cold material which might have caused significant tectonic episodes. Distinguishing these from plume events, or groups of plume events, might not be easy.

14.10.3 Alternatives to plates and consequences for thermal evolution

Although plate tectonics might proceed in a hot mantle, we have seen that it becomes less efficient at higher temperatures (Section 14.8.4, Figure 14.10). The consequences of this are illustrated in Figure 14.14. If the earth started hot, so that the upper mantle temperature was greater than the critical temperature at which plates are slowed, then the rate of heat loss would be less than the rate of heat generation by radioactivity and the mantle would get hotter. Even in a model with reduced radioactive heating (50% of present heat loss), starting at the quite modest temperature of 1400 °C results in a thermal runaway. Starting at 1300 °C, the heat loss rate initially rises, but then the temperature crosses the thresh-hold and the heat loss rate falls, yielding thermal runaway. Starting the model at 1200 °C just avoids thermal runaway, because of the

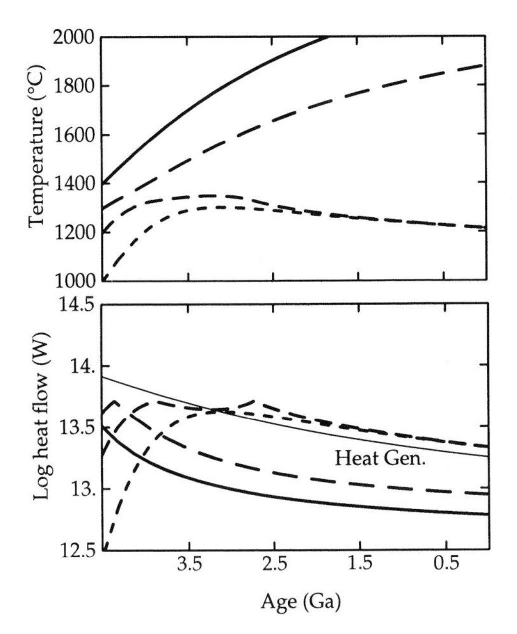

Figure 14.14. Hypothetical thermal evolutions resulting from the 'hindered plate' regime (Section 14.8.4) in which the heat loss rate declines at higher temperatures (Figure 14.10).

decline of radioactive heating. Starting at 1000 °C avoids the hindered plate regime entirely.

These models are not intended to be realistic. They demonstrate the kind of behaviour that this kind of model would yield. The starting temperatures are unrealistically low, and the radioactive heating rate is also rather low, as shown by the low final temperature (about 1200 °C) of the cooler models. It is interesting to note in passing how the final temperature is essentially independent of the initial temperature in the cooler models, but it suddenly switches to being a strong function of the initial temperature when thermal runaway occurs. Qualitatively similar behaviour can be expected for the subcrustal foundering modes of Section 14.8.5 (Figure 14.12).

In contrast, the very high efficiency of heat loss at high temperatures that would by accomplished by foundering melt residue (Section 14.8.6) would preclude any possibility of thermal runaway. Figure 14.15 shows the result of assuming that this mode operates as depicted in Figure 14.12, with plates operating normally at lower temperatures. The foundering residue mode is so efficient at high temperatures that the initial high temperature is reduced within a few million years to that required to balance the radiogenic heating.

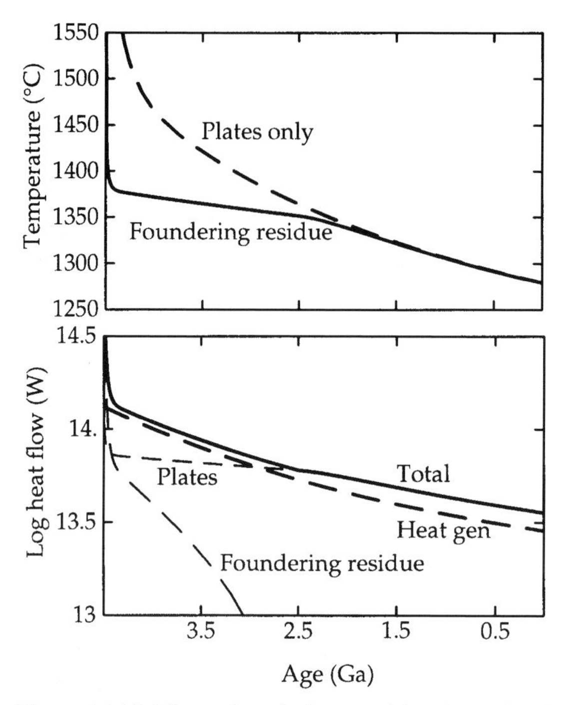

Figure 14.15. Thermal evolution resulting from the 'foundering residue' regime (Section 14.8.6) in which the heat loss rate increases rapidly at higher temperatures (Figure 14.12). A 'plates only' curve (like Figure 14.3) is included for comparison.

With the assumptions of this model, normal plate tectonics becomes dominant after about 2 Ga. A 'plates only' temperature curve is included for comparison (similar to that of Figure 14.3). This reveals how much more rapidly the initial cooling is with the foundering residue mode operating.

Although the foundering residue mode would serve admirably to cool the early earth, it seems unlikely based on present esimates of the relevant densities. It is included here to illustrate its potential importance, in order to encourage the acquisition of more accurate data, and to show the effect of a very efficient heat loss mode on the thermal evolution. The other modes, in which the buoyancy of the oceanic crust inhibits the ability of the top thermal boundary layer to sink, are not viable because they cannot explain geological evidence that mantle temperatures in the Archean were at least as high as at present. Thus although these modes may pass a dynamical test, in that the forces are sufficient to drive a mode of mantle convection, they fail the thermal test: they cannot cool the earth. We are forced to look for other possibilities.

14.10.4 Possible role of the basalt–eclogite transformation

It has been conjectured for some time that the high-pressure transformation of basalt to the denser mineral assemblage 'eclogite' would promote the subduction of plates and hence the viability of plate tectonics. In fact it was the hypothetical driving mechanism of Holmes' [46] qualitative mantle convection theory. The challenge is to show quantitatively how this might work, and what would be the character of the resulting mode of convection.

The character of a convection mode in which the basalt-eclogite transformation plays a key role is likely to depend strongly on whether the transformation is limited more by temperature or by pressure. The transformation requires a minimum pressure, but the reaction kinetics also require high temperature. If the rock is not very hot, then it may be carried metastably to higher pressures without transforming. Presumably at some high pressure the transformation will occur even if the rock is still cold, but this pressure is not known.

Now if the transformation is limited by temperature, then the volumetric rate at which basalt can transform will be limited by how fast it can be heated, and heat conduction occurs at slow and predictable rates at the scale of crustal bodies. The result is likely to be that the transformation rate, and hence the convection rate, is steady or only moderately fluctuating.

There are two circumstances in which the transformation might be limited by pressure rather than temperature. First, if the basalt remains hot after its formation, then only an increase of pressure will be required for it to transform. On the other hand, it might cool as it accumulates. Then if the heating rate is slow, the basalt might accumulate to large thickness, such that its base reaches a pressure at which it will transform even if it is cold. As basalt transforms into the denser assemblage, the pile will subside because buoyant basalt is replaced by negatively buoyant eclogite. This will carry more basalt through the critical pressure, which will also transform, and the process may accelerate into runaway. It will be limited either by the availability of basalt or by the rate at which basaltic crust can be pulled into the transforming region. In either case an episodic behaviour is likely, and the result is likely to produce large bodies of cold, dense eclogite that will sink into the mantle. Some aspects of this have been modelled numerically by Vlaar and others [47].

Which of these modes is more likely under particular conditions is not known. Nor is it known what would be the efficiency of heat transport of such modes, so it is not known whether the basalt–eclogite transformation provides an escape from the paradox illustrated in Figure 14.14.

14.10.5 Discriminating among the possibilities

Theories of the remote past of the earth must be tested against observations, but the connection between a putative dynamical mode of mantle convection 3.5 billion years ago and a rock at the earth's surface today may be a circuitous one involving multiple physical and chemical processes. Thus, on the one hand, we must search for testable implications of the theory and be ready to discard or modify a theory if it seems to be contradicted by observations. On the other hand, we must bear in mind that the prediction itself may be faulty, because of an overlooked complication in the path from model to presently observable consequence, and so models should not be too lightly discounted.

14.11 References

1. P. F. Hoffman and S. A. Bowring, Short-lived 1.9 Ga continental margin and its destruction, Wopmay orogen, northwest Canada, *Geology* **12**, 68–72, 1984.
2. A. W. Hofmann and W. M. White, Mantle plumes from ancient oceanic crust, *Earth Planet. Sci. Lett.* **57**, 421–36, 1982.
3. M. Stein and A. W. Hofmann, Mantle plumes and episodic crustal growth, *Nature* **372**, 63–8, 1994.
4. C. B. Agee, Petrology of the mantle transition zone, *Annu. Rev. Earth Planet. Sci.* **21**, 19–42, 1993.
5. G. F. Davies, Geophysical and isotopic constraints on mantle convection: an interim synthesis, *J. Geophys. Res.* **89**, 6017–40, 1984.
6. H. S. C. O'Neill and H. Palme, Composition of the silicate Earth: implications for accretion and core formation, in: *The Earth's Mantle: Composition, Structure and Evolution*, I. N. S. Jackson, ed., Cambridge University Press, Cambridge, 3–126, 1998.
7. F. D. Stacey and D. E. Loper, Thermal histories of the core and mantle, *Phys. Earth Planet. Interiors* **36**, 99–115, 1984.
8. G. F. Davies, Cooling the core and mantle by plume and plate flows, *Geophys. J. Int.* **115**, 132–46, 1993.
9. R. I. Hill, I. H. Campbell, G. F. Davies and R. W. Griffiths, Mantle plumes and continental tectonics, *Science* **256**, 186–93, 1992.
10. G. Gastil, The distribution of mineral dates in space and time, *Amer. J. Sci.* **258**, 1–35, 1960.
11. M. T. McCulloch and V. C. Bennett, Progressive growth of the Earth's continental crust and depleted mantle: geochemical constraints, *Geochim. Cosmochim. Acta* **58**, 4717–38, 1994.

12. S. R. Taylor and S. M. McLennan, *The Continental Crust: Its Composition and Evolution*, 312 pp., Blackwell, Oxford, 1985.
13. P. Machetel and P. Weber, Intermittent layered convection in a model mantle with an endothermic phase change at 670 km, *Nature* **350**, 55–7, 1991.
14. P. J. Tackley, D. J. Stevenson, G. A. Glatzmaier and G. Schubert, Effects of an endothermic phase transition at 670 km depth in a spherical model of convection in the earth's mantle, *Nature* **361**, 699–704, 1993.
15. G. F. Davies, Punctuated tectonic evolution of the earth, *Earth Planet. Sci. Lett.* **136**, 363–79, 1995.
16. A. Davaille and C. Jaupart, Transient high-Rayleigh-number thermal convection with large viscosity variations, *J. Fluid Mech.* **253**, 141–66, 1993.
17. D. P. McKenzie, Speculations on the consequences and causes of plate motions, *Geophys. J. R. Astron. Soc.* **18**, 1–32, 1969.
18. A. Holmes, Radioactivity and earth movements, *Geol. Soc. Glasgow, Trans.* **18**, 559–606, 1931.
19. P. F. Hoffman, Did the breakout of Laurentia turn Gondwanaland inside-out?, *Science* **252**, 1409–12, 1991.
20. M. Gurnis, Large-scale mantle convection and the aggregation and dispersal of supercontinents, *Nature* **332**, 695–9, 1988.
21. S. Zhong and M. Gurnis, Dynamic feedback between a continentlike raft and thermal convection, *J. Geophys. Res.* **98**, 12 219–32, 1993.
22. A. E. Ringwood and T. Irifune, Nature of the 650-km discontinuity: implications for mantle dynamics and differentiation, *Nature* **331**, 131–6, 1988.
23. A. E. Ringwood, Phase transformations and their bearing on the constitution and dynamics of the mantle, *Geochim. Cosmochim. Acta* **55**, 2083–110, 1991.
24. S. E. Kesson, J. D. Fitz Gerald and J. M. G. Shelley, Mineral chemistry and density of subducted basaltic crust at lower mantle pressures, *Nature* **372**, 767–9, 1994.
25. T. Irifune, Phase transformations in the earth's mantle and subducting slabs: Implications for their compositions, seismic velocity and density structures and dynamics, *The Island Arc* **2**, 55–71, 1993.
26. G. F. Davies, Penetration of plates and plumes through the mantle transition zone, *Earth Planet. Sci. Lett.* **133**, 507–16, 1995.
27. G. F. Davies, Mantle plumes, mantle stirring and hotspot chemistry, *Earth Planet Sci. Lett.* **99**, 94–109, 1990.
28. D. Loper and T. Lay, The core-mantle boundary region, *J. Geophys. Res.* **100**, 6379–420, 1995.
29. U. R. Christensen and A. W. Hofmann, Segregation of subducted oceanic crust in the convecting mantle, *J. Geophys. Res.* **99**, 19 867–84, 1994.
30. R. Jeanloz and E. Knittle, Density and composition of the lower mantle, *Philos. Trans. R. Soc. London Ser. A* **328**, 377–89, 1989.

31. L. H. Kellogg and S. D. King, Effect of mantle plumes on the growth of D″ by reaction between the core and the mantle, *Geophys. Res. Lett.* **20**, 379–82, 1993.
32. C. G. Farnetani, Excess temperature of mantle plumes: the role of chemical stratification across D″, *Geophys. Res. Lett.* **24**, 1583–6, 1996.
33. G. F. Davies and M. Gurnis, Interaction of mantle dregs with convection: lateral heterogeneity at the core-mantle boundary, *Geophys. Res. Lett.* **13**, 1517–20, 1986.
34. R. Boehler, A. Chopelas and A. Zerr, Temperature and chemistry of the core-mantle boundary, *Chemical Geology* **120**, 199–205, 1995.
35. I. H. Campbell and R. W. Griffiths, The changing nature of mantle hotspots through time: Implications for the chemical evolution of the mantle, *J. Geol.* **92**, 497–523, 1992.
36. G. F. Davies, On the emergence of plate tectonics, *Geology* **20**, 963–6, 1992.
37. R. White and D. McKenzie, Magmatism at rift zones: the generation of volcanic continental margins and flood basalts, *J. Geophys. Res.* **94**, 7685–730, 1989.
38. G. F. Davies, Conjectures on the thermal and tectonic evolution of the earth, *Lithos* **30**, 281–9, 1993.
39. M. J. Bickle, Implication of melting for stabilisation of the lithosphere and heat loss in the Archean, *Earth. Planet. Sci. Lett.* **80**, 314–24, 1986.
40. Y. Niu and R. Batiza, In situ densities of MORB melts and residual mantle: implications for buoyancy forces beneath ocean ridges, *J. Geol.* **99**, 767–75, 1991.
41. G. F. Davies, Heat and mass transport in the early earth, in: *Origin of the Earth*, H. E. Newsome and J. H. Jones, eds., Oxford University Press, New York, 175–94, 1990.
42. Y. Bottinga and D. F. Weill, The viscosity of magmatic silicate liquids: a model for calculation, *Amer. J. Sci.* **272**, 438–75, 1972.
43. Y. Bottinga and P. Richet, Thermodynamics of liquid silicates, a preliminary report, *Earth. Planet. Sci. Lett.* **40**, 382–400, 1978.
44. W. A. Duffield, A naturally occurring model of global plate tectonics, *J. Geophys. Res.* **77**, 2543–55, 1972.
45. W. B. Tonks and H. J. Melosh, The physics of crystal settling and suspension in a turbulent magma ocean, in: *Origin of the Earth*, H. E. Newsom and J. H. Jones, eds., Oxford University Press, New York, 151–74, 1990.
46. A. Holmes, *Principles of Physical Geology*, Thomas Nelson and Sons, 1944.
47. N. J. Vlaar, P. E. van Keken and A. P. van den Berb, Cooling of the earth in the Archean: Consequences of pressure-release melting in a hotter mantle, *Earth Planet. Sci. Lett.* **121**, 1–18, 1994.

APPENDICES

APPENDIX 1

Units and multiples

The units of some of the quantities used in the text are summarised here, particularly those less commonly encountered in other contexts. I also include their relation to the basic units of length, mass and time, their standard symbols and the various multiples and fractions used here.

Table A1.1. *Units.*

	Unit		
Quantity	Symbol	Name	Composition
Length	m	metre	—
Mass	kg	kilogram	—
Time	s	second	—
	a	year (année)	3.16×10^7 s
Force	N	newton	kg m/s^2
Stress	Pa	pascal	N/m^2
Viscosity	Pa s	pascal second	Ns/m^2
Elastic modulus			Pa
Heat	J	joule	Nm
Power (heat flow rate)	W	watt	J/s
Heat flux			W/m^2
Conductivity (thermal)			W/m °C
Thermal diffusivity			m^2/s
Specific heat			J/kg °C
Thermal expansion			°C^{-1}

Table A1.2. *Multiple and fractional units.*

Power of 10	Symbol	Name
12	T	tera
9	G	giga
6	M	mega
3	k	kilo
0	—	—
−3	m	milli
−6	μ	micro
−9	n	nano
−12	p	pico
−15	f	femto

APPENDIX 2

Specifications of numerical models

Parameters of the original numerical models shown in the text are summarised here, identified by the figure in which the model appears. The models are grouped into two tables for convenience. Parameters that are the same for all models in the table are given at the bottom of the table.

Several forms of temperature dependence of the viscosity were used in the models. These forms are specified by the following equations, and the appropriate equation is identified in the tables. The basic form is given by Equation (6.10.4), and the first form below is a version of this.

$$\mu = \mu_r \exp\left[q_A\left(\frac{T_m + T_z}{T + T_z} - 1\right)\right] \tag{A2.1}$$

where μ_r is the reference viscosity, T is dimensionless temperature and

$$q_A = \frac{E^*}{RT_r} \tag{A2.2}$$

is a measure of the activation energy, E^* (R is the gas constant and T_r is the reference temperature; see Section 6.10.2). T_m is either the maximum dimensionless temperature (if the bottom thermal boundary condition is a prescribed heat flux) or 1:

$$T_m = \begin{cases} T_{max}/T_r; \ (q) \\ 1; \ (T) \end{cases} \tag{A2.3}$$

$T_z = 0.21T_m$ is (approximately) the correction from °C to K (273/1300 = 0.21).

Table A2.1. *Parameters of subduction and related models.*

Quantity	(Defining equation)	Figure 8.4 left	8.4 right	10.1, 10.2	10.3	10.9	10.12	10.13
Reference viscosity		10^{22}	10^{22}	10^{22}	10^{22}	10^{22}	3×10^{20}	3×10^{20}
Rayleigh number								
Ra	(8.3.2)	6×10^{6}						
R_q	(8.6.1)		10^{8}	10^{8}	10^{8}	10^{8}	7.2×10^{9}	7.2×10^{9}
Peclet number	(8.3.5)	—	—	—	—	2000	5000	5000
Plate velocity (mm/a)		—	—	—	—	21	53	53
Viscosity								
T-equation		—	—	A2.1	—	A2.1	A2.1	A2.1
maximum		—	—	100	—	100	300	300
q_A	(A2.2)	—	—	20	—	20	20	20
activation energy (kJ/mol)		—	—	260	—	260	260	260
depth exponent		—	—	—	0	—	10	1–10
lower mantle		—	—	—	100	—	10	1–30
Internal heating		0.0	1.0	1.0	1.0	1.0	1.0	1.0
Bottom thermal boundary		T	q	q	q	q	q	q
Top boundary								
f, free slip; v, velocity		f	f	f	f	v	v	v
Side boundary								
m, mirror; p, periodic		m	m	m	m	m	p	p, m

Mantle depth 3000 km; reference temperature 1300 °C; numerical grids 256 × 64; Cartesian geometry.

For the illustration of the controls on the head and tail structure of the plume (Figure 11.7), the viscosity function is

$$\mu = \mu_r \exp\left[-q_A\left(\frac{T}{T_m} - 1\right)\right] \tag{A2.4}$$

with the same definitions.

The illustration of rifting a plume head (Figure 12.6) used an earlier function that approaches the maximum prescribed viscosity smoothly at low temperatures (whereas in the other cases the viscosity is simply truncated at the maximum value), while yielding a strong dependence on temperature in the interior of the model.

Table A2.2. *Parameters of plume and ridge models.*

Quantity	(Defining equation)	Figure 11.6	11.7	11.10	11.11	12.6	12.8
Reference viscosity		10^{22}	3×10^{21}	3×10^{20}	3×10^{20}	10^{21}	10^{21}
Internal heating		0.0	0.0	0.0	0.0	0.0	0.0
Bottom thermal boundary		T	T	T	T	T	q
Bottom temperature (dimensionless)		1.3	1.31	1.4	1.4	1.0	
Bottom heat flux (mW/m^2)							77
Reference temperature (°C)		1420	1300	1300	1300	1280	1400
Model depth (km)		3000	2900	3000	3000	2000	660
Thermal expansion ($\times 10^{-5}$ °C)		2	3	2	2	2	3
Model Rayleigh number							
Ra	(8.3.2)	3×10^6	1.24×10^7	10^8	10^8	8×10^6	
R_q	(8.6.1)						5.7×10^6
Bottom thermal boundary layer							
temperature jump (°C)		426	400	520	520	—	—
viscosity at reference temp.		10^{22}	3×10^{21}	6×10^{21}	6×10^{21}		
local R_T	(8.3.2)	9×10^5	3.8×10^6	2×10^6	2×10^6	—	—
Peclet number	(8.3.5)	0	0	0	0	2000	400
Plate velocity (mm/a)		0	0	0	0	32	20, (10, 30)
Viscosity							
T-equation		A2.1	A2.4	A2.1	A2.1	A2.5	A2.1, A2.6
maximum		1000	100	1000	1000	3000	30
q_A	(A2.2)	30	0, 11, 17.3	30	30	10	20
activation energy (kJ/mol)		420	0, 144, 225	390	390	130	280
depth exponent		—	—	10	10	—	1
upper layer		—	—	1	1	—	0.1
lower layer		—	—	20	20	—	1
Phase change							
model height		—	—	0.7	0.7	—	—
Clapeyron slope (MPa/K)		—	—	−2	−2, −2.5, −3	—	—
density jump (%)		—	—	10	10	—	—
Geometry							
cart, Cartesian; cyl, cylindrical		cyl	cyl	cyl	cyl	cart	cart
Numerical grid							
horizontal		128	128	128	128	256	512
vertical		128	128	128	128	128	64

Zero internal heating; velocity prescribed on top boundary; mirror side boundaries.

$$\mu = \mu_r \exp[v_m + (0.5q_A - 4v_m)x^6 - (0.5q_A - 3v_m)x^8] \qquad (A2.5)$$

where $v_m = \ln(\mu_{max})$ and $x = T/T_m$.

The upper mantle spreading centre model of Figure 12.8 used Equation (A2.1), but with T_m defined in terms of the average interior temperature away from the boundary layers, T_{av}:

$$T_m = T_{av}/T_r \qquad (A2.6)$$

For this model, the region over which T_{av} was defined excluded the upper and lower 1/4 of the box, and the left and right 1/8 of the box.

Index